New Chemistry and
New Opportunities from the
Expanding Protein Universe

23rd International Solvay Conference on Chemistry

New Chemistry and New Opportunities from the Expanding Protein Universe

Hotel Métropole, Brussels   16 – 19 October 2013

Scientific Editors

Kurt Wüthrich
The Scripps Research Institute, USA & ETH Zürich, Switzerland

Ian A. Wilson
The Scripps Research Institute, USA

Donald Hilvert
ETH Zürich, Switzerland

Dennis W. Wolan
The Scripps Research Institute, USA

International Solvay Institutes Editor

Anne De Wit
Université Libre de Bruxelles, Belgium

World Scientific

NEW JERSEY · LONDON · SINGAPORE · BEIJING · SHANGHAI · HONG KONG · TAIPEI · CHENNAI

Published by

World Scientific Publishing Co. Pte. Ltd.

5 Toh Tuck Link, Singapore 596224

USA office: 27 Warren Street, Suite 401-402, Hackensack, NJ 07601

UK office: 57 Shelton Street, Covent Garden, London WC2H 9HE

Library of Congress Cataloging-in-Publication Data
International Solvay Conference on Chemistry (23rd : 2013 : Brussels, Belgium)
  New chemistry and new opportunities from the expanding protein universe : proceedings of the 23rd International Solvay Conference on Chemistry, Hotel Métropole, Brussels 16–19 October 2013 / editor, Kurt Wüthrich, The Scripps Research Institute, USA & ETH Zürich, Switzerl [and 4 others].
    pages cm
  Includes bibliographical references and index.
    ISBN 978-9814603829 (hardcover : alk. paper) -- ISBN 9814603821 (hardcover : alk. paper) -- ISBN 978-9814616560 (pbk. : alk. paper) -- ISBN 9814616567 (pbk. : alk. paper)
  1. Proteins--Synthesis--Congresses. 2. Enzymes--Congresses. I. Wüthrich, Kurt, editor. II. Title.
  QP551.I5188 2013
  612'.01575--dc23
                                                              2014024092

British Library Cataloguing-in-Publication Data
A catalogue record for this book is available from the British Library.

Contents

The International Solvay Institutes

Honorary Members

Baron André Jaumotte
Honorary Director
Honorary Rector and Honorary President ULB

Mr. Jean-Marie Piret
Emeritus Attorney General of the Supreme Court of Appeal
and Honorary Principal Private Secretary to the King

Professor Jean-Louis Vanherweghem
Former Chairman of the Board of Directors of the ULB

Professor Irina Veretennicoff
Professor VUB

Guest Members

Professor Anne De Wit
Professor ULB and Scientific Secretary of the International
Committee for Chemistry

Professor Hervé Hasquin
Permanent Secretary of the Royal Academy of Sciences, Letters and
Fine Arts of Belgium

Professor Franklin Lambert
Professor VUB

Professor Alexander Sevrin
Deputy Director for Physics
Professor VUB and Scientific Secretary of the International
Committee for Physics

Professor Géry van Outryve d'Ydewalle
Permanent Secretary of the Royal Flemish Academy of Belgium for
Sciences and the Arts

Professor Lode Wyns
Deputy Director for Chemistry

Director

Professor Marc Henneaux
Professor ULB

Solvay Scientific Committee for Chemistry

Professor Kurt Wüthrich (Chair)
ETH Zürich, Switzerland and the Scripps Research Institute, La Jolla, USA

Professor Graham Fleming
University of California, Berkeley, USA

Professor Robert H. Grubbs,
California Institute of Technology, Pasadena, USA

Professor Roger Kornberg
Stanford University, USA

Professor Harold W. Kroto
University of Sussex, Brighton, UK

Professor Henk N.W. Lekkerkerker
Utrecht Universiteit, The Netherlands

Professor K.C. Nicolaou
The Scripps Research Institute, La Jolla, USA

Professor JoAnne Stubbe
Massachusetts Institute of Technology, Cambridge, USA

Professor George M. Whitesides
Harvard University, Cambridge, USA

Professor Ahmed Zewail
California Institute of Technology, Pasadena, USA

Professor Anne De Wit
Université Libre de Bruxelles, Belgium, Scientific Secretary

Acknowledgements

The organization of the 23rd Solvay Conference has been made possible thanks to the generous support of the Solvay Family, the Solvay Group, the Université Libre de Bruxelles, the Vrije Universiteit Brussel, the Belgian National Lottery, the Fédération Wallonie-Bruxelles, the Vlaamse Regering, the Brussels Region, BNP-Paribas-Fortis, the David and Alice Van Buuren Foundation, Belgacom and the Hôtel Métropole.

Participants

Markus	Aebi	ETH Zürich, Switzerland
Frances H.	Arnold	California Institute of Technology, USA
Nenad	Ban	ETH Zürich, Switzerland
Jason W.	Chin	MRC, Cambridge, UK
Aled M.	Edwards	University of Toronto, Canada
Joachim	Frank	Columbia University, USA
John A.	Gerlt	University of Illinois, USA
Rudolf	Glockshuber	ETH Zürich, Switzerland
Adam	Godzik	Sanford-Burnham Medical Research Institute, USA
Markus G.	Grütter	University of Zürich, Switzerland
Bernard	Henrissat	Aix-Marseille University, France
Donald	Hilvert	ETH Zürich, Switzerland
Kendall N.	Houk	University of California – Los Angeles, USA
Judith P.	Klinman	University of California – Berkeley, USA
Henk N.W.	Lekkerkerker	Utrecht Universiteit, The Netherlands
Richard M.	Lerner	The Scripps Research Institute, USA
Kaspar	Locher	ETH Zürich, Switzerland
Tom W.	Muir	Princeton University, USA
Dario	Neri	ETH Zürich, Switzerland
Harry F.	Noller	University of California – Santa Cruz, USA
Christine	Orengo	UCL-IRIS, London, UK
Jacques	Prost	ESPCI, Paris, France
Jean-Louis	Reymond	University of Berne, Switzerland
Marina V.	Rodnina	Max Planck Institute, Germany
Bryan L.	Roth	University of North Carolina, USA
Sachdev S.	Sidhu	University of Toronto, Canada
Gebhard	Schertler	Paul Scherrer Institute, Switzerland
Peter H.	Seeberger	Max Planck Institute, Germany
Arne	Skerra	Technische Universität München, Germany
Raymond C.	Stevens	The Scripps Research Institute, USA
Jan	Steyaert	VIB and Vrije Universiteit Brussel, Belgium
Hiroaki	Suga	The University of Tokyo, Japan
Craig A.	Townsend	Johns Hopkins University, USA
Wilfred A.	van der Donk	University of Illinois, USA
Gunnar	Von Heijne	Stockholm University, Sweden
Christopher T.	Walsh	Harvard Medical School, USA
Ian A.	Wilson	The Scripps Research Institute, USA

Dennis W.	Wolan	The Scripps Research Institute, USA
Chi-Huey	Wong	Academia Sinica, Taiwan
Kurt	Wüthrich	ETH Zürich, Switzerland; The Scripps Research Institute, USA
Lode	Wyns	Vrije Universiteit Brussel, Belgium

Auditors

Antoine	Amory	Solvay Group
Kristin	Bartik	Université Libre de Bruxelles, Belgium
Gilles	Bruylants	Université Libre de Bruxelles, Belgium
Nico	Callewaert	Universiteit Gent, Belgium
Jean-François	Collet	Université Libre de Bruxelles, Belgium
Stéphanie	Deroo	Université Libre de Bruxelles, Belgium
Anne	De Wit	Université Libre de Bruxelles, Belgium
Jean-Marie	Frère	Université de Liège, Belgium
Erik	Goormaghtigh	Université Libre de Bruxelles, Belgium
Cédric	Govaerts	Université Libre de Bruxelles, Belgium
Fabrice	Homble	Université Libre de Bruxelles, Belgium
Denis	Lafontaine	Université Libre de Bruxelles, Belgium
Anne	Lesage	GrayMatters Consulting, Belgium
Remy	Loris	Vrije Universiteit Brussel, Belgium
Philippe	Marion	Solvay Group
Serge	Muyldermans	Vrije Universiteit Brussel, Belgium
Marc	Parmentier	Université Libre de Bruxelles, Belgium
Vincent	Raussens	Université Libre de Bruxelles, Belgium
Han	Remaut	Vrije Universiteit Brussel, Belgium
Jean-Marie	Ruysschaert	Université Libre de Bruxelles, Belgium
Michel	Vandenbranden	Université Libre de Bruxelles, Belgium
Wim	Versées	Vrije Universiteit Brussel, Belgium

Opening Address by Professor Marc Henneaux

Dear colleagues, dear friends,

In the name of the International Solvay Institutes, it is my great pleasure to welcome all of you to the 23rd Solvay Conference on Chemistry, "New Chemistry and New Opportunities from the Expanding Protein Universe".

The Conference that starts today is part of a long and prestigious tradition and 2013 is a year that marks the anniversary of many important dates in the history of the Solvay Institutes.

The Institute of Chemistry was created in 1913 by Ernest Solvay. 2013 is therefore the year of its hundredth anniversary. [The Institute of Chemistry fused in 1970 with the Institute of Physics to become *The International Solvay Institutes for Physics and Chemistry.*]

We are celebrating also another anniversary, the 60th anniversary of the 9th Solvay Conference on Chemistry that took place in Brussels in 1953. This conference was quite remarkable because it is during this conference, on the 9th of April, that the original announcement of the double helix structure of DNA deduced by Watson and Crick, was made, prior to the publication of their paper in Nature on April 25, 1953.

The announcement was made by Bragg, who was rapporteur at the conference, and the director of the Cavendish Laboratory in Cambridge where Crick and Watson worked. Bragg had actually very closed ties with the Solvay Institutes since he was then the chair of the International Committee for Physics, in charge of the Solvay Conferences on Physics. The report by Bragg in the Proceedings contains a note by Watson and Crick, who were clearly aware of the importance of their discovery.

We are also celebrating the 30th anniversary of another important conference, the 18th Solvay Conference on Chemistry "Design and Synthesis of Organic Molecules Based on Molecular Recognition". In that conference, topics related to this year's conference were discussed. And indeed at least two of this year's participants were already participants in 1983. This was an excellent conference, and a significant fraction of the participants got the Nobel prize in the meantime!

If I have chosen to mention these two conferences, that of 1953 and that of 1983, it is, besides the fact that they both occurred a multiple of ten years ago, because proteins played a central role in them. Proteins are again the stars of this year's conference. This is not a topic that I know well — I am a physicist — but while reading the Proceedings of the previous conferences on the subject and preparing my speech, I have been impressed and amazed by the beauty and intricacy of the discipline. I am convinced that the research carried in the last sixty years has not only brought an enormous, overwhelming understanding of the growing protein universe, which might be exponentially expanding if I understand well, but also has led to new questions, new puzzles and new challenges which are as fascinating and intriguing as the challenges that existed in the fifties.

As you know, the Solvay Conferences are very special. These are elitist conferences by invitation-only, with a limited number of participants. There are few presentations but a lot of discussions. People come to the Solvay Conferences for the scientific interactions, which are indeed privileged, not for giving a talk. The format adopted this year to implement this philosophy and to initiate the discussions is an innovation with respect to the original format where each session started with one or two rapporteur talks. The rapporteur talks have been replaced by 5 to 6 short introductory statements.

For the discussions to be fruitful, a careful preparation is needed. Here is how it goes: the subject and scientific organizers of the Solvay Conferences are chosen by the Solvay International Scientific Committee for Chemistry, which has complete freedom in doing so. The organizers are then in charge of the invitations and the program, and again have complete "carte blanche" for achieving this task.

I would like to express our deepest thanks to the Solvay Scientific Committee for Chemistry, and in particular to its chair, Kurt Wüthrich, for choosing the subject of the 23rd Solvay Conference on Chemistry. I would also like to express our gratitude to Ian Wilson, Donald Hilvert and Kurt Wüthrich, for accepting to organize the Conference. I know that this means an enormous amount of work.

Since the discussions are important, they are included in the Proceedings. This is the tradition. We have an editorial committee that already remarkably worked since all the contributions have been received before the meeting. This is to be praised and I would like to thank the editorial committee, and in particular Anne De Wit and Dennis Wolan, for carrying the publication task. They are assisted for the transcriptions of the discussions by a scientific secretariat to whom we are also very grateful.

Thank you very much for your attention. I wish you a very fruitful meeting.

Marc Henneaux,
Director of the Solvay Institutes

Preface

On the occasion of the 150th anniversary of the Solvay Company and the centennial anniversary of the first Chemistry Council in 1913, the November/December 2013 issue of *Chemistry International* devoted its cover to "A Look Back at Ernest Solvay", and pages 4–11 to this same theme and to "The Solvay Chemistry Councils". It is thus highly fitting that a Solvay Conference on Chemistry was held in the same year, with the theme "New Chemistry and New Opportunities from the Expanding Protein Universe". This 23rd Solvay Conference on Chemistry assembled an illustrious group of invited scientists representing a diverse range of research disciplines, including chemistry, chemical biology, computational biology, structural biology, biochemistry, and cell biology. All of the attendees are at the forefront of their fields and have advanced our knowledge of the structure and function of biological macromolecules and their supramolecular assemblies, the chemistry of essential biological processes, and the development of new small-molecule drugs and biologicals.

The theme of this Solvay Conference was inspired by explorations into the expanding protein universe, using the tools of proteomics and structural genomics that have opened up new frontiers and opportunities for scientific investigation. The program was presented in six Sessions, and the Proceedings are organized accordingly. Session 1 on "New Chemistry in the Expanding Protein Universe" was devoted to innovative reports on discovering new chemistry associated with exploration of today's expanded protein space. Session 2 on "Exploring Enzyme Families and Enzyme Catalysis" was focused on recently identified families of enzymes and on studies of enzyme mechanisms. Session 3 on "Microbiomes and Carbohydrate Chemistry" was devoted on the one hand to the microbiota that are adapted to their particular environment in the human gut, and consequently have enhanced networks for carbohydrate metabolism, and on the other hand to recent advances in carbohydrate synthetic chemistry. Session 4 on "GPCRs and Transporters: Ligands, Cofactors, Drug Development" was dominated by exciting new results on GPCRs and membrane transporters from structural biology and biophysics, with a strong emphasis on biomedical applications. Session 5 on "Biologicals and Biosimilars" provided a perspective on the current state of development of protein-based drugs. The final Session 6 on "Proteins in Supramolecular Machines" focused on the impressive recent developments in understanding and dissecting complex macromolecular assemblies with the use of structural biology. The six Sessions thus

provided a spectacular overview of the present state of protein science and an outlook to anticipated developments during the coming years and in the more distant
future.

Considering the wide scope of present-day protein science, we decided on a
novel format for this Conference. Instead of having one "Rapporteur" providing a
one-hour introduction to each Session, we scheduled short "Personal Statements"
by the Session Chairperson and six to eight Panel Members at the outset of each
Session. These presentations segued into a discussion among the designated Panel
Members, which initiated lively exchanges with the Plenum. The Proceedings are
organized in such a way that for each Session the short "Personal Statements",
which had been submitted prior to the start of the Conference, are followed by
the transcription of the ensuing discussion. This Conference format was designed
to promote thought-provoking questions and answers that are not usually found
in formal presentations. The Conference participants were great in accepting this
rather unusual format. As a result, the combination of the "Personal Statements"
and the discussion transcriptions for each Session now present unique snapshots of
present-day protein science and outlooks to the future.

Protein science presented an ideal foundation for the organization of the "First
Solvay Roundtable on the State of Science & Present Challenges: Biologicals and
Biosimilars" on one of the afternoons of the Conference week. The Roundtable
assembled a group of industry leaders with Conference participants taken primarily
from among the speakers in Session 5. The following format was proposed by Jean-
Marie Solvay: "The aim of the roundtable is to enable an informal discussion of the
state-of-the-art on biologicals and biosimilars, making it possible for outstanding
scientists participating in the 23rd Solvay Conference on Chemistry to share their
views with interested business leaders". The envisaged aim was achieved due to the
mutual interest in the theme for the Roundtable, giving rise to intense discussions
during the afternoon and thereafter over dinner.

The culmination of the Conference week was the Public Event held by the International Solvay Institutes at the Flagey Cultural Center. The well-attended event
began with the Solvay Award Ceremony for young scientists working at local uni-
versities in the fields of physics, chemistry and engineering. This was followed by
lectures of Professors Joachim Frank (Columbia University, USA) and Jason Chin
(University of Cambridge, UK) on their current research. The afternoon concluded
with a debate prompted by a wide range of excellent questions posed by the audience to a panel of Conference attendees. It was a pleasure for me to chair this panel
of Professors Jason Chin (MRC, Cambridge, UK), Joachim Frank (Columbia University, USA), Gunnar von Heijne (Stockholm University, Sweden), Donald Hilvert
(ETH Zürich, Switzerland), Christopher Walsh (Harvard Medical School, USA) and
Ian A. Wilson (The Scripps Research Institute, USA). I would like to thank Prof.
Franklin Lambert for his role as bilingual moderator, and the Solvay Institutes for
hosting this well-received and successful event of public outreach.

It was a privilege for me to chair this Solvay Conference on Chemistry, which gave me and my colleagues, Prof. Donald Hilvert and Prof. Ian A. Wilson, the opportunity to select a stellar group of scientists from different disciplines for a truly integrative discussion on timely subjects. We greatly appreciate the generous support provided by the Solvay Institutes. I was impressed by the commitment of the Solvay family, with Jean-Marie Solvay participating in the scientific part of the Conference and being joined by Mrs. Solvay at the social functions. We want to thank the Solvay Institutes represented by Prof. Marc Henneaux. Special thanks go to Dominique Bogaerts and Isabelle Van Geet from the Solvay Institutes, for their commitment and dedicated assistance that made the logistical running of the conference perfect, to Prof. Anne De Wit, who arranged the recording and transcription of the discussions during the meeting, and to the local scientists who aided with the transcriptions.

Kurt Wüthrich
Chair of the Conference

Session 1

New Chemistry in the Expanding Protein Universe

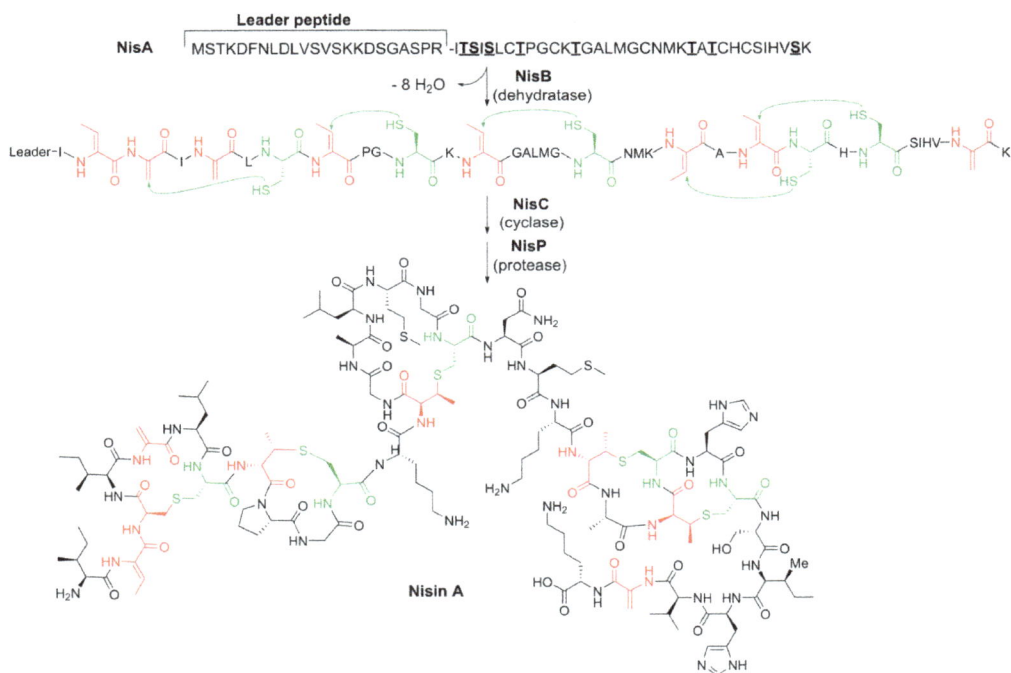

NisA

Leader peptide

MSTKDFNLDLVSVSKKDSGASPR -ITSISLCTPGCKTGALMGCNMKTATCHCSIHVSK

- 8 H₂O NisB (dehydratase)

NisC (cyclase)

NisP (protease)

Nisin A

Biosynthesis of nisin A, a lantibiotic that has been used for nearly 50 years in the food industry to combat food-borne pathogens. See Figure 2 contributed by Wilfred van der Donk on page 8.

NOVEL CHEMISTRY STILL TO BE FOUND IN NATURE

CHRISTOPHER T. WALSH

Biochemistry & Molecular Pharmacology Department
Harvard Medical School, Boston, MA 02115, USA

My view of the present state of research on new chemistry in the expanding protein universe

New chemical transformations catalyzed by proteins continue to be discovered. Most of the novel scaffolds in small molecule frameworks emerge from studies on microbes, prokaryotes and single cell eukaryotes, from underexplored niches with both anaerobic and aerobic (oxidative) metabolic transformations prominent in bond-breaking and bond-making steps. Bioinformatic analysis of protein superfamilies may highlight particular family members for novel catalytic activities. Post-translational modification of ribosomally generated proteins has also been implicated in the morphing of peptide frameworks into complex architectures. *De novo* design and protein evolution activity can also create novel chemical transformations, including reactions not previously seen in Nature.

My recent contributions to new chemistry in the expanding protein universe

Research from my group has focused on the morphing and maturation of peptide scaffolds into rigidified, compact scaffolds by the two major biosynthetic strategies for peptide bond formation: ribosomal and non-ribosomal assembly lines. A hallmark of the RNA-independent nonribosomal peptide synthetases has been the use of nonproteinogenic amino acid building blocks in place of the 20 canonical proteinogenic amino acids. As an example, we have examined how a subset of mononuclear nonheme iron oxygenases act instead as halogenating catalysts to provide γ-chloro amino acid and cyclopropyl amino acid building blocks to NRPS assembly lines. The nonproteinogenic β-amino acid anthranilate is a building block for a series of fungal alkaloids ranging from bicyclic to octacyclic scaffolds, put together by bi- and trimodular NRPS assembly lines, followed by action of dedicated tailoring enzymes.

Complexity generation in peptide scaffolds can also be achieved from ribosomally generated nascent proteins by a series of post-translational modifications (PTMs). A remarkable cascade of more than a dozen PTMs occur as a 14 residue C-terminal peptide region of a 52mer is morphed into the trithiazolylpyridine core of thiazolyl peptide antibiotics in the thiocillin and nosiheptide class of antibiotics. These involve conversion of cysteines to thiazoles, threonines to methyloxazoles and partitioning of serine residues down two distinct PTM pathways. One route is to the corresponding oxazole by cyclization, dehydration, and aromatization; the other route is net dehydration to dehydroalanines. A pair of dehydroalanines can undergo condensation and dehydrative aromatization to yield the core pyridine ring of the 2,4,6-trithiazolyl pyridine core at the center of these mature antibiotic scaffolds.

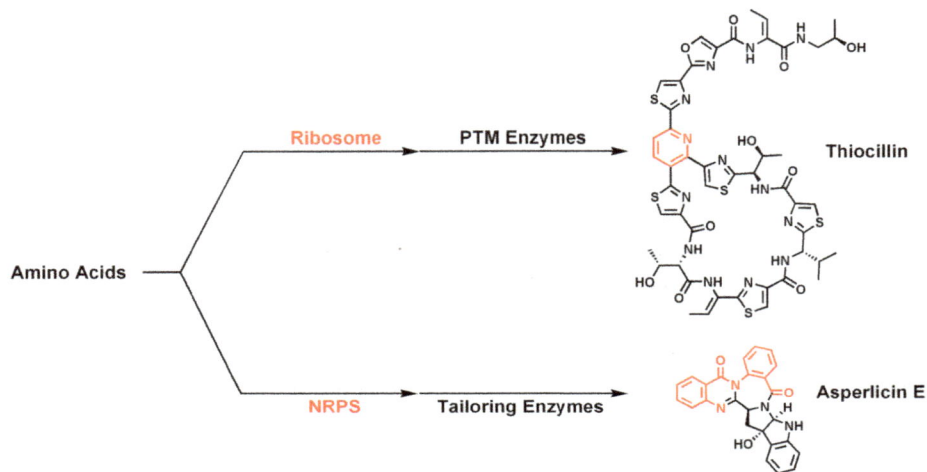

Fig. 1. Complexity generation in peptide scaffolds: two strategies.

Outlook to future developments of research in the chemistry of the expanding protein universe

As bioinformatics, structural genomics and proteomics continue to define the existing protein universe and guide protein engineers in search of new kinds of catalysts, the future is bright for novel chemistry to continue to emerge. On one end of the catalytic spectrum the large superfamily of radical S-adenosyl methionine (SAM) enzymes holds particular promise for catalytic diversity and novelty of chemical transformations: the reactions catalyzed by such famly members as riibonucleotide reductase, lipoyl and biotin synthases, ThiiC, and the tRNA and mRNA C-methyltransferases are likely to be the tip of the iceberg in the chemical capacity of this superfamily. In a distinct superfamily, the hemeprotein cyochrome P450 oxygenases, protein engineering has recently led to evolution of synthetically useful carbene chemistry. *De novo* protein design should enable completely abiotic chemistry to move into the realm of protein catalysis.

References

1. C. T. Walsh, R. O'Brien, C. Khosla, *Angew. Chem. Int. Ed.* **52**, 7098 (2013).

2. C. T. Walsh, S. Haynes, B. Ames, X. Gao, Y. Tang, *ACS Chem. Biol.* **8**, 1366 (2013).
3. C. T. Walsh, S. Malcolmson, T. Young, *ACS Chem. Biol.* **7**, 429 (2012).
4. P. S. Coelho, J. Wang, M. E. Ener, S. A. Baril, A. Kannan *et al.*, *Nat. Chem. Biol.* **9**, 485 (2013).

NATURAL PRODUCT BIOSYNTHESIS IN THE GENOMIC AGE

WILFRED A. VAN DER DONK

Department of Chemistry, University of Illinois at Urbana-Champaign and the Howard Hughes Medical Institute, 600 South Mathews Ave, Urbana, IL 61801, USA

My view of the present state of research on new chemistry in the expanding protein universe

Natural products (NPs) have featured prominently in the development of pharmaceuticals and as tools to study biology. However, since the turn of the century, many natural product discovery platforms in industry have been dissembled. Several explanations have been given for the withdrawal of pharmaceutical companies from NP discovery. For antibiotics, small projected profits have been a major driving force. But when considering *e.g.*, antitumor agents, the move away from NPs has been motivated by other factors including high rediscovery rates of known compounds and the difficulty to perform medicinal chemistry because of complex structures. At the same time, the available sequenced genomes have demonstrated that the number of NP biosynthetic gene clusters in a typical microorganism far exceeds the number of compounds it produces under laboratory conditions. Based on their sequences, the overwhelming majority of these "silent" gene clusters are expected to encode new NPs. As a result, genome mining has been widely touted as a potentially efficient route to new NPs. An important bonus of investigating NP biosynthetic pathways with respect to the theme of this panel is their richness in novel biochemical transformations.

My recent research contributions to new chemistry in the expanding protein universe

The genome sequencing efforts have revealed that ribosomally synthesized and post-translationally modified peptides (RiPPs) form a much larger class of NPs than anticipated [1]. The extensive post-translational modifications endow these peptides with greatly expanded chemical structures, restricted conformational flexibility for target recognition, and increased metabolic stability. In retrospect, it may not be surprising that RiPP biosynthetic pathways are so widespread because they offer a unique advantage to their producing organisms: high evolvability.

Nearly all RiPPs are initially synthesized as a longer precursor peptide, with a leader peptide appended to the N-terminus of the core peptide, which will be converted to the final NP (Fig. 1(a)). The leader peptide is important for recognition by many of the post-translational modification enzymes [2]. This leader peptide-guided strategy results in highly evolvable pathways because the post-translational processing enzymes can be intrinsically permissive with respect to mutations in the core peptide. The leader peptide-guided biosynthesis also renders RiPPs particularly well-suited to both genome mining and synthetic biology approaches because their substrates are DNA encoded and the biosynthetic enzymes have intrinsically relaxed substrate specificity [3].

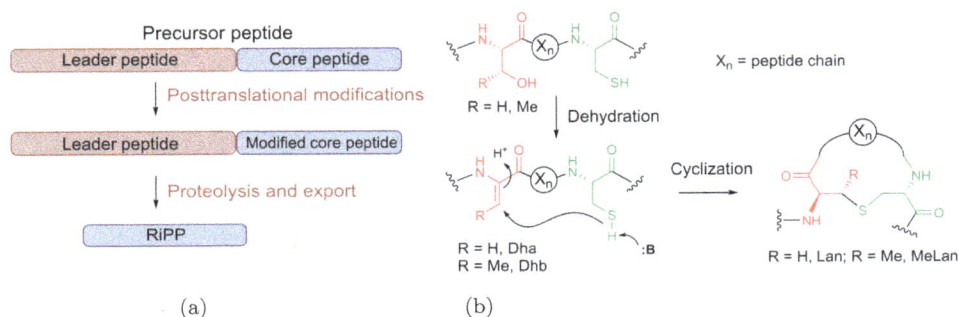

Fig. 1.   Biosynthesis of RiPPs. (a) General biosynthetic pathway of RiPPs. (b) General biosynthetic strategy that results in the formation of thioether corsslinks named lanthionine and methyllanthionine.

Based on the available genomes, lanthionine-containing peptides (lanthipeptides) are the most abundant class of RiPPs. Currently known lanthipeptides have a range of activities including antimicrobial (called lantibiotics), morphogenetic, antiviral, and antiallodynic [4]. Lanthionine (Lan) consists of two alanine residues crosslinked via a thioether linkage that connects their β-carbons; 3-methyllanthionine (MeLan) contains one additional methyl group (Fig. 1(b)). These structures are installed by a dehydratase that eliminates water from Ser and Thr residues to generate dehydroalanine (Dha) and dehydrobutyrine (Dhb), respectively, and a cyclase that catalyzes the subsequent addition of thiols of Cys residues to the dehydro amino acids to generate Lan and MeLan, respectively. How the enzymes coordinate these complex chemical transformations in which the substrate peptide structure is continuously changing is unknown (*e.g.*, Fig. 2). Furthermore, most lanthipeptide biosynthetic enzymes do not have homology with non-lanthipeptide proteins, and hence many questions remain about their evolutionary origin.

At least four different pathways to these polycyclic natural products have evolved [4], reflecting the high efficiency and evolvability of a post-translational modification route to generate conformationally constrained peptides. Recent studies have shown that three of the four pathways involve Ser/Thr phosphorylation and subsequence phosphate elimination to generate the dehydro amino acids [6–8]. The fourth route unexpectedly involves Ser/Thr glutamylation [9]. As can be seen in Fig. 2, the dehydratase and cyclase enzymes involved in lanthipeptide biosynthesis

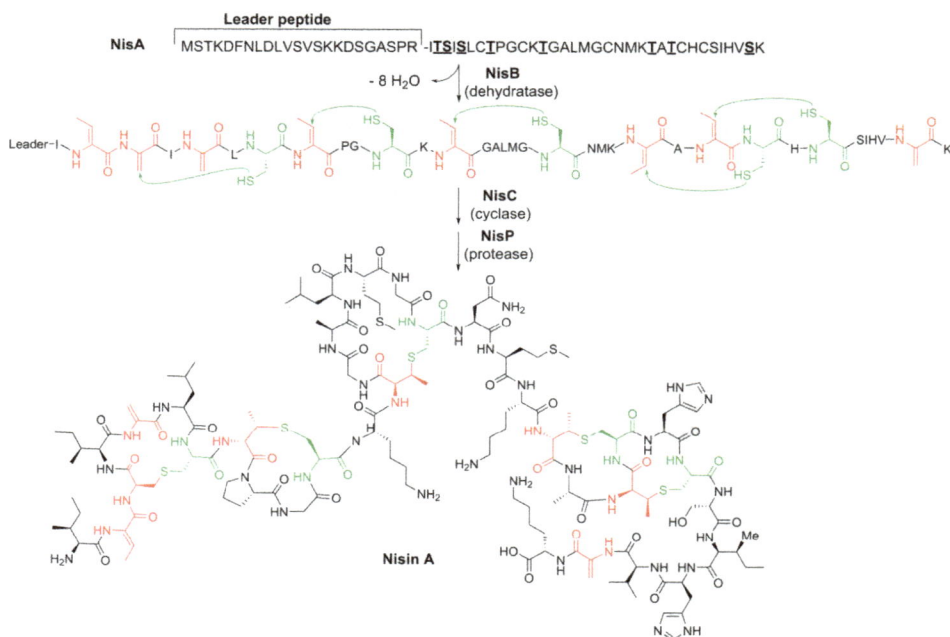

Fig. 2. Biosynthesis of nisin A, a lantibiotic that has been used for nearly 50 years in the food industry to combat food-borne pathogens. Note the challenge for the cyclase to regioselectively connect five Cys nucleophiles with the correct dehydro amino acids, a process that can generate more than 6,500 different constitutional isomers [5]. In this process, five rings that differ greatly in size and amino acid sequence are formed by one enzyme.

act on residues that are located in diverse sequence contexts. These lanthionine synthetases appear to have retained the low substrate specificity of their primitive progenitors but acquired a dependence on a leader peptide. The exact role of the leader peptide has been widely debated, and no definitive answer has been provided thus far. Our studies suggest that the leader peptide plays an allosteric role [5], but other roles cannot be ruled out and it is possible that the leader peptides have different functions in different classes of RiPPs.

A recent genome mining exercise revealed a stunning example of natural combinatorial biosynthesis, illustrating the considerable potential of RiPP biosynthesis for synthetic biology. The genome of a strain of *Prochlorococcus*, a planktonic marine photosynthetic cyanobacterium, encodes a single class II lanthionine synthetase (ProcM) but no less than 29 different putative substrate peptides. These substrates have highly conserved leader peptides but hypervariable core peptides (not a single homologous pair). We cloned the enzyme and a subset of its putative substrates and showed that all 17 peptides tested were indeed substrates for ProcM [10]. In a follow-up study, we determined the structure of a subset of the resulting compounds termed prochlorosins (Pcn), and demonstrated that their ring topology is highly diverse [11]. These findings open up a large number of intriguing questions, including (1) how can one enzyme make 29 very different polycyclic structures, (2) what is the function of these cyclic peptides, and (3) can the ProcM-activity be harnessed for synthetic purposes?

Outlook to future developments of research on new chemistry in the expanding protein universe

Genome mining efforts are highly likely to uncover many new NPs. However, genome mining approaches have their own challenges. Most importantly, unlike phenotypic screens that provide compounds with a desired activity, tools to predict the type of bioactivity of a putative new compound identified by genome mining are lacking. Hence, for genome mining to be a viable approach to new compounds with desirable activities, new predictive tools need to be developed and/or high-throughput production platforms must be engineered such that the odds of finding molecules with desired activities are increased. Efficient production methods could either be based on production in heterologous hosts or on new strategies to elicit production from the original organisms [12]. RiPPs lend themselves particularly well for heterologous expression as the 20 amino acids are common building blocks that are present in any heterologous hosts and because their substrates are gene encoded. In addition, synthetic biology approaches may take advantage of the high substrate tolerance of RiPP biosynthetic enzymes to generate cyclic peptide libraries that can be selected for specific properties.

Acknowledgments

This research was supported by the National Institutes of Health (RO1 GM 58822) and the Howard Hughes Medical Institute.

References

1. P. G. Arnison, M. J. Bibb, G. Bierbaum, A. A. Bowers, T. S. Bugni *et al.*, *Nat. Prod. Rep.* **30**, 108 (2013).
2. T. J. Oman, W. A. van der Donk, *Nat. Chem. Biol.* **6**, 9 (2010).
3. J. E. Velásquez, W. A. van der Donk, *Curr. Opin. Chem. Biol.* **15**, 11 (2011).
4. P. J. Knerr, W. A. van der Donk, *Annu. Rev. Biochem.* **81**, 479 (2012).
5. X. Yang, W. A. van der Donk, *Chem. Eur. J.* **19**, 7662 (2013).
6. C. Chatterjee, L. M. Miller, Y. L. Leung, L. Xie, M. Yi *et al.*, *J. Am. Chem. Soc.* **127**, 15332 (2005).
7. W. M. Müller, T. Schmiederer, P. Ensle, R. D. Süssmuth, *Angew. Chem. Int. Ed.* **49**, 2436 (2010).
8. Y. Goto, B. Li, J. Claesen, Y. Shi, M. J. Bibb *et al.*, *PLoS Biol.* **8**, e1000339 (2010).
9. N. Garg, L. M. Salazar-Ocampo, W. A. van der Donk, *Proc. Natl. Acad. Sci. USA* **110**, 7258 (2013).

10. B. Li, D. Sher, L. Kelly, Y. Shi, K. Huang *et al.*, *Proc. Natl. Acad. Sci. USA* **107**, 10430 (2010).
11. W. Tang, W. A. van der Donk, *Biochemistry* **51**, 4271 (2012).
12. K. Scherlach, C. Hertweck, *Org. Biomol. Chem.* **7**, 1753 (2009).

PEPTIDE DENDRIMERS AND POLYCYCLIC PEPTIDES

JEAN-LOUIS REYMOND

Department of Chemistry and Biochemistry, University of Bern, Freiestrasse 3
3012-Bern, Switzerland

My view of the present state of research on new chemistry in the expanding protein universe

Proteins are classically understood as the products of genes as found in Nature. Advances in protein engineering such as mutagenesis, directed evolution, and non-natural amino acid mutagenesis have allowed to not only better understand how proteins work, but also to produce new proteins with new functions. The bulk of these advances were made possible by the techniques of molecular biology. Synthetic chemistry follows closely behind in exploring the protein universe using solid-phase peptide synthesis (SPPS) [1] and native chemical ligation (NCL) [2], and contributes by allowing the chemical synthesis of proteins as well as the introduction of targeted modifications that cannot be produced in a cell.

My recent research contributions to new chemistry in the expanding protein universe

Our approach to *de novo* protein design focuses on rethinking peptide topologies from first principles, intentionally departing from Nature's models in search for new protein-like macromolecules with unexpected properties. In the hierarchical organization of organic matter, atoms assemble in topologically diverse patterns of covalent bonds (graphs) to form millions of different molecules [3]. At the next level amino acids assemble through peptide bonds to form peptides and proteins, however in this case topological diversity is limited almost exclusively to linear peptide chains as encoded in genes and translated by the ribosome. Proteins acquire their diverse shapes and functions not through topologically diverse peptide bond networks but rather by folding. One advantage of chemical synthesis is that one is not limited to linear chains when assembling a polypeptide, and one can also access different topologies by introducing multiple branching points in form of diamino acids or amino diacids. Capitalizing on this additional degree of freedom, we have developed SPPS protocols to prepare artificial proteins with non-natural topologies.

In one such approach inspired by the field of organic dendrimers [4], we are investigating peptide dendrimers [5] as artificial proteins [6]. Peptide dendrimers are symmetrically multi-branched peptides not accessible by biosynthetic routes. We

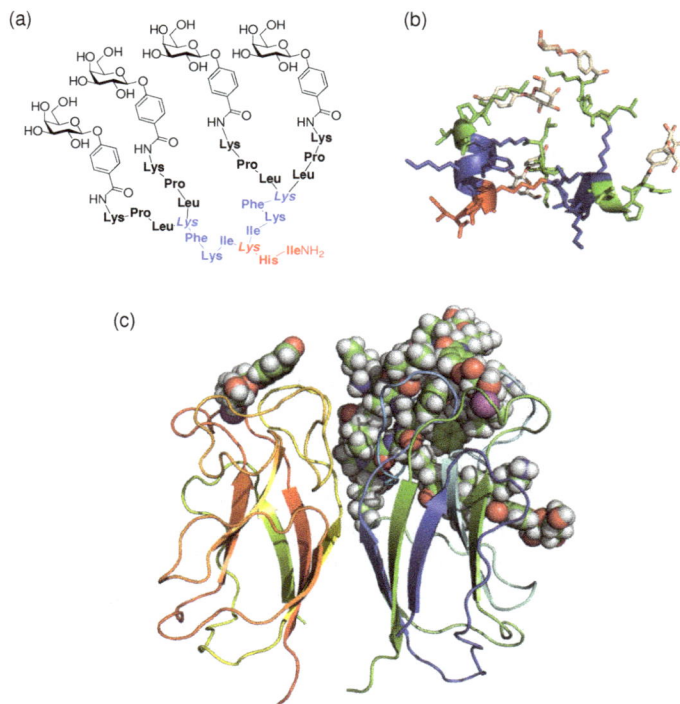

Fig. 1.   The glycopeptide dendrimer *P. aeruginosa* biofilm inhibitor **GalAG2** is shown in (a) chemical structure, (b) MD minimized model and (c) modelled complex with its target LecA.

install branching points every 2nd, 3rd or 4th amino acid along the growing peptide chain in form of diamino-acids (lysine or 2,3-diaminopropanoic acid) resulting in G1 (one branching point, two terminal branches), G2 (three branching points, four terminal branches), or G3 (seven branching points, eight terminal branches) peptide dendrimers with variable amino acid positions throughout the dendrimer branches. This divergent dendrimer SPPS delivers products in excellent yields and purities for sequences of up to approximately fourty residues in total, corresponding to twelve consecutive coupling steps. The method is amenable to combinatorial split-and-mix synthesis to produce libraries of thousands of different dendrimers which can be screened for function by on-bead or off-bead assays [7, 8]. We have discovered peptide dendrimers with a variety of functions including enzyme-like catalysis (esterase [9], aldolase [10], cobalamin binding proteins [11], peroxydase [12]), antimicrobial activity [8], and biofilm inhibition using glycopeptide dendrimers lectin inhibitors (Fig. 1) [13]. We also found that certain polycationic/lipophilic peptide dendrimers acts as highly efficient yet non-toxic gene transfer reagents [14], while others can deliver small molecule cargos such as cytotoxic agents into cells [15, 16].

One remarkable feature of peptide dendrimers is their increased proteolytic stability compared to linear peptides, probably caused by steric crowding around the peptide bonds [17]. Contrary to linear peptides which are frequently prone to aggregation, peptide dendrimers are generally well behaved in solution and do not aggregate. Structural investigations by CD, FTIR and NMR spectroscopy and by molecular dynamics simulations show that peptide dendrimers are conformationally flexible and adopt a globular molten globule like structure with relatively few peptide backbone H-bonds and consequently few secondary structure elements, although α-helical peptide dendrimers can be obtained by design [18, 19].

The direct dendrimer SPPS yields pure products up to G3 dendrimers and ca. 5 kDal in size. We have explored chemical ligation methods to prepare higher generation, protein sized dendrimers through convergent assembly of G2 and G3 dendrimers. While NCL and click reactions were unsuccessful in our hands, protein-sized dendrimers were obtained by the coupling of multiple copies of a dendrimer carrying a cysteine residue at its core with another dendrimer substituted with four or eight chloroacetyl groups at its N-termini. This so-called ClAc (Chloroacetyl thioether) ligation allowed us to prepare peptide dendrimers up to 340 amino acids characterized as enzyme models [10].

In a comparable approach to artificial proteins, we are investigating the synthesis and properties of topologically diverse polycyclic peptides designed as peptide analogs of saturated alkane graphs [20]. We have established an SPPS protocol comprising on-bead cyclisation, purification of the cyclic peptide, and closing of the second ring in solution and used the approach to prepare bicyclic peptide analogs of "non-zero bridged" bicyclic alkanes such as norbornane (Fig. 2).

Bicyclic peptides define a very large structural family. For instance, there are 25 possible bicyclic peptide scaffolds up to 10 amino acids forming $3.3.10^{12}$ possible products using 20 different amino acids at the variable non-bridgehead positions.

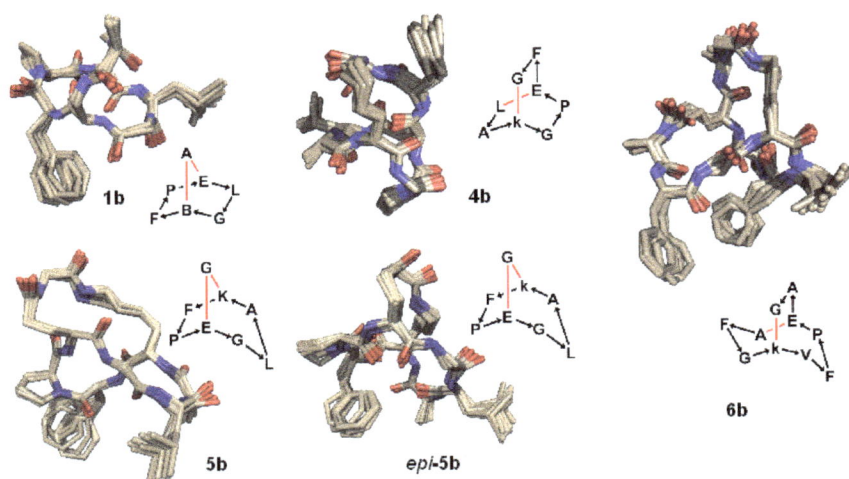

Fig. 2.   Experimental MR structures of selected bicyclic peptides.

Analysis of molecular shape shows that bicyclic peptides populate a distinct area of peptide shape space. Thus, while linear peptides in extended or β-sheet conformation are rod-like and monocyclic peptides are disk-like, "non-zero bridged" bicyclic peptides have a significant third molecular dimension indicative of a globular shape, and "zero-bridged" bicyclic peptides are elliptical. The shapes of bicyclic peptides resemble those of short α-helical segments extracted from proteins in the PDB. Bicyclic peptides are conformationally rigid and thus overcome the major challenge encounterd sofar in our peptide dendrimer work. Bicyclic peptide have the potential for selective protein binding, and are one of the current focus of our research.

Outlook to future developments of research on new chemistry in the expanding protein universe

The chemical synthesis of peptides by SPPS is sufficiently versatile and practical to allow the rapid exploration of functional sequence space, in particular through the use of parallel or combinatorial synthesis. We are exploiting this technique to explore non-natural topologies defining entirely new classes of macromolecules, yet with the same building block composition as proteins and therefore a particular propensity for biocompatibility. In the near future we are particularly interested to integrate computational modelling for the design of peptide dendrimers and polycyclic peptides, and thus replace untrained combinatorial screening with educated guesses and the synthesis of focused series. If modelling delivers on its promise this approach should allow a more focused exploration of protein sequence and topological space outside of what Nature has realized sofar.

Acknowledgments

This work was supported financially by the University of Berne, the Swiss National Science Foundation, and the EU ITN BioChemLig.

References

1. B. Merrifield, *Meth. Enzymol.* **289**, 3 (1997).
2. S. B. Kent, *Chem. Soc. Rev.* **38**, 338 (2009).
3. L. Ruddigkeit, R. Van Deursen, L. C. Blum, J. L. Reymond, *J. Chem. Inf. Model.* **52**, 2864 (2012).
4. D. Astruc, E. Boisselier, C. Ornelas, *Chem. Rev.* **110**, 1857 (2010).
5. L. Crespo, G. Sanclimens, M. Pons, E. Giralt, M. Royo *et al.*, *Chem. Rev.* **105**, 1663 (2005).
6. J.-L. Reymond, T. Darbre, *Org. Biomol. Chem.* **10**, 1483 (2012).
7. N. Maillard, A. Clouet, T. Darbre, J. L. Reymond, *Nat. Protoc.* **4**, 132 (2009).
8. M. Stach, N. Maillard, R. U. Kadam, D. Kalbermatter, M. Meury *et al.*, *MedChemComm.* **3**, 86 (2012).

9. R. Biswas, N. Maillard, J. Kofoed, J. L. Reymond, *Chem. Commun.* **46**, 8746 (2010).

10. N. A. Uhlich, T. Darbre, J.-L. Reymond, *Org. Biomol. Chem.* **9**, 7071 (2011).

11. N. A. Uhlich, A. Natalello, R. U. Kadam, S. M. Doglia, J. L. Reymond *et al.*, *ChemBioChem.* **11**, 358 (2010).

12. P. Geotti-Bianchini, T. Darbre, J. L. Reymond, *Org. Biomol. Chem.* **11**, 344 (2013).

13. J. L. Reymond, M. Bergmann, T. Darbre, *Chem. Soc. Rev.* **42**, 4814 (2013).

14. A. Kwok, G. A. Eggimann, J. L. Reymond, T. Darbre, F. Hollfelder, *ACS Nano* **7**, 4668 (2013).

15. E. M. Johansson, J. Dubois, T. Darbre, J. L. Reymond, *Bioorg. Med. Chem.* **18**, 6589 (2010).

16. G. A. Eggimann, S. Buschor, T. Darbre, J. L. Reymond, *Org. Biomol. Chem.* **11**, 6717 (2013).

17. P. Sommer, V. S. Fluxa, T. Darbre, J. L. Reymond, *ChemBioChem.* **10**, 1527 (2009).

18. S. Javor, A. Natalello, S. M. Doglia, J. L. Reymond, *J. Amer. Chem. Soc.* **130**, 17248 (2008).

19. H. K. Ravi, M. Stach, T. A. Soares, T. Darbre, J. L. Reymond *et al.*, *Chem. Commun.* **49**, 8821 (2013).

20. M. Bartoloni, R. U. Kadam, J. Schwartz, J. Furrer, T. Darbre *et al.*, *Chem. Commun.* **47**, 12634 (2011).

WHAT CAN COMPARATIVE GENOMICS REVEAL ABOUT THE MECHANISMS OF PROTEIN FUNCTION EVOLUTION?

NATALIE L. DAWSON*, ROMAIN STUDER*, NICK FURNHAM†, DAVID LEES*,
SAYONI DAS*, JANET THORNTON† and CHRISTINE ORENGO*

*University College London, London, UK
†European Bioinformatics Institute, Hinxton, UK

My view of new chemistry in the expanding protein universe

Over the last decade, international genomics initiatives have a brought a flood of new protein sequence and structural data, increasing our knowledge of the protein family and functional repertoire across the tree of life, particularly for microbial organisms. In parallel with major developments in sequencing technologies, the worldwide structural genomics initiatives have brought improved strategies for solving the 3D structures of proteins and have expanded our knowledge of the universe of protein folds [1]. Whilst this data is still much more sparse than the sequence data, methods for mapping sequences to known structures have advanced and have facilitated the comparison of functional sites between related proteins. Currently, there are over a thousand evolutionary protein enzyme families characterized both structurally and by sequence, of which ∼ 100 have been very widely exploited during evolution [2]. Since the 3D locations of the active sites are frequently conserved across a family, by comparing the differences in physico-chemical characteristics of residues in the active sites of relatives it is possible to recognize those which have evolved novel chemistries. Metagenomic studies are revealing great diversity in the chemical repertoires of bacteria which should help in understanding enzyme mechanisms and designing novel catalysts.

Recent research contributions to new protein chemistry

Some domain superfamilies are very ancient and highly populated [3]. Duplication and divergence in these superfamilies has played a major role in expanding the functional repertoires of species. My group has been analysing, computationally, the ways in which divergence in sequence and structure can modify protein function.

Identifying functionally distinct relatives within a protein domain superfamily

Identifying functionally distinct homologues from the sequence data is a challenging computational problem as extensive residue insertions between very diverse relatives means that it is often hard to align all sequences in the robust manner needed to derive a phylogenetic tree. FunFHMMer which classifies functional families in our CATH classification [4, 5] uses sensitive profile-profile based methods to detect conserved sequence patterns highly likely to be shared by functionally similar homologues. Classification of 2,500 completed genomes and UniProt (> 25 million sequences in total), gives $\sim 50,000$ different functional families, 30,000 of which have experimentally characterised functions.

Using structural data to derive phylogenetic trees that reveal the evolution of new functions in superfamilies

Because the 3D structures of proteins are much more conserved than their sequences, structural data can guide the alignment of relatives to generate phylogenetic trees. Together with Janet Thornton at the EBI, we are using these trees to study the emergence of new functions in enzyme superamilies. The FunTree resource developed by Nick Furnham [6] integrates phylogenetic information with a host of data collected by the Thornton group on the substrates, chemistry and reaction mechanisms performed by each relative, together with information on the catalytic and binding residues involved. This treasure trove of functional data can help reveal the mechanisms leading to functional changes; insights that can be valuable for protein design.

To what extent can function diverge within a protein superfamily

Using FunTree and stepping down through the evolutionary tree in each superfamily, from the root to the leaves, it is clear that dramatic changes in the chemical class of the reaction performed are rare, although possible [6].

Changes in active site residues can modify the chemistry performed

Dramatic changes in chemistry are likely to be mediated by variations in the physiochemical characteristics of residues lining the active site pockets, which are either directly involved in the reaction mechanism or contribute to an electrostatic environment that promotes certain types of reactions. We used structural alignments to identify equivalent positions in the active sites of diverse functional relatives within a superfamily. Figure 1 shows the extent to which relatives can diverge in physicochemical properties of residues at these sites and also shows the number of different enzyme chemistries identified for relatives in these superfamilies.

In some superfamilies, there are clearly positions in the structure where mutations more frequently lead to functional change, whilst other superfamilies show plasticity across the whole active site pocket [7]. A large proportion of enzyme superfamilies adopt TIM and Rossmann folds, which have structural frameworks that support huge sequence diversity.

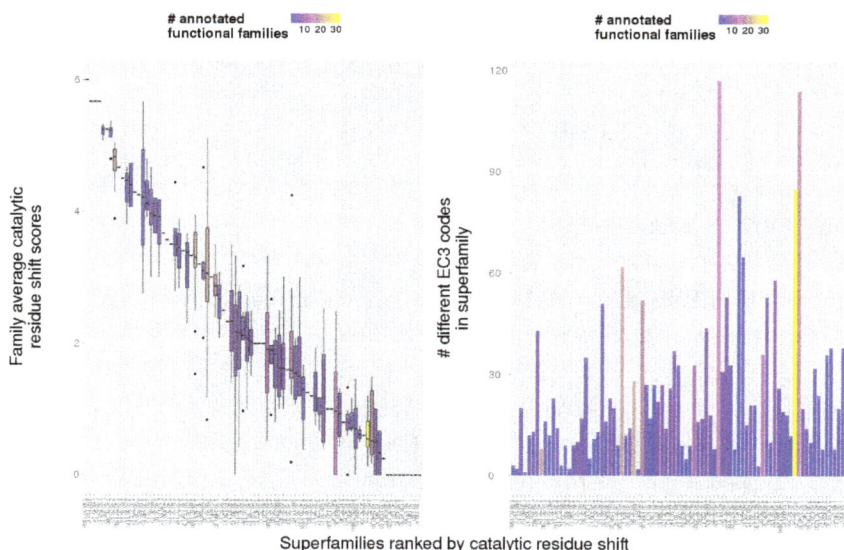

Fig. 1. Plot showing the average similarity in physicochemical properties (measured by the McLachlan substitution matrix) of residues in catalytic sites, for 101 well-annotated enzyme superfamilies. Superfamilies are ranked by similarity in catalytic residues.

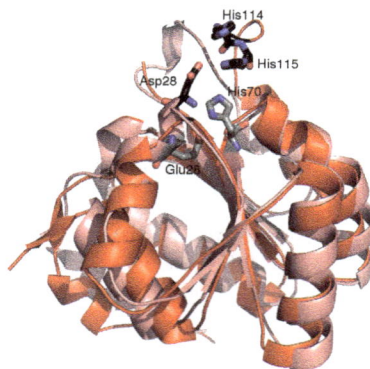

Fig. 2. Superposition of two relatives from the CATH TPP superfamily. Pyruvate decarboxylase is in red and benzoylformate decarboxylase in pink with their catalytic residues shown in black and grey.

Convergent evolution of functions in superfamilies

In some superfamilies, similar chemistries have been enabled by multiple evolutionary routes. Figure 2 shows homologous proteins which perform the same reaction but differ in their catalytic machinery.

In the RuBisCo family, most plant species contain the C3 form, which is rather inefficient. Convergent evolution to a more efficient C4 form has occurred multiple times during evolution. By reconstructing the ancestral tree and building 3D models we found several stabilising mutations leading up to the transition between the two forms (C3/C4), whilst the transition itself is accompanied by significantly more destabilising mutations. By accumulating stabilising mutations the enzyme can support the occurrence of destabilising mutations that have beneficial impacts on the function. The transition to C4 is immediately followed by compensatory mutations which restore global stability.

Outlook to future developments

An exciting new pool of sequence diversity and novel functional repertoires is emerging from the international metagenome studies. These initiatives are capturing sequence information from microbial communities found in a wide range of environments (*e.g.*, human gut, spoil, deep sea). Preliminary studies have already revealed many new relatives in known enzyme superfamilies and the tools developed by our group (*e.g.*, the CATH functional family profiles) and those of many other groups classifying protein superfamilies and functional families will shed light on the nature of changes in these proteins. As mentioned already the common active site exploited by relatives facilitates comparison of sequence changes in new relatives. The challenge for the future will be to predict the nature of the shifts in function induced by these residue mutations. This will enable a better understanding and modelling of the active site environment and the mechanisms by which chemistries can be modified and diverse substrates utilised.

Acknowledgments

NLD thanks the MRC for funding. RST thanks NSF for funding. DL thanks NIH for funding and NF and SD thank the Wellcome Trust for funding.

References

1. B. H. Dessailly, R. Nair, L. Jaroszewski, J. E. Fajardo, A. Kouranov *et al.*, *Structure* **17**, 869 (2009).
2. C. Orengo, J. Thornton, *Annu. Rev. Biochem.* **74**, 867 (2005).
3. A. Cuff, O. C. Redfern, L. Greene, I. Sillitoe, T. Lewis *et al.*, *Structure* **17**, 1051 (2009).
4. D. A. Lee, R. Rentzsch, C. Orengo, *Nucl. Acids Res.* **38**, 720 (2010).
5. R. Rentzsch, C. Orengo, *BMC Bioinform.* **14**, S5 (2013).
6. N. Furnham, I. Sillitoe, G. L. Holliday, A. L. Cuff, R. A. Laskowski *et al.*, *PLoS Comput. Biol.* **8**, e1002403 (2012).
7. B. H. Dessailly, N. L. Dawson, K. Mizuguchi, C. A. Orengo, *Biochim. Biophys. Acta* **1834**, 874 (2013).

EXPLORING CHROMATIN BIOLOGY USING PROTEIN CHEMISTRY

TOM W. MUIR

Department of Chemistry, Princeton University, Princeton, NJ, USA

My view of the present state of research in protein chemistry

Recent decades have witnessed astonishing progress in our ability to analyze native proteomes. Technologies centered on mass spectrometry have led to the recognition that proteins are decorated with post-translational modifications (PTMs) to an extent hitherto unimaginable. Indeed, it seems likely that every protein in a eukaryotic cell is modified, perhaps in manifold ways, at some point during its lifetime. The number and type of PTMs found on proteins is ever expanding and the resulting complexity of proteomes greatly complicates our ability to understand cellular biology at the systems level.

Perhaps nowhere is this complexity more evident than in chromatin, the nucleoprotein complex that leads to spatial organization of genomic DNA in the eukaryotic nucleus. The fundamental repeating unit of chromatin is the nucleosome which comprises an octamer of four histone proteins (forming a proteinaceous spool) around which is wound the DNA. Histone proteins are modified in an astonishing number of ways; over 100 discrete PTMs have been reported at the time of writing this piece, and because histones form a hetero-octamer, the potential for combinatorial PTM signatures at the nucleosome level is enormous. Indeed, the last few years have seen a flood of *genomic* and *epigenomic* information that, among other things, reveal the existence of 'chromatin states' containing characteristic patterns of DNA and histone modifications that are correlated with different patterns of gene expression [1]. Such 'cataloging' is a critical first step towards an operational understanding of epigenetic control mechanisms, however, it does not in itself yield direct information on how these PTM patterns are installed, maintained and, most importantly, what they are actually doing at the molecular level. For this, we still require more-directed biochemical and cell biological methods, which provide the quantitative data needed to make testable statements on mechanism. Current biochemical approaches in the chromatin area are, however, very low throughput and so there is an ever-growing disconnect between the flood of *"omics"* data sets and our ability to follow up the numerous mechanistic hypotheses they generate. Thus, I believe that one of the grand challenges, and most urgent needs, in the chromatin area is the development of *high-throughput protein biochemistry* methods that will allow us to acquire a more comprehensive molecular understanding of epigenetic control mechanisms.

My recent research contributions to protein chemistry

My research group has for many years focused on the development of methods that allow chemical access to proteins [2]. These protein semisynthesis approaches, along with complementary unnatural amino acid mutagenesis technologies [3], now make it possible to manipulate the entire covalent structure of proteins using the tools of synthetic organic chemistry. Among the very many things now possible, is the direct generation of proteins bearing PTMs at user-defined positions. Such proteins are exceedingly hard to come by using classical biochemistry or molecular biology methods and, thus, represent a powerful application of these modern protein chemistries. For our part, we have largely focused on the histone proteins, which as noted earlier are elaborated with an astonishing range of PTMs. Over the last 4-5 years, we have put in place modular synthetic routes that allow the generation of histones carrying various PTMs, or combinations thereof, in biochemically useful quantities [1]. Modifications accessible to this approach include, methylation (at lysine and arginine), acetylation (lysine), ubiquitination (lysine), phosphorylation (histidine, serine, threonine and tyrosine), sumoylation (lysine) and ADP-ribosylation (glutamic acid). With these modified proteins as raw materials, we can, in principle, reconstitute any 'chromatin state' for detailed biochemical study. This reconstitution approach has begun to reveal the logic by which some of these modifications are installed (*e.g.*, crosstalk mechanisms [4]) and readout (*e.g.*, multi-valent interactions [5]). Moreover, access to these reagents has yielded insight into how breakdown of epigenetic control mechanisms can lead to human disease [6]. Nonetheless, we have come to realize that our current modus operandi is too artisanal in Nature, meaning that it just takes too long to manufacture and test the chromatin complexes. The backlog of hypotheses we would wish to test, generated by the ongoing 'omics' initiatives, is simply overwhelming. I see this as the major bottleneck in the field generally, assuming of course that one still buys into the notion that a mechanistic understanding of biochemical processes is a useful thing! Thus, I believe that the future in this area must fundamentally address this bottleneck, and that one productive approach to this would be to devise *high-throughput protein biochemistry* methods that accelerate the manufacture and functional analysis of these 'designer chromatin' molecules and the factors that act on them. Achieving this goal will have significant challenges that will surely require innovative ideas from chemistry, biology and engineering.

Outlook to future developments of research on protein chemistry

Remarkable progress has been made in the field of protein chemistry in recent years. I think we have transitioned from asking whether or not a modified protein can be made, to perhaps the more exciting question of what can we learn about biology given that we probably can make it (certainly true for modified histones). This is not to say that we can synthesize any protein imaginable, and certainly there is still room for improved and streamlined methods. However, I would suggest that in thinking about such innovations, we should do so with an eye towards the even bigger problem, namely the very large number of PTM-modified proteins found in proteomes and how to functionally characterize them all. It seems likely to me that tackling this problem will have to involve the parallel manufacture of hundreds, perhaps thousands, of chemically defined, modified proteins and that this process will also have to incorporate a solution to the issue of keeping track of all these molecules thereby allowing one to perform massively parallel quantitative biochemistry. This is a complex problem, but one that I think represents the next chapter in the field of protein chemistry.

Acknowledgments

I am grateful to the US National Institutes of Health for supporting the research in my laboratory.

References

1. B. Fierz, T. W. Muir, *Nat. Chem. Biol.* **8**, 417 (2012).
2. M. Vila-Perelló, T. W. Muir, *Cell* **143**, 191 (2010).
3. L. Wang, J. Xie, P. G. Schultz, *Annu. Rev. Biophys. Biomol. Struct.* **35**, 225 (2006).
4. R. K. McGinty, J. Kim, C. Chatterjee, R. G. Roeder, T. W. Muir, *Nature* **453**, 812 (2008).
5. A. J. Ruthenburg, H. Li, T. A. Milne, S. Dewell, R. K. McGinty *et al.*, *Cell* **145**, 692 (2011).
6. P. W. Lewis, M. M. Müller, M. S. Koletsky, F. Cordero, S. Lin *et al.*, *Science* **340**, 857 (2013).

OUR EXPANDING PROTEIN UNIVERSE

ADAM GODZIK

Program in Bioinformatics and Systems Biology, Sanford-Burnham Medical Research Institute
10901 N. Torrey Pines Rd, La Jolla, CA 92037, USA

My view of the present state of research on new chemistry in the expanding protein universe

Last few years witnessed an unprecedented explosion of information on protein coding genes from individual species (genomics) and environments (metagenomics), many of them coding for enzymes with unknown functions and specificities.

We have now amassed sequences of billions of protein coding genes from over 200 eukaryotes, thousands of bacteria and archea as well as from microbial communities (microbiomes) from thousands of environments, sampling from a vast "protein universe" of all existing protein [1]. Actually, over 99% of all known proteins are coming from the latter group and for the most part are not associated with specific organisms nor with specific neighborhoods on their genomes. This makes the function prediction for them very difficult. Based on ratios of different types of proteins in bacterial genomes, we can estimate that between 30 and 40% of these genes code for enzymes, for some of which we can predict general functions (hydrolases, proteases, kinases, etc.) but as much as 40% remain completely unknown. We can only speculate that many of them would be involved in metabolism of various natural compounds.

My recent research contributions to new chemistry in the expanding protein universe

Over last several years in our group we have collected and analyzed genomic and metagenomic data sets from different sources. Public protein databases, such as translations of the GeneBank [2] coding sequences and UniProt [3], collecting protein sequences from genomic studies and sequencing of individual genes, undergo a steady, exponential growth (see Fig. 1, blue rombi) and toward the end of 2013 their size approached over 45M non-redundant protein sequences (http://uniprot.org). However, in last few years, the biggest impact on the growth of protein sequence databases came from metagenomic studies, where DNA sequencing is applied directly to microbiomes, complex microbial communities that occupy almost every imaginable environmental niche on Earth. By skipping the traditional step of growing microbes in pure cultures, we are now able to sample DNA of organisms that

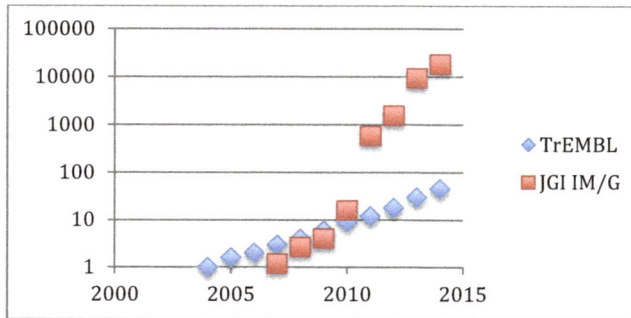

Fig. 1. Growth of protein sequence datasets, UniProt TrEMBL (blue rhombi) and JGI IM/G (red squares). The Y axis scale is in million (10^6) sequences.

don't grow in such cultures, at least not easily. Recently, developments in the study of a human metagenome, dubbed by some a second human genome [4], garnered a lot of attention. With several large studies around the world, from NIH Human Microbiome Project [5, 6] to European/Chinese metaHIT project [7], already identified 100M of predicted protein sequences, doubling the size of traditional datasets. However, studies of environmental microbiomes, such as present in soil and marine environments dwarf all previous efforts, providing to date over 16B (10^9) predicted protein sequences (Fig. 1, red squares) in only one of several metagenomic resources, IMG/M at the DOE Joint Genomics Institute [8]. Our protein universe is now over 17B large! Even if expected redundancies and wrong ORF calls currently inherent in metagenomics data would reduce this number, the pace of discovery of new genes would not stop and it is clear that early estimates of the size of the protein universe were incorrect and a much larger part of protein space of all possible protein sequences is actually explored in Nature.

These developments bring the question of what it the internal structure of the newly discovered part of protein universe. The part studied so far, for the most part can be classified into a hierarchical system of protein families and superfamilies. Information about such families is collected in databases such as Pfam [9], SMART [10] or CCD [11] with the number of defined protein families counted in tens of thousands, for instance there are 14,831 protein families defined in the most recent Pfam release (27.0). As shown in Fig. 2 (lowest panel), over 85% of proteins in a set of 220 fully sequenced genomes representing all phyla with sequenced representatives can be matched to a Pfam family. This coverage drops to little over 60% for metagenomics sets from the human metagenome focused metaHIT [7] and type 2 diabetes studies [12] (top and middle panels in Fig. 2, respectively). Similar level of coverage is seen in other metagenomics datasets (data not shown), suggesting that while the diversity and novelty of genes found in these microbial communities is significant, the core is formed by known and defined protein families. At the same time, its worth noting that even well studied microbial genomes still contain

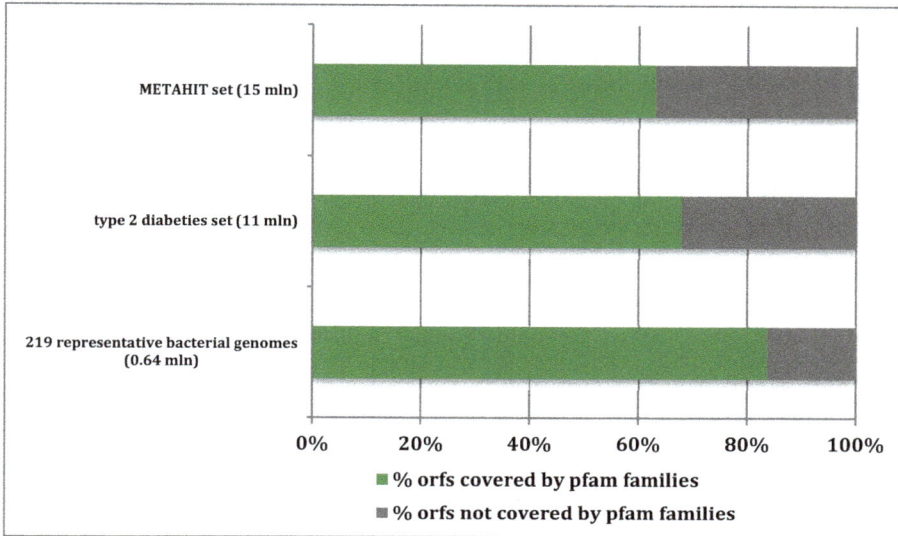

Fig. 2.   Coverage of different subsets of protein space with currently defined protein families.

a significant number, around 15%, of proteins that cannot be classified into families using current resources. Another comment here is that classification of a protein into a known protein family is not synonymous to assigning function to this protein, as many protein families in Pfam and other similar databases do not have any verified functional annotations and thus are known as domains of unknown function (DUF) families.

Since first metagenomic datasets became available, our group focused on identifying novel protein families in these datasets, with special emphasis on families of secreted proteins from microbes present in the human gut microbiome [13]. As of today, close to 400 such families have been identified and deposited in the Pfam database. In collaboration with the Joint Center for Structural Genomics, an NIH Protein Structure Initiative center for high-throughput structural determination, we have systematically targeted these families for structure determination, solving structures of first representatives of over 40 of these families.

However limited in scope to a specific segment of protein universe, this exploration brought rather unexpected results that give us an interesting perspective of what can we expect from the future, more extensive studies of novel proteins from new regions of the expanding protein universe. Most if not all families identified in the human gut microbiome represented divergent branches of already known families, with differences on the sequence level that often made it impossible to recognize the homology, but with structural similarities so strong that the couldn't be explained by convergent evolution. While functions of protein in many of these new families remain unknown, we have identified novel families of glycosyl hydrolases and other carbohydrate metabolic enzymes [14, 15] — as expected in the gut environment, rich in polysaccharide based plant fibers such as starch, and multiple families of proteins involved in cell-cell interactions [16, 17], again expected in the dense environment of the gut.

Outlook to future developments of research on new chemistry in the expanding protein universe

We can expect even faster growth of protein sequence databases and explorations of novel environments, resulting in the known protein universe becoming even bigger and more diverse. With the help of other high-throughput techniques such a proteomics and metabolomics we expect that integration of these multiple sources of information would be both a main challenge, but also a main source of discovery of new functions, new chemistry and new processes in the new environmental niches. The expanding protein universe is going to catch up with equally extensive world of natural products, which are all after all synthetized and metabolized by protein-based catalysis.

Acknowledgment

Research described here was funded by the NIH grant U54 GM094586 (JCSG).

References

1. I. Ladunga, *J. Mol. Evol.* **34**, 358 (1992).
2. D. A. Benson, M. Cavanaugh, K. Clark, I. Karsch-Mizrachi, D. J. Lipman *et al.*, *Nucl. Acids Res.* **41**, D36 (2013).
3. The UniProt Consortium, *Nucl. Acids Res.* **41**, D43 (2013).
4. E. A. Grice, J. A. Segre, *Annu. Rev. Genom. Hum. Genet.* **13**, 151 (2012).
5. L. M. Proctor, *Cell Host Microbe* **10**, 287 (2011).
6. P. J. Turnbaugh, R. E. Ley, M. Hamady, C. M. Fraser-Liggett, R. Knight *et al.*, *Nature* **449**, 804 (2007).
7. J. Qin, R. Li, J. Raes, M. Arumugam, K. S. Burgdorf *et al.*, *Nature* **464**, 59 (2010).
8. V. M. Markowitz, I. M. Chen, K. Chu, E. Szeto, K. Palaniappan *et al.*, *Nucl. Acids Res.* **40**, D123 (2012).
9. M. Punta, P. C. Coggill, R. Y. Eberhardt, J. Mistry, J. Tate *et al.*, *Nucl. Acids Res.* **40**, D290 (2012).
10. I. Letunic, T. Doerks, P. Bork, *Nucl. Acids Res.* **40**, D302 (2012).
11. A. Marchler-Bauer, S. Lu, J. B. Anderson, F. Chitsaz, M. K. Derbyshire *et al.*, *Nucl. Acids Res.* **39**, D225 (2011).
12. J. Qin, Y. Li, Z. Cai, S. Li, J. Zhu *et al.*, *Nature* **490**, 55 (2012).
13. K. Ellrott, L. Jaroszewski, W. Li, J. C. Wooley, A. Godzik *et al.*, *PLoS Comput. Biol.* **6**, e1000798 (2010).

14. D. Das, M. Herva, M. A. Elsliger, R. U. Kadam, J. C. Grant *et al.*, *J. Bacteriol.* **195**, 5555 (2013).
15. A. P. Yeh, P. Abdubek, T. Astakhova, H. L. Axelrod, C. Bakolitsa *et al.*, *Acta Crystallogr. Sect. F Struct. Biol. Cryst. Commun.* **66**, 1287 (2010).
16. Q. Xu, B. Christen, H. J. Chiu, L. Jaroszewski, H. E. Klock *et al.*, *Mol. Microbiol.* **83**, 712 (2012).
17. Q. Xu, P. Abdubek, T. Astakhova, H. L. Axelrod, C. Bakolitsa *et al.*, *Acta Crystallogr. Sect. F Struct. Biol. Cryst. Commun.* **66**, 1281 (2010).

THE SCIENTIFIC IMPACT OF FREELY AVAILABLE CHEMICAL PROBES

ALED M. EDWARDS

Structural Genomics Consortium, University of Toronto
101 College Street, MaRS South Tower, Suite 706
Toronto, ON M5G 1L7, Canada

Enabling research efforts on the proteome

The human genome sequence provided a near complete list of human genes and proteins. A comprehensive understanding of human normal and pathobiology will come only after the role(s) of each human gene and protein, and their inter-relationships, are well understood. However, the anticipated impact of having our genome sequence is not being quickly realized. Bibliometric analysis of research activity on each human gene/protein revealed a clustering of activity on a relatively small subset of human genes/proteins, highly enriched in those that had been the subject of research efforts for decades [1, 2]. Progress toward a deeper understanding of human biology and the discovery of new medicines is being limited by this behaviour.

Additional bibliometric analysis also pointed to a partial solution. Historically, the availability of high quality, small molecule inhibitors of proteins provided impetus for the research community to initiate research on proteins that previously had garnered little attention, provided the small molecule inhibitor was made available without restriction on use. This raised the hypothesis that research activity could be increased on proteins residing among the less studied by making chemical research tools readily available. To test this hypothesis, we have engaged in a large-scale collaboration with academic and pharmaceutical medicinal chemists with the aim to produce, and make available, cell-active small molecule inhibitors of proteins and enzymes that regulate chromatin biology.

Our recent research contributions

Our project is focused on families of human proteins that regulate protein acetylation and methylation. Specifically our objective is to generate inhibitors of enzymes that acetylate and methylate other proteins, enzymes that demethylate methylated proteins, and the suite of proteins that bind preferentially to acetylated or methylated side chains [3, 4]. These modifications regulate, among things, proteins that control gene expression. The aim of the project is to increase research activity on these protein families, and in so doing, to define new links between their members and human disease.

The project revolves around the technology platform within the Structural Genomics Consortium (SGC). Within the aforementioned protein families, for most family members, the SGC has generated purified proteins, their 3D structures and has built suites of biophysical and biochemical assays to monitor protein function. This core resource enables a large number of medicinal chemistry groups to identify weak inhibitors and, using structure as a guide, to advance these to become potent, selective and cell-active molecules.

With nine pharmaceutical companies and a network of academics involved, the project has produced "chemical probes" for a number of proteins, some of which have been used to uncover new biological insights [5–9]. Importantly, the more general influences on academic research activity are beginning to appear. The papers reporting the chemical probes for the G9a methyltransferase [6], the JMJD3 histone demethylase [7] and the BET family of bromodomains [5] have become the top cited papers on those proteins since their publication (Edwards, unpublished). In turn, the research activity on these proteins, compared to other family members, has increased significantly (Fig. 1). With these early indications in hand, it does appear that the availability of chemical probes is sufficient to induce the research community to focus on "novel" proteins.

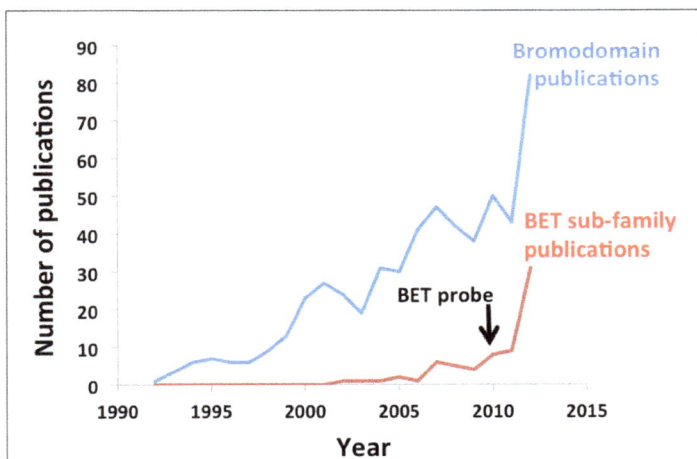

Fig. 1.   Publication trends on bromodomains-containing proteins. Annual publications naming bromodomains-containing proteins in the title, abstract or key words were quantified using the Scopus database (blue line). Papers referencing the BET subfamily of bromodomains are indicated in the red line. The black arrow indicates when the BET chemical probe papers were published.

Outlook to future developments

The availability of the human genome sequence places biomedical research in a unique situation among all sciences — it provides us a map of the known unknowns. However, even given this map, the research funding system appears fearful to go exploring. Although there is strong justification to continue studying proteins of known importance or biomedical interest, there is equal justification to encourage more research into the unknown. But this is difficult to accomplish within the current global research funding framework, in which funding is allocated along national lines and according to the opinions of a group of peers. It is my opinion that scientists would be willing explorers if the granting, tenure and peer-recognition systems permitted. But of course, these systems were created and are endorsed by the research community — we have met the enemy and it is us.

Acknowledgments

The Structural Genomics Consortium is a registered charity (number 1097737) that receives funds from AbbVie, Boehringer Ingelheim, Genome Canada, GlaxoSmith-Kline, Eli Lilly Canada, Janssen, the Novartis Research Foundation, Pfizer, Takeda, the Canada Foundation for Innovation, the Ontario Ministry of Economic Development and Innovation, and the Wellcome Trust.

References

1. A. M. Edwards, R. Isserlin, G. D. Bader, S. V. Frye, T. M. Willson *et al.*, *Nature* **470**, 163 (2011).
2. http://baderlab.org/Data/RoadsNotTaken
3. D. J. Patel, Z. Wang, *Annu. Rev. Biochem.* **82**, 81 (2013).
4. C. H. Arrowsmith, C. Bountra, P. V. Fish, K. Lee, M. Schapira, *Nat. Rev. Drug Discov.* **11**, 384 (2012).
5. P. Filippakopoulos, J. Qi, S. Picaud, Y. Shen, W. B. Smith *et al.*, *Nature* **468**, 1067 (2010).
6. M. Vedadi, D. Barsyte-Lovejoy, F. Liu, S. Rival-Gervier, A. Allali-Hassani *et al.*, *Nat. Chem. Biol.* **7**, 566 (2011).
7. L. Kruidenier, C. W. Chung, Z. Cheng, J. Liddle, K. Che *et al.*, *Nature* **488**, 404 (2012).
8. W. Yu, E. J. Chory, A. K. Wernimont, W. Tempel, A. Scopton *et al.*, *Nat. Commun.* **3**, 1288 (2012).
9. L. I. James, D. Barsyte-Lovejoy, N. Zhong, L. Krichevsky, V. K. Korboukh *et al.*, *Nat. Chem. Biol.* **9**, 184 (2013).

SESSION 1: NEW CHEMISTRY IN THE EXPANDING PROTEIN UNIVERSE

CHAIR: CHRISTOPHER T. WALSH

AUDITORS: K. BARTIK[1], E. GOORMAGHTIGH[2]

(1) Engineering of Molecular Nanosystems, Université Libre de Bruxelles, CP 165/64, 50 av. F.D. Roosevelt, 1050 Brussels, Belgium

(2) S.F.M.B., Université Libre de Bruxelles, CP206/2, Campus Plaine, Bd du Triomphe, B-1050 Brussels, Belgium

Discussion among panel members

Christopher Walsh to Aled Edwards: Aled, it's not as if these molecules are pheromones for humans. How did you decide what probes to make? Suppose I don't care about kinases and methyltransferases from all domains? You are still looking under a lamppost in a sense.

Aled Edwards: Absolutely! For us, it was pragmatic. If you work on protein families, the chemistry is shared and you don't have to invest so much in chemistry because you have common libraries that provide starting points. I think we just had to try and show that the model works first and then hopefully it will be copied.

Aled Edwards to Adam Godzik and Christine Orengo: You have been at these meetings for a decade and everyone is talking about the domains of unknown function. To your knowledge, has there been any concerted effort around the world, anywhere, to say, "let's start looking at these" because the probability of finding out interesting or novel science is higher there? To my knowledge, there is none!

Adam Godzik: Surprisingly, as far as I know, nobody other than us, nobody has picked it up, but it's certainly coming. What we noticed is that, when you publish papers or describe them, in time, after 2 or 3 years, there is a peak of interest. For us, the catalyst of interest is publishing papers on them.

Christine Orengo: I think the problem is, as you say, that people are quite nervous about studying proteins for which there is no information at all and I suppose that's why we try to organize things so that we can group the novel proteins so that at least we can say what they are most similar to. So there are some clues there as to what their functions might be, or where they might have diverged from. You then have a starting point and a few clues. I think that if you are talking about some of these completely novel families we actually have no insight at all and you are talking about a lot of experimental work. You have got to do a vast amount of screening to get any idea what they could be doing. So, I think at the moment the way people are using this data, is to take sequences that have at least some similarities to a family for which something is known about the active site and see what the changes might be. From that, some deductions can be made about which ligands might bind, and then they've got a clue. In this way it is not such an expensive operation. I think what you're proposing is very exciting but it is also hugely expensive.

Wilfred van der Donk to Aled Edwards: I was a little bit surprised that RNAi didn't seem to have made much of a dent. I would have thought that it would level the playing field much more than it seems to have.

Aled Edwards: This is an interesting observation. When you get the RNAi knocked down, if you do it *en masse*, then you have to think "What do I do next?" I need a cell line, I need a reagent to prove that the protein was knocked down, I need a whole bunch of stuff that costs me, or my students, a lot of time and effort to get to the point where I can actually do an experiment. And if we have a five-year grant or a three-year PhD student, often it is proving not enough.

Jean-Louis Reymond to Aled Edwards: Counting papers is one thing but you have to measure the depth of insight, so I think it's only natural that there are different levels of insight. The question is: "Have you measured whether the additional papers add anything significant or can one measure the learning level of an additional publication?"

Aled Edwards: It's a hard experiment but you would imagine that, because we continue to publish in those areas, because the funding system now is quite austere and only good, great grants get funded, that a lot of the papers on the old chestnut proteins are important. It's not that they are not important proteins, it's just that as a scientific institution, maybe we don't have the balance right between them. You can discover interesting things about these proteins. As Tom said, they are all probably modified by weird and wonderful things and they all have interesting activities, and so it's not as if the science is poor or is yesterday's science, it is just that we tend to focus on what we have always done. As Roger Kornberg said, we like to "fondle our problems".

Christine Orengo to Adam Godzik: You're analysing in much more detail some of these very novel families and as you say, it is something we find as well, that quite often they turn out to be remote homologues of what we have seen already. And from our sort of preliminary scans of the meta-genome data, we see that a lot of the new enzymes that we pick up are still relatives of some of the very common favourable structural scaffolds like the TIM barrel or the Rossmann fold. We see them over and over again appearing in this dataset and I just wondered, because you have done a larger scan, whether you see some other types of structural scaffolds also emerging and being exploited quite frequently by these new sequences you are analysing?

Adam Godzik: Yes, we did and the result was sort of depressing in the sense it looks like the most popular folds are getting even more popular. So, in answer to my question on whether this new environment would exploit these new scaffolds, they don't. They actually increase the popularity of the most popular fold. So yes, in genomes TIM barrels represent 10%, in meta-genomes, they represent 15% and so the distribution is getting even more skewed.

Tom Muir to Chris Walsh and Wilfred van der Donk: You spoke about these amazing peptides and natural products. I wondered though where we are in terms of understanding of what all these things are doing. Do we even know in many cases which organisms they are targeting in the ecology of the soil or wherever they come from? Does informatics have any role to play in predicting any of this stuff?

Wilfred van der Donk: In most cases, we don't. As you pointed out, we know which organism produces them in most cases but we don't necessarily know enough about the ecology, where they're living, who they are targeting in their natural environment. We can test them against a panel of organisms that we are interested in as humans, and that's what most people do. Most people don't really test them against the organisms they are probably living with. Can we predict activities? To some extent, but I think to a low-level, you can sometimes say "well this looks like something we have seen before" and we can predict the activity but as soon as it goes away from something that we know and have seen before, we essentially have no idea. And I would say this is actually a problem for genome mining because a lot of the molecules we have made by genome mining. When we have the molecule we can see that indeed it is the molecule the organism also produces. After we do some tricks to try have it produce the compound, then we see no bioactivity whatsoever. And we have no idea really to predict what clusters to go after! And hence, as you said in your talk about your field, we have to do this in scale in order to really get back to where we were in the 60's and 70's where simply by a large number of natural products being available a number of them made it to pharmaceuticals.

Tom Muir: How much interaction is there between chemists, such as yourself, and microbiologists or ecologists who actually are thinking about bacterial communities in the soil, in the ocean or in our gut? Do you guys talk to each other to try to think about a concept?

Wilfred van der Donk: Yes, I collaborate with several microbiologists both at Illinois and elsewhere to talk about the chemical ecology. I am mostly interested in the chemistry, but yes we have to find out what these molecules are doing. As you probably know, there is even an hypothesis out there, that what we call "antibiotics" are actually not made in the environment to kill other organisms but are more signalling molecules. And so yes, we do. I don't know if Chris wants to add something to both of your questions?

<u>Chris Walsh</u>: First I would make a distinction between eukaryotes, prokaryotes and multicellular eukaryotes where I think there isn't much interesting new chemistry that has been discovered. That may be a red flag for somebody like Aled Edwards. Are nuclear hormone receptors actually interesting from a point of view of new chemistry? I'll come back to that. It is meant to be an interesting but not antagonistic statement. Let me go to fungi which are a fascinating set of organisms which make many of these unusual molecules and almost nothing is known about the natural physiology of what these molecules do. I suspect part of that is the historical view that you can figure out the function of conditional metabolites in laboratory-based cultures. Almost clearly the missing links are the fact that you have to look at ecologic communities. It has been estimated that in the kinds of molecules I or Wilfred or others study, maybe there are 30 gene clusters in the typical fungal cells or *Actinomyces* and only 5 to 10% of them have been turned on, transcribed under laboratory conditions. So the other 90% of the conditional metabolite pools are just not being exploited in the physiology that we look at. To me, it is about looking under the lamppost and that's what our community does. I'm hard pressed to find a paper for example on kinases which really is interesting or novel in terms of chemistry, that is a different question from the health relevance and the applied importance. It is worth factoring those two things though.

<u>Aled Edwards to Adam Godzik</u>: I have a question about anaerobic and aerobic biology. If you parse your datasets by strict anaerobes and aerobes, do you see a difference in the sort of patterns of protein family that are in each?

<u>Adam Godzik</u>: Yes we do. We haven't really followed on that, but yes there are definitely differences. We even have a theory that the distribution of folds in anaerobes is sort of more ancient because they map to families, which form certain specific parts of the metabolic network, which people argue are older. So we thought we saw a pattern in this distribution.

<u>Aled Edwards to Chris Walsh</u>: The anaerobic biochemistry is really hard, you've got to go under hoods and they have cofactors...

<u>Chris Walsh</u>: It's not hard for the anaerobes!

<u>Aled Edwards</u>: Exactly, and they have cofactors, different corrinoids. So if it's a source of great chemistry, where are we going to get the biochemists that can do it because we can probably count on one hand the people who are really skilled in the art of anaerobic biochemistry.

<u>Chris Walsh</u>: I think that in the general discussion later on and in the session this afternoon, we'll hear about the "enzyme function initiative" people and they have just built the anaerobic biochemistry in line with several stations from, doing assays, to doing purifications, to large throughput crystallography. So it will take

those kinds of activities to reduce the barriers. So it's not just "Are chemicals available?" but also getting through the technology. Just a comment — one of my first comments on my first slide was that I thought anaerobes were interesting and I'll just give you one example: the reduction of benzene to cyclohexane can be done by anaerobes but not by aerobes. It's a sort of interesting Birch type reduction chemistry. As far as I know, reported in the literature 25 years ago but never studied in any detail.

Wilfred van der Donk: I would like to add to the discussion. Much of metallo-biochemistry requires the same type of very controlled environment. There is actually a really large research community that is set up to study these enzymes. So, whether it is from an anaerobe or an aerobe, you often run into oxygen-sensitive enzymes and so there are a fairly large number of people that are set up to do that.

Aled Edwards: Yes, some of the problems though are that you cannot make them recombinantly because the hosts sometimes don't have the appropriate cofactors that their host organisms would make. So you need to grow them, but they double every two weeks and you just have to wait. It's hard to do bucket-biochemistry on a thing that is barely alive as you know.

Christine Orengo to Adam Godzik: My question is about the orphan families that he's finding, whether he has done any further analysis on the sort of characteristics they have? Some studies we have done suggest that they're perhaps more disordered, or have higher proportions of disorder, so it's harder to spot any relationship with other known families. Some of the relationships may have been obscured. So I just wondered, have you looked at the features of those orphan families? Perhaps they might be less tractable for structural determination as well?

Adam Godzik: No, we didn't do it in this way. What we were interested in is to see how quickly they can be de-orphanized. So what we noticed is that calling something an orphan, meaning that has no homologue, is very time-dependent because most of the orphans known at a given time would find homologues once the dataset gets larger. But then more orphans would be introduced so it's like you're doing a cycle. So in some sense, them being orphans is nothing special about them. We just happened to see the top of the mountain first and then we looked deeper and we saw the rest. Our thought is that there is nothing special about them — it is just random events. These were not very large families and we just found just one representative, we looked again and we see more and more. They are not special anyway. By random chance, we found one example first and then, in next dataset, we found plenty more.

Christine Orengo: So do you think it's reasonable to suppose that, as the data accumulate the information we have for each family grows and actually the sequence patterns we derive become much more sensitive and so we can recognize these very distinct homologues? Rather than the universe exploding if you like, actually it will start to collapse again. We will be able to bring these relatives back into their correct superfamily. At that point, what becomes interesting is the ways in which the different domains combine because that seems to me where you do have a huge explosion of data that is going to be much harder to understand. At that point, you have to try to understand what the individual domain is contributing to the function of the multi-domain assembly. That's a different problem that we have barely started to tackle. Once we have organized the sequences into their constituent domains we have the central building blocks of these proteins. Then we have to start to understand how, when you combined them in different ways, you're again starting to modify their functions. I think that is something we have barely started doing.

Adam Godzik: Two questions in one. We tried to do both. The first one we believed it would happen, that we would start to saturate at some point and that we would become so good in recognizing families that all these orphans would be "folded back". It doesn't seem to be happening yet. Yes it happens for some orphans but then for every orphan we can add to a family we have 10,000 more coming. We are still in a completely linear growth curve in terms of orphans. So we can all speculate when it will happen but we haven't yet seen the traces of such rating. For the second question, I agree with you. We tried to work on this too but I think this is much more important for eukaryotes. Most bacterial and microbial proteins are single domain and we don't have so many problems or so many issues understanding how domains contribute, unless you go into operons. The diversity in eukaryotes is mostly built by reshuffling already known domains. We definitely have two very different methods of how, bacteria or microbes versus multicellular eukaryotes, deal with a complex world. One of them is diversifying into the sequence space and the designing of more and more homologues, the others are actually shrinking their existing domains by recombining them into new patterns. It's definitely different paths and I think you mentioned that in yeasts, for instance, there are definitely different patterns of diversification, of finding new chemicals they need or new challenges by mixing domains rather than inventing new ones.

Chris Walsh: I want to go back to Adam's comment. So far, it looks like there are 1300 or so protein folds and you estimate it might be possible to have 5000 by, I think you said graph theory or perhaps I misunderstood. At what point does it not become worthwhile looking for new folds, that it's a diminishing return and one should be doing something else? Or is that just coming along for the ride and it's not a frontier problem anymore?

Adam Godzik: We are not looking for new folds specifically. We cannot really predict whether the protein will have a new fold. But we're looking for new families and increasingly we're finding that even if they look novel, they have an old fold.

The fact that the number of folds seems to be saturating is a result, it is not an assumption. It's surprising because we thought it would keep growing and it is definitely on the level of saturation. We very rarely find new folds.

Christine Orengo: If I can contribute to that. Yes in fact I think that the number of folds is reducing as we get more information. I think it is rather difficult how to define a fold. It's a quite controversial discussion about what we mean by fold! If you think about the core of proteins, the core structural scaffold, I think really there only seems to be a thousand of those arrangements seen over and over again. During the 20 years that we have been classifying protein sequences and structures into families, the number of folds has actually started to diminish as we define these core structures more clearly. I completely agree with Adam and I don't think we are likely to see many more new folds. It's the variation, it's the sort of peripheral decoration to these folds that are interesting and we are going to see many more of those.

John Gerlt to Chris Walsh: I think every time Chris Walsh says something I have to add a slide to my talk this afternoon. I will make a comment. I was at a workshop a couple of years ago with DOE and, Claire Frasier who was at the workshop, made the comment that maybe what we should do is stop sequencing and maybe figure out exactly how we decipher what's already in the database. That is, how do we go about figuring out what the functions are of the proteins that we have now? The problem is only getting worse, and I don't think Claire was terribly serious about stopping sequencing, but the problem is overwhelming. I think, and I'm going to talk about this a bit this afternoon, that one needs not just bioinformatics (I don't mean to insult bioinformatics) but really a multidisciplinary approach to figure out what all this sequence information means. Is there some way to decipher function from the information? I think that's the real problem, from my perspective, that is confronting this community. If we want to discover new chemistry, we are going to have at some point to stop and figure out what that chemistry is and how do we go about doing that.

Chris Walsh to Tom Muir: You said in passing Tom, regarding histone modifications in chromatin, that essentially every combination of acetylation, methylation, phosphorylation, ubiquitylation on a residue that could be modified, can be seen. Maybe I'm putting words in your mouth. What is the chance that those are random and not useful? Could they represent promiscuous activity of post-translational modifications enzyme sets and one gets a statistical or a bio-statistical set? And some of the combinations which you or others may make are not going to be physiologically relevant. How is one ever going to sort that out? And I pick that because you pointed out there are so many combinations, many of which you've actually seen, in contrast to the antibiotic case that van der Donk mentioned where you only see a very small fraction. What is the chance those are either unplanned, not purposeful, mistakes or maybe they are evolving from new functions? What do you think about that?

Tom Muir: It is a great question. I guess to rephrase: how much collateral damage is really on these proteins, very abundant proteins? To answer your question in short we really don't know...

Chris Walsh: Maybe they're just things for acylating and methylating chemically active side chains in proteins...

Tom Muir: I mean essentially anything that can be put on CoA is now being found on histone tails. So how much of that really has function versus it just being collateral damage? We really have no idea. This is the problem. The more sensitive the analytical techniques become (to dig into the noise), the more of these things you will find and so, there lies the challenge...

Chris Walsh: That's the curse of modern mass spectrometry!

Tom Muir: That's actually true but it doesn't mean that it is not incredibly interesting of course! But one has to think long and hard before one undertakes a 2-year expedition to make these things and try to study their biochemistry. If you're wrong, if it really is noise, which I see as being one of the biggest problem here, we need ways to speed this up because there is going to be noise some of the times. We have to rethink how we approach this problem.

Aled Edwards to all chemists: I went to a plant meta-genome meeting and there are two things I want to mention. The first, they asked, "Why aren't there any soil microbiologists"? They figure that they don't chat with one another at this meeting anyway which was kind of interesting. And there are no plant people here I notice, I don't think. There is a lot of chemistry in plants you know. The second thing is that most of the talks in the plant meeting were on carbon and nitrogen, and trace metals are essential for microbes. No one worked on halogen cycles, trace metals. Is there a chemical myopia to the kinases where oxygen and sulphur, and carbon, and nitrogen are on this side, and equally important for life molecules are understudied? I mean I don't know the literature but it sounds like there is.

Wilfred van der Donk: I think you are right and I think that, at least in the area of natural products, it is obvious why plants are so much harder, why they are not being looked at. The genes are not clustered and so you don't get these very nice packages of genes that you can look at for chemistry. In respect to metals in plants, I'm not qualified to comment as to why. First of all, whether people do or do not study them, of course things like the photosystem is very well studied. But you could argue that it is easier studied in bacteria even, then in plants. But yet, I think for plants, as you pointed out in your answer to my question about RNAi, the overhead in order to set up plans in your lab is much much larger than working on microbes, single organisms and hence I think most chemists shy away from it.

Arne Skerra: At the Technical University of Munich, we have a strong Plant Science Department and I have the feeling that metabolism in plants is much better explored than the enzymology behind it. In addition, the genomic efforts in the plant area are also under-developed. So I would foresee rather fruitful research in this area, finding new enzyme functions, not only in plants but also in fungi.

Wilfred van der Donk to Jean-Louis Reymond: You showed that your artificial enzymes or enzyme analogues can mimic certain reactions that we find in Nature. I was wondering whether you or others had looked at the chemistry that you don't find in Nature and that you might be able to catalyse with these structures?

Jean-Louis Reymond: Our structures are not functioning as proteins but to get new functions, the idea is just to figure out what you want and put it inside. We know so much about chemistry that I think you can actually design things into proteins by importing things from outside. There are all the efforts to create metallo-enzymes using metals that don't appear in natural proteins. So we can do all of that in a completely synthetic structure like we do, or in the natural ones. As to our approach to dendrimers, it reveals that you can do things very simply, so certain things are simple to plant in the structure. Except that we cannot denature these things. We are not that far yet.

Chris Walsh: To follow-up maybe. When I think of Nature's dendrimer equivalent in proteins, I think of the carbohydrate chains on glycoproteins because they are the branched versions although the backbones of proteins are not. Have you made any mixed peptide oligosaccharide dendrimers to look at questions of different kind of ligand recognition?

Jean-Louis Reymond: Well we've played the usual game of attaching single carbohydrates at the end of the dendrimers so you get this multivalency effect. It's very clear that you have a recognition effect that is achieved by multivalency. We can obtain this effect also in the synthetic systems. Whether the branch nature of the carbohydrate exists because of that need for multivalency, I don't think it is that clear at this stage. I mean if you look at the complex oligosaccharides, they have probably used the opportunity to branch that is offered by the carbohydrates structure; Nature evolved because it could do it, because the sugar is a branching structure. The dendrimers just offer the branching but doesn't answer the question of whether the branching is naturally necessary for the function.

Chris Walsh: In the spirit of dendrimers, maybe I can ask Jason Chin. Are you or your colleagues actually building dendrimer or branched proteins using multifunctional unnatural amino acids in site-specific ways?

Jason Chin: So, other than carbohydrates, another example that you might imagine from literature is of course ubiquitin chains where ubiquitin is attached by an isopeptide bond built from a lysine to the C-terminal of the next protein in the chain. Ubiquitin itself has multiple lysines so you can build a number of different topologies. There are a number of different lysines within ubiquitin so you can build dendrimers like structures from ubiquitin proteins and from similar and other homologues. Other people have to use amino acids that allow you to synthetically build ubiquitin just like Tom done, some of his work...

Chris Walsh: I meant dendrimers, not ubiquitin.

Jason Chin: We are not working on building dendrimers *per se*.

Tom Muir to Jean-Louis Reymond: I have a question about topology. We are very good at making peptides at this point and have some beautiful examples of dendrimers and polycyclic compounds. Is there a topology that you would love to access using chemistry that perhaps Nature accesses, that is currently off limits? That we just don't have the synthetic know-how to do it yet?

Jean-Louis Reymond: It depends on how you define your topology. Clearly the example of these polycyclic natural products made from peptides, already have this polycyclic topology, and that it is possible and very simple. It's what you learn from graphs. And then of course Nature has other thing like these lasso-peptides where you actually have additional dimensions of topologies with interlocked rings, where I would say supra-molecular chemistry is much more at home. That is another area of possible completely new functions that might exist when you make tied knots and things like that.

Gebhard Schertler to Jean-Louis Reymond: In your chemical space, there is a lot of empty space. Is it actually really principally empty or is it again evolution that has been chosen in biology or the chemistry, when we have reactions to make polymers, that actually then reinforces certain branches to be very strong? I mean the DNA branches are very strong; the peptides branch is very strong. But we saw for example when we have these cyclic peptides, then we get rigidity. If you really ask for new chemistry shouldn't it be in this empty space?

Jean-Louis Reymond: Of course you can do this. The pretty picture I showed is of course a representation of chemical space. We have this front page of the conference booklet with the painting "ceci n'est pas une pipe" so this is all we do also in our representations. It all depends on how you describe it. But clearly, for example if we map the bicyclic peptide on the map, they clearly move to the right so we actually fill a completely new area of chemical space. So clearly there are a lot of opportunities in studying molecules that have different properties by the nature of how they are made. Then you have to think carefully because I think the wonder with proteins is for example that they have just the right balance for very high diverse functions. You might also just take the lesson from Nature by going in that direction but again, hybrid molecules are highly interesting as well and it is a change for synthetic chemistry I would say.

Discussion among all attendees

Gebhard Schertler to Tom Muir: In GPCRs, we have phosphorylation on the C-terminus, sometimes in the loop and there was a proposal of a kind of barcode for this modification. It would be established by different kinases possibly putting different modifications onto the GPCR. Now if we look at the modification field in general, is this barcode maybe a wider idea and how would we figure it out? Could we just do something which people are doing in computational sciences now, look for patterns? So if we have a method to look at modifications in a large scale like we do sequencing, could we then look for patterns and could we, for example, change the signaling state and see how these patterns change? And could we actually see if there are any systematic changes in these modifications?

Tom Muir: I think that is an interesting idea that there might be patterns. There are some analytical challenges to identifying them. Unless they are right next to each other in sequence and you can get them on a single peptide when you do mass spectrometry, when they are distal in a protein sequence and you are averaging over a population it is hard to know whether or not they're all coming from one discreet polypeptide or whether they are from different members in a population. So this is certainly an analytical challenge to identify them in the first place. But assuming one could do that in some fashion, the question would be "Do they actually signify some defined output and how do we figure out what it is?" Is that essentially what you're asking? If one were able to correlate those patterns with some phenotype in the cell, coming back to chromatin where we work, if it correlates with a particular type of gene transcription, we label the gene transcript. Maybe you can infer that a pattern does that, which is certainly part of what people are doing in that area. It doesn't tell you anything about the mechanism with which that might happen of course. This again brings us back to this business of getting hold of reagents that actually have that pattern, to study the behaviour by structural methods or by biochemical methods or however you may want to do it. You need to get beyond correlations and I think for that you need good old-fashioned biochemistry and you need to be able to make the reagents and that's a challenge.

Christopher Walsh to Kurt Wüthrich: Just a comment, a thought. The C-terminal tail of RNA polymerase is phosphorylated by different kinases and there is very good evidence in yeasts at least, from genetics and biochemistry, that in fact the multiple phosphorylations at specific serines dramatically affects the behaviour of the polymerase. So that might be a very good place to start analyses or calculations.

Kurt Wüthrich: Well, I grew up in an environment where peptide synthesis was king, with Professors Robert Schwyzer and Josef Rudinger, at a time when all the Swiss pharmaceutical companies had big teams working on peptide synthesis. ACTH, oxytocin calcitonin and growth hormone were big names, and for us it was glucagon. Then peptide chemistry literally disappeared, all the peptide synthesis laboratories were closed in the pharmaceutical industry in Basel. Today, we are witnessing a rebirth. In the old days it used to be that the synthesis of cyclic peptides was a particular challenge, but now we see that with both biosynthetic and chemical methods, it seems to be daily routine that you make mutlicyclic peptides. I have 2 questions. One is: "How do you compare the potential of synthetic and biosynthetic methods to generate these complex peptides?" And then a question for the bioinformaticians: "How do you handle peptides, and in particular cyclic peptides?"

Tom Muir: When comparing the synthetic approaches and the biosynthetic approaches, obviously with the synthetic chemistry one is not constrained by the natural amino acids so you can use unnatural residues and unnatural backbones and unusual topologies that we heard about earlier. There have been great strides though in the biosynthetic, even ribosomally synthesized, cyclic peptides. It is now possible to generate these using various protein chemistry tricks involving proteins called inteins that might work that some of you may be familiar with. You can generate quite large libraries of cyclic peptides, which are genetically encoded using these sort of reagents. This obviously then leads to a selection and all the fun things that are associate. So the biosynthetic approaches have the advantage of ease of synthesis, it's easy to encode a library of cyclic peptides at the DNA level, but you are constrained by what you can put in. You simply wouldn't be able to make easily some of the wonderful architectures that Jean-Louis or Wilfred, can generate using chemistry.

Wilfred van der Donk: I think that summarizes pretty much my thoughts as well. Perhaps one thing to add is that using biosynthetic methods, I would also call out Hiroaki Suga approaches to bring in non-natural amino acids. Those are perhaps the best to get lead compounds because you can make libraries, but then synthetic chemistry can be used to optimize those much better because you have a larger structure or functional space that you can explore with synthetic chemistry. So I think maybe a combination of both is the best of both worlds.

Chris Walsh: To go back to your comments historically about the early days of peptide commercial triumphs, that was the low-hanging fruit. Those were small peptides known to be active as hormones, ACTH a little bit bigger. One can look at the field and say very few peptides of any size beyond 20 residues have been commercialized, it's an interesting problem. I do sit on the board of Ironwood Pharmaceuticals, they have commercialized their 17mer for GI and of course the

Roche-Trimeris group had a 37 residues anti-HIV peptide but it didn't sell effectively. Hard to tell whether cyclic peptides and the new generations will turn out to be a commercially interesting set of targets for the pharma industry. I would be interested by what other people think. Maybe we can go to the bioinformatics.

Adam Godzik: The bioinformatics of peptides is way behind the bioinformatics of full proteins. Not much is known about these peptides. They could just be an overlooked part of human genome. I read recently some papers which started from environmental proteomics where numbers of peptides secreted by the human host is going up to the hundreds and some of them are not even coded in the genome. Our knowledge about this is very scattered, we know few families and they're characterizing hundreds of them now. So it's a completely new area, which I think is not nearly as developed as everything. There are no databases for them, there are no good tools because we have problems with recognition of similarity between such peptides. I think it's a huge area to be explored but not much has been done and not much research exists yet.

Christine Orengo: I completely agree with Adam. I can't think of any resources and it seems to me an ideal opportunity for young chemists or young chemical informaticians to develop something like this. I would think if the problem was more tractable in some way, than what we have to do with protein structures and sequences, I could imagine the tools being somewhat easier to design to compare these compounds. I mean there are good methods for comparing similarities.

Adam Godzik: We struggled with this a little. I was part of a project of meta-proteomics on human skin and human gut and we discovered a lot of peptides, which we can track back to being produced by humans. But we realized there was no catalogue of them and for many of them we don't know whether they are functional peptides or just pieces of proteins, which happened to be in the neighbourhood. Some results suggest they're actually functional peptides. More is known about this in flies because I remember there was a big project where they went and identified peptides secreted by flies. Many of them have antimicrobial properties for instance.

Tom Muir: There is some interesting new data coming out on so-called "small open-reading frames" or SORF's where these are not processed from larger proteins but there are the small genes that maybe have been under the radar in terms of the size. The analytical tools, the mass spectrometry, being used to detect these in cells are being developed. A lot of these peptides do seem to be around and persist for quite some time in cells, implying that maybe they have some function but the functional aspect of what these things are doing is somewhat embryonic.

Jason Chin to Wilfred van der Donk and Chris Walsh: We thought about how to catalogue and categorize and how to understand mechanisms. We can do that endlessly but one question I have, just as a thought, is where are we with really understanding the level of functions? For example in the natural product area, can we recreate ecosystems and understand what the functions of these are? Is that something that is worth investing time?

Wilfred van der Donk: I think that knowing mode of action is of course something we are always looking at not only for the reason that you pointed out but also because any type of target that you identify could be potentially an interesting target for other molecules, not necessarily your molecule that might not have great properties. It's not so easy to identify targets and we have found that the hard way. You often get many possible targets or leads to each target. For some molecules you can look at, you can already guess what the target will be based on what we know, but the great majority of molecules we know that are antimicrobial, how they work, we often don't know.

Chris Walsh: I just want to relate one example. There is a molecule called surfactin, a lipoheptapeptide produced by *Bacillus subtilis*, which was identified by its ability to lower the surface tension dramatically of fluids like water. So it is called surfactin and for years its mode of action was not known, it is a long chain acylheptapeptide. Two of my Harvard colleagues were studying the interaction of this particular *Bacillus* strains with *Streptomyces* and they observed that the *Bacillus* put out a peptide, which blocks the growth of *Streptomyces*. It turned out it was a surfactant only turned on under specific circumstances. They also realised it functions as a potassium channel and so, this potassium signalling had turned on certain kinase genes. The *Bacillus* caused the differentiation pattern. So it's clear that if you want to study the behaviour of genes that are not normally turned on, that is are conditional, one should not study them in pure cultures in a laboratory. It's still an open-research question to figure out what kinds of microenvironments you want to reproduce. That is a sort of a nice example. Surfactin was used actually to reduce the viscosity of oil in oil wells; you can pump up crude oil for decades before the physiological functions was detected.

Richard Lerner: Chris, if you get a mixture of peptide, how do you know it is not just junk, some sort of proteolysis? I think a very good initial cut would be those peptides that are C-terminally amidated because I don't know of any situation where Nature is going to do work to C-terminally amidate a peptide that isn't in some sense being important. I remember in the early days of the Amylin Pharmaceutical company, when during the dark days, when people wondered if amylin was real, or some artifact. The share price was 25 cents. I said if it's C-terminally amidated it must be important and I don't know any company or any research group that has done the sort of the modern and analytical world a screen for C-terminally amidated peptides.

Chris Walsh: To your point Richard, I think almost all of that post-translational amidation reflects the processing of precursor proteins. So these really are not primary transcripts. I think Tom Muir was also referring to work for example of Jonathan Weissman. Ribosome profiling showed you the prediction of primary proteins which are anywhere from 40 to 100 residues which had been missed in sequencing and whose functions are being worked out one at a time. But I can go with, most of the C-terminally amidations come from hydroxylation and then loss of the last residue. So it's a purposeful post-translational processing.

Richard Lerner: A lot of people have tried to make an antibody which could be a great tool to pull out interesting peptides from a mixture. It has not been successful.

Brian Roth and Don Hilvert: I'd like to expand on that. We thought a lot about deconstructing the activities of peptides and small molecules that are in the brain for instance to find their molecular targets and what we decided was that two things were needed. The first was basically a robust genome-wide way to interrogate the targets that is scalable. We now can basically look across entire families of receptors or kinases or whatever in a very robust and relatively cheap fashion. But the problem, the stumbling block that we have is to get compounds that are likely to have biological activity and with the peptide, it's a very difficult problem. So you can imagine if you do a mass spectrometry based approach, you might find thousands of these potentially biologically active peptides that are secreted from a particular neuron. And you don't know, I mean some of them may be amidated and that may be a way to pull them out but if you want to do this from a unbiased approach, basically to screen the entire home of peptides against all of the potential targets that they might interact with, you need to have some ways, cheap and robust way of making these all in quantities that are sufficient to interrogate their biological activities. We've tried to do this over the years with peptide chemists that basically are mining genomic data and coming up with what they think are biologically active peptides identified from genomic and mass spec data. The stumbling block is always to get sufficient amount of material to actually test. Each one of those can be a relatively difficult project in itself. So if there is some way of basically robustifying this so that it can be done in an unbiased way, I think we could really push things forward. The same relates to small molecules, we're now realizing that many of the known and unknown intermediary metabolites are actually ligands for receptors or can modify enzymatic activity. Again the problem is not identifying them which can be done relatively simply with the available analytical techniques but to obtain sufficient quantities of them to actually test. This is where chemistry comes in. We can imagine one of those can be a project but if you want to make thousands of them, it becomes a really monumental task.

Don Hilvert: One potential problem with the bioinformatics approach to identify bioactive proteins is the extensive post-translational post-synthetic modifications that we have heard about. In that context, to what extent are these post-translational post-synthetic events deterministic? Does the first event dictate the next and so one? Obviously in the case of the histones, if you acylate a particular lysine, you can't then methylate it, but there are other sites that may be influenced. That may also be true in smaller peptide systems as well.

Tom Muir: The idea that one modification becomes another one is pretty well established, at least in the chromatin area. Something that is called crosstalk and it's clear that the enzymes that install these things require initiating modifications to work. They are either not recruited properly or not allosterically activated properly. Either way, they will not work on unmodified template. That crosstalk is a very important part of regulation in these systems and very common. Similarly for removing modifications that is just an installation taking the right ones off, there needs to be other constellations or other modifications to do it properly at the right place. So that is an extra layer of complexity to the whole business.

Wilfred van der Donk: So with the biosynthesis of these natural products, you find both. We find cases where chlorination will only happen if the molecule is already cyclized and otherwise the chlorinase doesn't touch it. We find other examples where a tailoring enzyme works on the linear peptide just as well as on the already cyclized peptide. So it seems to be sometimes required and in other cases not.

Chris Walsh: In the kinase field, there are many examples where the first phosphorylation could be stochastic. There might be 500 possible substrates in the cells but the following phosphorylation may require that first one to have been installed. So there are cases where it can become more deterministic as one goes to a second and a third post-translational modification step. Of course there is also the issue that it's very rare for a site to be fully modified in a protein, that is to say it is modified in a given protein but if there are thousands of such protein molecules in the cell, there will be a gradation in terms of mole fractions phosphorylated or acylated which both complicates things and explains stochasticity for example.

Tom Muir: To add to that, there is an important role for informatics here because many of the proteins that install modifications contain domains whether they can recognize the same or different modifications and you can identify these domains from sequences. If you take a step further, obviously many proteins are part of large complexes, but then those complexes are maybe domains that recognize particular PTMs. Based on that information, you can come up with hypotheses that certain patterns may be necessary for a given complex or enzyme complex to work.

Jason Chin to Wilfred van der Donk and Tom Muir: So the idea that, in Wilfred's case, maybe doing highly parallel biosensors and, in Tom's case, highly parallel high-throughput biochemistry to either get a functional to make a collection of natural products. I'm curious if both of you would be able to talk a bit more about how in practice you think that is going to be achievable?

Wilfred van der Donk: So there was a call by NIH this last year for a merge of synthetic biology and natural products biosynthesis. The goal was simply to scale up what we are doing now in genome mining and not to one molecule at a time but doing it hundreds if not thousands at a time. And so that has brought together a large number of synthetic biologists with natural product biochemists. We also put in a proposal as a team. So yes the idea would be to essentially use the merging tools of synthetic biology and expressing proteins in non-natural hosts and doing that in a robotic way. And that's what we are applying right now to our genome mining efforts.

Tom Muir: Our focus has been more on *in vitro* biochemistry in this regard. So we know we can make a lot of these proteins in bulk. The question is "how do you do it in parallel?" I think a key part of this is going to be integrating the synthesis with microfluidics, which is something my lab is doing quite a bit. It is a matter of doing it in parallel and scaling down. Engineering a proteic-fluidic approach is going to be an important part of generating all these things in parallel. Of course, you get smaller amounts of materials, so then you have to start to build at the back end of this, analytical approaches that allow you to manipulate and analyse tiny amounts of proteins, which is another part of the puzzle. But certainly in terms of the manufacture of these things, fluidics I think it's an important part of the solution.

Jason Chin: Can I follow up and ask whether and to what extent there might be scope for doing your sort of work using, or not using hosts, essentially, as *in vitro* experiments and how realistic that is?

Wilfred van der Donk: We do all of our mechanistic work *in vitro* but when it comes to genome mining, we do other things, mostly in *E. coli*, a little bit in yeasts, and a little bit in *Streptomyces*, simply because you need quantity. At the end of the day, once you get your new natural product, you still have to determine the structure and actually we find that that might well be the biggest hurdle to scale this up, to determine the structure of the final product and to make sure that you actually have the final product because it might be just an advanced intermediate. Those are hurdle steps we would face.

Lode Wyns to Christine Orengo: We are overwhelmed by the explosion of sequences, up to a point that we should maybe hold the sequencing. Now there are so many proteins looking for a function, I wonder what the reciprocal exercise will do? We know about functions and there are many enzymes doing particular functions. Statistically, can you do the inverse problem looking for a function and see which folds, which families are connected to it and if so, what is coming out of such an exercise?

Christine Orengo: I think it's difficult answering that. When we analyse families, and when we analyse the types of functions they are able to support, we can see that there are many more families than the functions we observed. So in other words the same function can be performed by multiple families in different ways, with different geometries in the active site. We could ask the question for this function "what are the various different ways in which you can engineer this function, and what are the different ways in which Nature has done that? I think that would be an interesting use of the data. You've alluded to this explosion of sequence data. I think I disagree with the comment that is overwhelming. I think that actually it's very easy for us to organize quite a large proportion of it. Maybe 80 or 90% of the sequences that Adam has alluded to as well are distant homologues of what we have seen already. We can very easily catalogue this new information. So I don't think it's overwhelming in that sense, I think it's actually very exciting because having done that, we can then start to see what are the changes that we are getting in the active site for this particular scaffold. So we can look for positions that are hot spots, we can look for context in the active site where we can see conserved or varying electrostatic and things like that. So we can ask lots of questions very easily from these data. Having more information does give us a lot more insights into ways in which you can change the chemical environment for particular scaffold. So I hope they continue with the sequencing rather than stopping it. We are not finding too much of them at the moment.

Judith Klinman to Christine Orengo: I wanted to comment on your search engine where you focused on functional groups that can be easily identified to play a role in catalysis (acids, bases, catalytic triads), whatever it is that would define a particular enzyme reaction because so much more of the protein is going to be involved in catalysis. And so, going beyond this very small and constrained region and then moving out by degree — I don't know to what extent you've tried that... We know so much now about the impact of remote mutations on catalysis, the role of protein dynamics and other kinds of features. And I think that is also interesting in the context of what you just said about convergent evolution, where you have different scaffolds coming together to give you the same activity. What are the roles of rates in these different families and can you use that information to guide you in terms of structural boundaries and where you look? So, it's a question and comment at the same time.

Christine Orengo: It's an interesting question and you're right. It's not always the active site that you're interested in. And the example I showed of RuBisCo where we used all the sequence information we had in the evolutionary tree, it allowed us to find these sites under positive selection where we could see that frequently there were mutations in these sites that were altering the efficiency of the enzyme. Very few actually are directly in the active sites, they are in a sort of next shell around the active site ensuring that conformation or adjusting the conformation or geometry around the active site. They were also in the interface between some subunits as well, in fact affecting the allostery. What was interesting was we could see multiple routes to achieving this efficient enzyme. So Nature found many ways of adjusting and tuning these positions that could have some influence on the chemistry ultimately.

Judith Klinman: Yes I think the importance of subunit interactions is something that needs more attention for sure, and how that is directed into the active site. I did have a question for you, when you talked about the change in efficiency of RuBisCo, does that relate to the ability to discriminate between CO_2 and O_2? Is that what you mean or is it the turnover number.

Christine Orengo: Yes it is quite complex because there is this competition between oxygen and carbon dioxide for respiration and photosynthesis. And these plants that have become more efficient, it's because they had actually reduced their affinity for carbon dioxide. And by doing so, that actually increased the turnover, it's inversely proportional to their affinities. So that's really what I meant, that was the effect that we saw of these mutations.

Richard Lerner to Adam Godzik: Two practical questions maybe to any of the panel. So in evolution we talk about when that disappeared and when that appeared, and so forth. Do we know when folds appeared? And is there a single case where we actually know what the order was of acquisition of folds?

Adam Godzik: I think that folds appeared in the time we cannot really probe with tools for evolutionary analysis. They all appeared somewhere around when the first cellular life probably appeared. So when we try to track evolutionary folds, it's very hard to say anything about it because it is really too far for us to say anything. They are some folds, which appeared later but I think there are very few. When we ask exactly this question, I think we found like 5 or 6 folds, which are specific to eukaryotes.

Richard Lerner: But you know proteins that used to be something, pick up one more fold and become something else?

Adam Godzik: But we see... for instance especially multi-domains proteins acquiring certain folds so we go track back to the core enzyme.

Richard Lerner: No but very specifically, is there an example of a protein that picked up a new fold and became something else? One fold?

Christine Orengo: Perhaps I could just add one thing to what Adam said. I think if you use the phylogenetic trees, if you trace back, you do see that you can identify perhaps what you would consider to be the most ancient folds and they're often the most highly populated folds. There are lots of superfamilies that adopt these kinds of structure scaffolds. If you look at them, they make sense in some ways in that they are highly regular structure arrangements with quite a lot of hydrogen bonds. Here for example, the β-barrels or the α/β-barrels. You can understand why these structure arrangements have great plasticity to sequence change, so they can support wide range of sequences. In fact you can theoretically predict that these types of structure arrangements would be able to support much more sequence divergence than some of the more irregular structure arrangements. People who have looked at the emergence of new folds using the phylogenetic trees, have shown that, to a certain extent, these new folds are less ordered. They have a more irregular arrangement in secondary structures. More recent ones also seem to be more unique to species, so they have not been so widely used across the tree of life. I don't think there is much more information on that at the moment. The problem is that the structural data is still very sparse compared to the sequence data, which makes it a bit difficult to answer these questions using phylogenetic information, because you can't build very good phylogenetic trees just with a structural data.

Gebhard Schertler to Christine Orengo: Can I make just a little comment to this? So you are looking at sequences but if you look at folds, kind of three-dimensional context become very important. There is a lot of small amino acids (glycine, lysine, leucine), they pack and if something is hydrophobic, they look conserved. But somehow, they are not fully conserved so this is difficult to analyse. In GPCRs, we have used a kind of network analysis to actually get on this, it's in a Nature review recently. And what we can do with this, it's actually we can kind of remove us a little bit from the sequence but work with proximity. And I think if you would look at fold evolution, so how do you transform from one fold to the other, you would need the kind of tool like that and the sequence might not be the ideal thing because many things are held together by this hydrophobic core of the proteins which is made out of these small amino acids, which are really not so distinctive in sequence analyses. So when you reduce the protein to kind of contact and abstract it to this, you are actually looking at this space in a different way and you don't need every structure, you can use your structure and the sequence variability and use that in the network analysis. And I actually hope that brings more structural insights in the comparison. That's why we started doing that as an idea but I don't think we did it very well.

Christine Orengo: If I could briefly respond to that. I think you are right and you are sort of translating your sequence code into a set of properties that you put then against structure scaffold. There are other people in the room involved in protein design and I know there was a competition at one point and I don't know if it still runs to try to convert one structure into a completely different structure which uses the idea of an alphabet of residue properties and switching those properties dramatically to try to engineer a completely different fold. There might be people in the room who know about those competitions?

Bernard Henrissat, Adam Godzik and Christine Orengo: I would like to say that I agree very much with what Christine Orengo said about the fact that the number of sequences is not so much of a problem. There are bioinformatics methods that are really designed to handle them. Where I have a problem is with the storage of the knowledge, of functional knowledge. There is no way that I know where this is done in a systematic manner to capture this information and to store it just by analogy to what is done with three-dimensional structures. There is a PDB with some rules that we can follow easily and we must follow if we want to publish. And for functional data, it's not so good, it's not good at all. And we have some systems that have changed of purpose. I suggest the EC numbers that were not designed for bioinformatics and they are being used now for bioinformatics. And these EC numbers however imperfect they are, they're only for enzymes and proteins are not only enzymes. We are missing also descriptors for binding relationships between a protein and a ligand.

Adam Godzik: Overall you're right but it's not because of lack of trying. So there are efforts such as gene ontology, which tries to map literature information into formalized function description. I think EC system was also an inspiration for the gene ontology. There are several groups working seriously but I agree that overall, as compared to the amount of structural information captured in PDB, the amount of functional information available in literature captured in this classification is relatively small. But there is also a problem with distribution of it. So I agree with you, we find routinely papers from ten years ago, which defined the functions of proteins that were not captured anywhere. So there are efforts to do this, but I agree that this is way behind our capturing structure information.

Christine Orengo: I think a lot of money has been put in to the gene ontology classification. They are trying to expand it even further to reach out to communities working on particular pathways of chemistry and get them to actually annotate the proteins within this resource. But obviously, that is quite a big initiative and it's quite slow, but another quite significant problem is with querying the data as well, it's much harder to query than the EC information. So that's another problem, it's not just organizing it, but organizing in the way that you can query easily is another challenge.

Ian Wilson to Adam Godzik and Chris Walsh: So given that there appears to be a lot less folds than initially anticipated. Adam, you talked about 1300 we currently know but I think you backed off on that in a previous question saying there is probably not that many more folds to find than we already know. So when is it that we're going to be able predict from the sequence alone, from any sequence, what fold a structure is going to adopt from the sequence?

Adam Godzik: First let me clarify that there are differences. So there are 1300 or so, plus/minus a hundred folds we have found. There are several thousands, which are mathematically possible, and to rephrase what David Baker did, they designed a sequence to fold into a new fold, which is not seen in Nature. They synthesized it and they did. So for many of these folds, which are technically or theoretically possible, there are no physiochemical reasons for them not to exist. Nature simply hasn't found it. And this is one of the puzzles here, why there are so many millions of sequences and there are still possibilities of how they can shape and they were not used. One of the possibilities here is that Nature simply was happy with what it found. The second part of the question is again, there are big efforts to go there and David Baker's group and few other competing groups have really made big progress. So at this point, there are sequences for which we can predict structure. When it could be possible for all the sequences, . . .

Chris Walsh: about 90%. . .

Christine Orengo: I think that's possible. . .

Adam Godzik: Basically the problem is that folds are very often defined by a delicate balance between different competing forces and getting this balance exactly right requires to both contribution to be calculated very accurately. So as I've just quoted David Baker, he can predict easily exaggerated folds — so folds, which are dominated by one type of force. This can be easily predicted. But especially proteins that are really on the bridge between two opposing forces are very difficult to get correct.

Richard Lerner: But isn't it a problem that you're looking at the thermodynamic outcome and you have no idea whether any fold would be kinetically possible in a context of the thermodynamically finished protein?

Christine Orengo: I think that also brings to another issue with the kinetics and why we maybe don't see so many of the folds that we predict should be stable and accessible to Nature. And some of the ones that we see very very frequently when you analyse their topologies, you can understand perhaps why they would fold very rapidly. They've got a lot of local contacts between secondary structures and things like that. So I think kinetics is another important consideration in perhaps explaining why we have this limited fold set.

Rudolf Glockshuber to Wilfred van der Donk: I would like to come back to the bioactive peptides and I also have a question to the bioinformaticians. So we heard that many biosynthesis pathways of these peptides are known nowadays. The enzymes introduce unusual modifications unknowingly. So do you think that there is a potential in the long term for bioinformatics to predict the covalent structures of all bioactive peptides that an organism produces from only knowing its genome sequence?

Wilfred van der Donk: I'm not a bioinformatician but currently, certainly not. An enzyme that hydroxylates an amino acid, no idea which amino acid, at which carbon, at which position and so whereas we can say from the enzymes, ok these are probably the post-translational modifications that are going to take place. Where they will take place? And when you have cyclic, polycyclic compounds what their topology would be? I cannot predict it right now but maybe if we have enough sequences and enough examples of solved structures, and that, I think, is going to be required as well, not of the enzymes but of the natural products. Maybe at some point we will get close but currently I would say no. But again, I am not in bioinformatician.

Adam Godzik: I don't know but I think we are looking at a sort of exponential, an explosion of possible permutations. So you can list all the possible mutations and you can calculate how many different possibilities there are, which of them will be realised. It's much more difficult to understand and you have to know how they are regulated and this is something which we don't know at this point.

Don Hilvert to Christine Orengo: I was really fascinated by the example of the carboxylases that you mentioned, that have the same scaffold, catalyse the same reaction but use a completely different functionality. With those, was the activity identified before the sequence analysis or did it emerge from the sequence analysis? Secondly where did this rare case arise evolutionarily? Can one trace its lineage? Did it come from a completely different activity and then regain the carboxylase activity?

Christine Orengo: So the function information came from the literature. So we've combined all of our family information with experimentally annotated functions that we have from either the EC or from the gene ontology. So that's were the function information came from. And then it's a difficult question as regards to the phylogeny because some of the families are so diverse and we don't have enough of the intermediates sequences to be able to trace exactly what the steps are. It is a difficult question as to whether it is convergent evolution. These arrangements have a reason and have been efficient, have been preserved within an organism. Whether you have a case of a where you have mutational drifts and it lost some of its chemical activity, it reduced efficiency but then that is subsequently recovered by further mutations in different sites in the pocket.

Don Hilvert: It went through an intermediate with completely different activity.

Christine Orengo: Exactly! So I mean that is why for me I would like many more sequences because then it's easier for us to trace what the steps might be in these different routes to different answers or how you can facilitate this chemistry given this sort of structure arrangement. But we actually do need more sequences and actually, more importantly we need more structural data as well. That's why the structure initiative in genomics in the States was so exciting because they were allowing us to answer some of these questions and try to solve structures with some of the intermediates sequences.

Jason Chin to Chris Walsh: So Chris, I think one of the things that you mentioned in your talk was the radical SAM superfamily having on the order to the 104 open-reading frames and this might be a good place to look for new activity or new chemistry. I wondered with all of your experience whether there is chemistry that you imagine might be in biology that you haven't seen experimentally and things like transformations that you're particularly interested in looking for? In that family or any other family more probably.

Chris Walsh: I believe oxidation chemistry is amongst the most interesting that happens biotically or abiotically. I think we'll here from Frances Arnold about some P450-mediated chemistry that is novel. And I guess I would just go back and say I think most of the sort of nucleophilic-electrophilic reaction manifolds, that happen in Nature are well understood and predictable. The electron radical carbon chemistry I think it has been very slow to come up in part because that happens in microbes that are anaerobic under simple biological niches and because the enzymes have been oxygen-sensitive and so not amenable to study under aerobic conditions. So I put my money on oxidative transformations that are yet to be worked out but I'm sure Frances and others may, during the course of this afternoon come back to that question.

Tom Muir to Chris Walsh: What about the fluorinationfor example? So everybody has a little chemist favourite modifications at the moment right, or maybe forever. To my knowledge there is only one fluorinase known right, the SAM-dependent fluorinating enzymes. Maybe that's out of date but that they are not so many, right? Is that surprising?

Chris Walsh: The comment is about fluorination and let me put it in a category of halogenation more generally. In fluoride, it's the most difficult of the halogens to incorporate into organic scaffolds, in part because it's so nucleophilic and you have solvation energy problem to get it de-solvated. So yes, as far as I know, there is only one class of fluorination that has been discovered. By contrast, there are 4 different chlorination families that have evolved and have used different kind of cofactors. So I think it's the difficulty of de-solvating fluoride ion that is the problem there. There are brominations very common in marine organisms because of the amount of bromide in sea water and I think there are only about a dozen iodo compounds, iodotyrosine and iodothymidine of course being the most famous that are found. So there are some interesting niches there perhaps in halogenation. But last comment, chlorination most of it is oxidative and you cannot oxidize fluoride ion under biological conditions.

Don Hilvert: It might be worth mentioning Michelle Chang's recent work incorporating fluorine using the non-ribosomal peptide synthesis or polyketide synthesis, which I think is really beautiful. Clearly, it needs a lot of work but I think it's a very very nice study and shows where we can go in this area.

Chris Walsh: Other comments?

Judith Klinman: Just this idea of aerobes and anaerobes, radical SAM. Someone mentioned the possibility of doing bioinformatics in the division between the anaerobes and aerobes. There are radical SAM enzymes that appear in aerobes and no one really mentioned that and how they succeed because these are so rare and intolerant, whether there is any information through bioinformatics that could give us a clue. It's kind of a crossover between the anaerobic and the aerobic world. I just put that out as a curiosity, I don't really have any insight to say beyond there but it is a puzzle and something to just put out.

Chris Walsh: There are many microaerophilic bacteria that practice anaerobic chemistry so to speak. All the non-heme mononuclear iron enzymes when isolated are very oxygen-sensitive. Some of them are turning over only 2 to the 5 times before they kill themselves. And yet presumably in microaerophilic organisms, they run thousands of turnovers. So it's a compartmentalisation and control of oxygen tension in the organism maybe.

Aled Edwards: I finally give some perspective just on the protein universe. I think the first meeting we held on it was probably 1998 when we were thinking about the PSI at the very beginning as we've tried to classify. Then the DOE had a meeting I went to in 2002 about the unannotated proteins in every genome that they were sequencing and the message was we need more and more experiments but it's too expensive. In 2006, I went to another meeting talking about the unannotated proteins every new genome has — the proteins we don't know what they do, we should do something but it's too expensive. And in 2010, the same, the PSI went for new folds, it was fantastic but too expensive, stop doing that and do some real biology. And now in 2013, we are all talking about the same problem, about the same fraction of the genome. So I think what we can do actually is to figure out how to get it done instead of talking about these unannotated ones.

Chris Walsh: John Gerlt, you better figure that out in the next hour and a half.

Session 2

Exploring Enzyme Families and Enzyme Catalysis

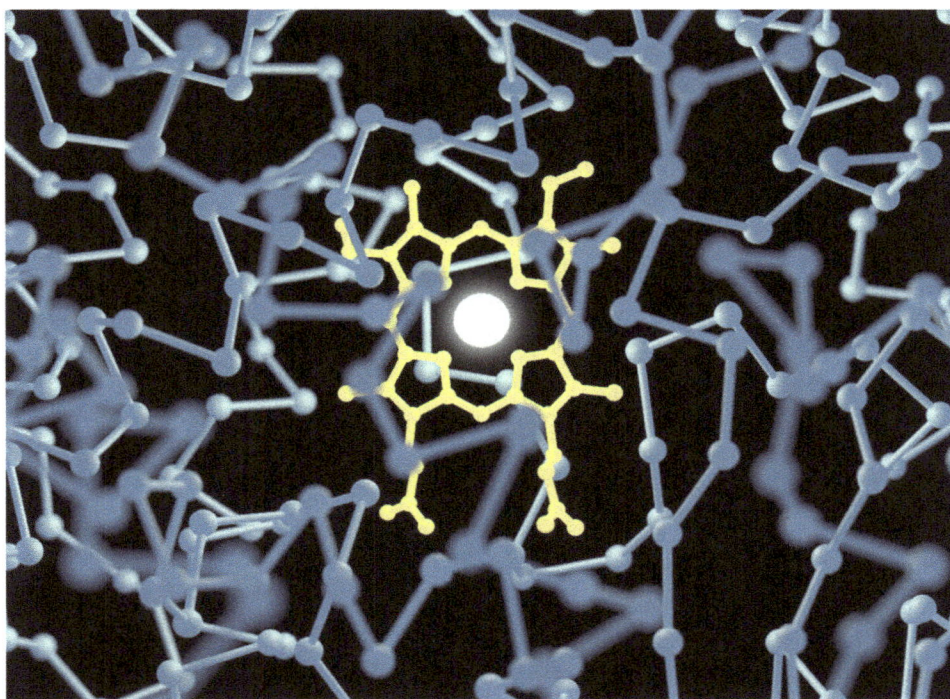

Illustration of the iron-heme center of a cytochrome P450 enzyme engineered by the Arnold laboratory at California Institute of Technology to catalyze non-natural reactions.

MECHANISTIC ENZYMOLOGY AND CATALYST DESIGN

DONALD HILVERT

Laboratory of Organic Chemistry, ETH Zürich
CH-8093 Zürich, Switzerland

My view of the present state of research on exploring enzyme families and enzyme catalysis

Enzymology is a mature field of study. Over decades, thousands of enzymes have been catalogued and characterized; thanks to intrepid biochemists and microbiologists and recent progress in genome mining, novel activities continue to emerge apace. Sustained interest in this area reflects the importance of enzymes as the catalytic engines of the cell and as targets for therapeutic intervention. Technological advances have also made biocatalysts available for novel applications inside and outside of the cell, paving the way for synthetic biology [1] and environmentally friendly synthetic chemistry [2].

The basic workings of many enzymes are now known. Detailed structural and mechanistic studies have elucidated the sequence of bond-making and bond-breaking events, the identity of key reaction intermediates, and the role of critical functional groups. Nevertheless, explaining the origins of catalytic proficiency in quantitative terms has proved difficult [3]. A full accounting of enzyme efficiency is problematic owing to the complexity of these systems and the non-additivity of the contributing factors. As a consequence, fundamental concepts such as the roles of electrostatics [4] and dynamics [5–7] or the relative importance of transition state versus ground state effects [8–10] are still controversially discussed.

Given our incomplete understanding of protein structure-function relationships, expanding the universe of enzymes by design constitutes a formidable challenge. Not only is the number of theoretically possible protein sequences astronomically large, functional sequences are also rare. First steps toward fully automated *de novo* design of artificial enzymes are being taken [11], but adapting natural protein scaffolds for new tasks is usually simpler than starting from scratch. Diverse strategies exist for tailoring enzyme properties [12]. Both targeted and random mutagenesis can be employed, for example, to increase stability, alter specificity or create completely new function. Once identified, low levels of activity are reliably optimizable by directed evolution [13]. Chemical guidance, provided either by structural data or computational design, can greatly enhance the chances of success of such endeavors by increasing the frequency of variants possessing interesting catalytic activities in the sample population [14]. Nascent enzymes for several abiological reactions have been created in this way [15, 16].

Design and evolutionary optimization of artificial enzymes has the potential to bring our understanding of basic enzyme chemistry into sharper focus. The general principles learned are likely to be transferable to increasingly complex systems and thus support wide-ranging applications in research, industry and medicine.

My recent research contributions to exploring enzyme families and enzyme catalysis

My group utilizes chemical synthesis, immunology, molecular biology, and genetics to investigate enzyme catalysis. The functional modification of existing active sites, epitomized by conversion of a PLP-dependent racemase into an aldolase through a single active site mutation [17], is one focus. Enzymes that catalyze pericyclic reactions, including chorismate mutase and various pyruvate lyases, are another. Exploitation of macrophomate synthase, a putative Diels-Alderase, as a promiscuous aldolase for the stereoselective synthesis of a wide range of 3-deoxyketo sugars highlights the practical opportunities made possible by such research [18].

Laboratory evolution — iterative rounds of mutagenesis and screening or selection — has become a powerful tool for studying and modifying proteins. We have exploited this approach to probe catalytic mechanism, investigate the determinants of protein structure, redesign protein topology, and tailor supramolecular structures (Fig. 1) [14]. The evolution of a highly active monomeric chorismate mutase with all the features of a molten globule — a dynamic ensemble of poorly packed and rapidly interconverting conformers — illustrates the surprises that can emerge from such studies [19, 20]. That laboratory evolution is not limited to catalysis is exemplified by protein capsids engineered to encapsulate and sequester toxic cargo [21]. We are further extending this technology to customize protein containers into nanoscale reaction chambers and delivery vehicles.

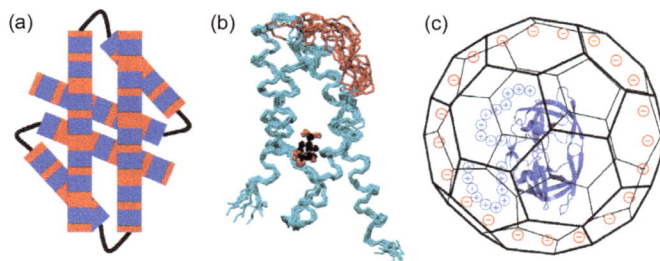

Fig. 1.   Evolved catalysts and constructs. (a) A binary-patterned chorismate mutase constructed from a limited library of 9 amino acids [22]. (b) A molten globular enzyme [19, 20]. (c) Encapsulation of toxic HIV protease in an engineered protein container [21].

Reactive functional groups, metal ions and organic cofactors can be employed to expand the capabilities of proteins [12]. For example, novel spectroscopic and redox properties of the element selenium render it a valuable mechanistic probe and catalytic group. In biological systems, selenium is incorporated into proteins as selenocysteine, sometimes termed the twenty-first proteinogenic amino acid. In the lab, we have engineered artificial selenoenzymes by chemical, semi-synthetic, and recombinant methods. In cytochrome P450cam, for instance, recombinant replacement of the proximal heme ligand by selenocysteine afforded a unique mechanistic and spectroscopic handle on reactive intermediates (Fig. 2(a)) [23]. To access an even wider range of chemical activities, coupling the unique reactivity of transition metals with the selectivity conferred by specific protein scaffolds represents an attractive strategy [24]. In one case, we attached a Grubbs–Hoveyda type catalyst to a protein framework to yield an artificial metalloenzyme that promotes ring-closing metathesis [25]. As an extension of efforts to create abiological enzyme mimics, we are also developing catalysts based on other sequence-defined polymers, such as β-peptides and mixed α/β-peptides, which can fold into well-defined secondary and tertiary structures [26]. Such molecules are useful for testing the tenets of molecular recognition and catalysis.

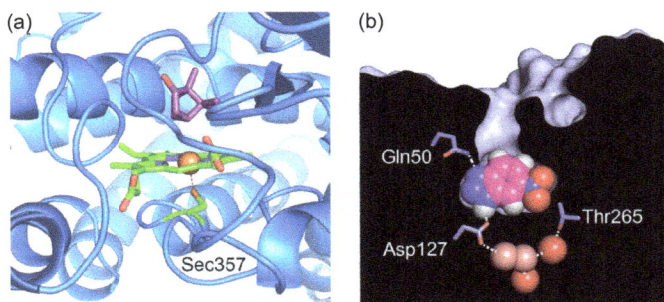

Fig. 2. Artificial enzymes. (a) A variant of cytochrome P450cam in which the axial cysteine ligand was replaced by selenocysteine [23]. (b) A computationally designed and experimentally optimized Kemp eliminase showing unusually high efficiency [30].

Exploiting the promise of computation, our group collaborates with David Baker (University of Washington, Seattle), Ken Houk (UCLA) and Steve Mayo (Caltech) on the design of enzymes for abiological reactions. Computational design is conceptually similar to the creation of catalytic antibodies [27, 28], but because it can explore protein sequence space more thoroughly, it has the potential to go far beyond. This approach has already yielded catalysts for simple hydrolytic and proton transfer reactions, a bimolecular Diels-Alder cycloaddition, and multistep retro-aldol reaction reactions [15, 16]. Although the starting designs typically exhibit modest activities, they are readily improved by directed evolution. Optimization of the originally programmed mechanism is the norm, but the unexpected does occur, as evidenced by extensive active site remodeling and a catalytic residue switch observed for a model retro-aldolase [29]. Evolution of a computationally designed enzyme that catalyzes the Kemp elimination, a simple proton transfer reaction, with a k_{cat} of 700 s^{-1} and a k_{cat}/K_M of 230,000 M^{-1}s^{-1} shows that very high rates can be achieved in favorable cases [30]; the 6×10^8 fold rate acceleration exhibited by this catalyst approaches the exceptional efficiency of highly optimized natural enzymes like triosephosphate isomerase. Structural studies are important to chronicle how shape complementarity and precisely placed catalytic groups give rise to such effects (Fig. 2(b)).

Outlook to future developments of research on exploring enzyme families and enzyme catalysis

As with many contemporary scientific challenges, protein chemistry is amenable to integrated, interdisciplinary approaches. An array of chemical, biological and biophysical tools is opening a multiplicity of pathways to a better-nuanced understanding of protein structure-function relationships, enzyme function and evolution, and the engineering of entirely new properties.

De novo design of enzymes for any conceivable reaction remains a distant prospect, yet there is reason for optimism that the protein universe will continue to expand in exciting new directions. Artificial enzymes are attractive alternatives to natural enzymes for elucidating basic principles of catalysis and as practical catalysts in their own right. Bottom-up production of these agents, as opposed to top-down dissection of naturally evolved systems, promises to enhance our understanding of the origins of catalytic efficiency. Emerging technologies, particularly the combination of computation and laboratory evolution, can be expected to facilitate exploration of varied catalytic motifs and starting scaffolds with respect to activity and specificity, on the one hand, and further diversification on the other.

Beyond protein catalysts for relatively simple chemical transformations await more complex assemblages, including non-biological polymers and molecular machines. The creation of orthogonal genetic materials, complex signal transduction systems, biosynthetic pathways for unnatural products, artificial organelles and even whole cells represent exciting prospects for the future. Modern challenges in human health, energy, and sustainability will certainly benefit from progress toward these aims.

Acknowledgments

The author is indebted to the ETH Zürich, the Swiss National Science Foundation and DARPA for generous support.

References

1. P. E. Purnick, R. Weiss, *Nat. Rev. Mol. Cell Biol.* **10**, 410 (2009).
2. U. T. Bornscheuer, G. W. Huisman, R. J. Kazlauskas, S. Lutz, J. C. Moore *et al.*, *Nature* **485**, 185 (2012).
3. D. Herschlag, A. Natarajan, *Biochemistry* **52**, 2050 (2013).

4. A. Warshel, P. K. Sharma, M. Kato, Y. Xiang, H. Liu *et al.*, *Chem. Rev.* **106**, 3210 (2006).
5. Z. D. Nagel, J. P. Klinman, *Nat. Chem. Biol.* **5**, 543 (2009).
6. V. L. Schramm, *Annu. Rev. Biochem.* **80**, 703 (2011).
7. G. G. Hammes, S. J. Benkovic, S. Hammes-Schiffer, *Biochemistry* **50**, 10422 (2011).
8. T. C. Bruice, S. J. Benkovic, *Biochemistry* **39**, 6267 (2000).
9. M. Štrajbl, A. Shurki, M. Kato, A. Warshel, *J. Amer. Chem. Soc.* **125**, 10228 (2003).
10. M. Garcia-Viloca, J. Gao, M. Karplus, D. G. Truhlar, *Science* **303**, 186 (2004).
11. B. Kuhlman, G. Dantas, G. C. Ireton, G. Varani, B. L. Stoddard *et al.*, *Science* **302**, 1364 (2003).
12. M. D. Toscano, K. J. Woycechowsky, D. Hilvert, *Angew. Chem. Int. Ed.* **46**, 3212 (2007).
13. F. H. Arnold, *Acc. Chem. Res.* **31**, 125 (1998).
14. C. Jäckel, P. Kast, D. Hilvert, *Annu. Rev. Biophys.* **37**, 153 (2008).
15. G. Kiss, N. Çelebi-Ölçüm, R. Moretti, D. Baker, K. N. Houk, *Angew. Chem. Int. Ed.* **52**, 5700 (2013).
16. D. Hilvert, *Annu. Rev. Biochem.* **82**, 447 (2013).
17. F. P. Seebeck, D. Hilvert, *J. Amer. Chem. Soc.* **125**, 10158 (2003).
18. D. G. Gillingham, P. Stallforth, A. Adibekian, P. H. Seeberger, D. Hilvert, *Nat. Chem.* **2**, 102 (2010).
19. K. Vamvaca, B. Vögeli, P. Kast, K. Pervushin, D. Hilvert, *Proc. Natl. Acad. Sci. USA* **101**, 12860 (2004).
20. K. Pervushin, K. Vamvaca, B. Vogeli, D. Hilvert, *Nat. Struct. Mol. Biol.* **14**, 1202 (2007).
21. B. Wörsdörfer, K. J. Woycechowsky, D. Hilvert, *Science* **331**, 589 (2011).
22. K. U. Walter, K. Vamvaca, D. Hilvert, *J. Biol. Chem.* **280**, 37742 (2005).
23. C. Aldag, I. A. Gromov, I. Garcia-Rubio, K. von Koenig, I. Schlichting *et al.*, *Proc. Natl. Acad. Sci. USA* **106**, 5481 (2009).
24. T. Heinisch, T. R. Ward, *Curr. Opin. Chem. Biol.* **14**, 184 (2010).
25. C. Mayer, D. G. Gillingham, T. R. Ward, D. Hilvert, *Chem. Commun.* **47**, 12068 (2011).
26. M. M. Müller, M. A. Windsor, W. C. Pomerantz, S. H. Gellman, D. Hilvert, *Angew. Chem. Int. Ed.* **48**, 922 (2009).
27. R. A. Lerner, S. J. Benkovic, P. G. Schultz, *Science* **252**, 659 (1991).
28. D. Hilvert, *Annu. Rev. Biochem.* **69**, 751 (2000).
29. L. Giger, S. Caner, R. Obexer, P. Kast, D. Baker *et al.*, *Nat. Chem. Biol.* **9**, 494 (2013).
30. R. Blomberg, H. Kries, D. M. Pinkas, P. R. E. Mittl, M. G. Grütter *et al.*, *Nature* **503**, 418 (2013).

LOOKING IN NEW DIRECTIONS FOR THE ORIGINS OF ENZYMATIC RATE ACCELERATIONS

JUDITH P. KLINMAN

Department of Chemistry, Department of Molecular and Cell Biology, and California Institute for Quantitative Biosciences, University of California Berkeley, CA 94720-3220, USA

Historical overview of enzyme catalysis

As the most powerful catalysts in Nature, enzymes are capable of accelerating rates of reaction by up to 10^{25}-fold [1]. A satisfactory explanation for the origins of these enormous rate accelerations remains a major challenge. Structural studies provide a robust context for our visualization of enzyme active sites. It is immediately clear that the "molecularity" of enzyme reactions has no counterpart in small molecule chemistry, with a multitude of side chains positioned close to the substrate-binding pocket to provide nucleophilic, acid/base and electrostatic catalysis. When this is combined with the action of active site bound organic cofactors and metal ions, the uniqueness and catalytic power of an evolved enzyme active site becomes apparent.

Historically, one of the enduring treatments of enzyme rate acceleration has been in the context of transition state theory, with catalysis emerging, by definition, as a result of a reduced free energy barrier height. This had led to an almost exclusive focus on "enhanced transition state binding" as the physical origin of enzyme catalysis. Many investigators have pondered how subtle changes in reactant bond lengths and angles that occur in the transition state can produce a binding enhancement by up to 10^{25}-fold. Additionally, the neglect of protein motions in earlier theories now seems untenable. Increasingly, investigators have been arriving at the view that the total protein structure will play a role in the tuning of active site structure and reactivity. Motions of the protein backbone and side chains have long been recognized as important for the entry and egress of substrate and product from an enzyme active site, together with the movement of protein domains and loop closures that bring key active site residues into proximity with bound substrate(s). More recently, another role for protein motion has emerged as an important element in catalysis. This is proposed to involve a stochastic sampling among multiple ground state structures to obtain the protein configurations that are optimal for catalysis (*e.g.*, [2]).

Evidence for active site compaction in enzyme catalysis

The present discussion will be focused on two widespread classes of biologically important reactions involving hydrogen and methyl transfer, respectively. *While the details of each study differ greatly, a similar picture is emerging in which barrier compaction, controlled in part by the size of strategically placed amino acid side chains, exerts a dramatic impact on the rate of the enzymatic chemical transformation step(s).*

C–H Activation and Hydrogen Tunneling. Studies on enzyme-catalyzed C–H activation reactions have indicated the importance of through-the-barrier, hydrogen tunneling behavior (cf. [3]). The ability to stably incorporate isotopes of hydrogen at the reactive bond and to measure kinetic isotope effects (KIEs) has played a key role in characterizing these systems. A number of distinctive features have emerged in relation to similar measurements for small molecules in solution [4]. First, almost uniformly for native enzymes operating under optimal conditions, the temperature dependence of the KIEs is found to be very small, often undetectable, Fig. 1(A, top). Second, deviations away from optimal conditions (*e.g.*, by mutagenesis or alteration in reaction conditions) are found to increase the barrier for reaction of the heavy isotopes of hydrogen much more than for protium itself, producing the pattern of temperature-dependent KIEs, Fig. 1(A, bottom).

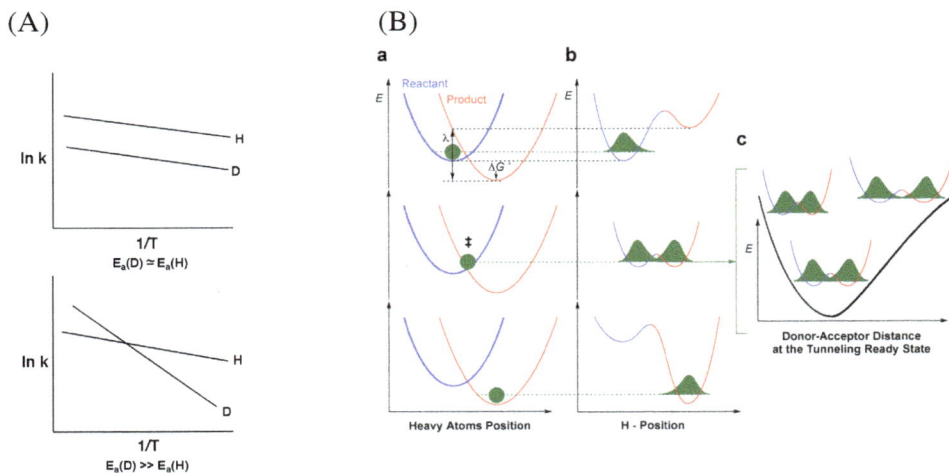

Fig. 1. (A) Depiction of the two extreme temperature dependencies of KIEs. Top is the pattern characteristic of native proteins, and bottom represents the behavior under non-optimal catalytic conditions. (B) The multi-dimensional nature of H-tunneling depends on the adjustment of ground state electrostatics (a), wave function overlap (b), and donor-acceptor distance samplings (c), [4].

The behavior in Fig. 1(A) has now been observed for a large number of enzyme reactions that differ with regard to substrate structure, active site cofactor and the nature of the hydrogen transfer step (hydrogen atom vs. hydride ion or proton).

Experimentally, mutagenesis of hydrophobic side chains has proven very informative, allowing a systematic analysis of the impact of the creation of protein packing defects on the generation of temperature-dependent KIEs.

Many theoretical models have been advanced in an effort to explain the properties of enzymatic C–H activation reactions. A physical model, capable of representing the behavior of Fig. 1(A), is illustrated in Fig. 1(B) [4]. The movement of hydrogen from donor to acceptor is treated fully quantum mechanically, with a mass dependence for the rate that comes from a differential efficiency of wave function overlap, $H > D > T$, Fig. 1(Bb). The extent of this mass dependence is linked to the distance over which the particle moves, with longer donor-acceptor distances (DADs) favoring the more diffuse wave property of H (giving rise to larger KIEs); shorter distances diminish the gap in reactivity among isotopes, producing smaller KIEs. Of considerable importance for interpreting the experimentally observed trends in KIEs, wave function overlap is independent of temperature.

The temperature dependence of the rate comes solely from heavy atom motions that are required to tune the active site environment for successful H-transfer, and these represent the activation barrier. A key coordinate describes the thermal activation necessary to achieve transient degeneracy of vibrational modes in the reactant and product wells (essential for all reactions where the equilibrium constant for inter-conversion of substrate and product is different from unity). This term is analogous to the Marcus description for electron tunneling [5], and depends on two primary parameters: λ, the environmental reorganization and ΔG°, the reaction driving force, Fig. 1(Ba). Separation of wave function overlap, Fig. 1(Bb), from the motions of the heavy atom environment, Fig. 1(Ba), results in a near absence of cross-terms for mass and temperature-dependent phenomena as required from the observations of temperature-independent KIEs.

One feature that is omitted from the above description is the magnitude of the initial DAD. In virtually all theoretical treatments of hydrogen tunneling, the distance between donor and acceptor atoms must approach ca. 2.7 Å for protium tunneling and even less than this value for the movement of the heavier isotopes D and T. The reactive internuclear distance is, thus, more than 0.5 Å reduced from an estimated van der Waals distance of ca. 3.2 Å. An additional class of thermally-activated protein is generally invoked to create short, tunneling-appropriate DADs. These motions are referred to as conformational sampling and though introduced here in the context of H-tunneling, are likely an essential component of all enzyme reactions [2].

SLO-1 offers a classical example of enzymatic tunneling behavior, illustrating how mutation at a single conserved hydrophobic side leads to a progression from $E_a(D) - E_a(H)$ close to zero for wild-type enzyme to 5.3 kcal/mol for Ile[553]Gly [6]. The incorporation of this behavior into the physical model for enzymatic H-tunneling is achieved via the introduction of a third coordinate to the tunneling process. As shown in Fig. 1(Bc), this involves a DAD sampling term that becomes a necessity when an enzyme loses its capacity to achieve the compact active site configurations of native protein. In contrast to the behavior of Figs. 1(Ba) and 1(Bb), Fig. 1(Bc) represents a thermally active motion that is significantly dependent on the mass of the transferred particle. The latter follows from the need to reach shorter DADs for tunneling of the isotopically heavy particles: once the controlled tuning of active site geometries within native enzymes has been disrupted, a behavior more characteristic of solution reactions begins to take over.

One question is how far an enzyme active site structure can be disrupted before it loses its capacity to approximate the behavior of the WT protein. The simple act of mutating pairs of hydrophobic side chains in SLO-1 has turned out to be incredibly informative in this regard. New data have recently emerged for a double mutant of SLO-1, Leu^{546}Ala/Leu^{754}Ala. A combination of steady-state and single turnover kinetics, X-ray crystallography and theoretical modeling on this variant provide a powerful test/confirmation of the model outlined above for enzymatic C–H activation [7].

Enzyme-Catalyzed Methyl Transfer Reactions. In order to ascertain the generality of reduced barrier width in native enzyme catalysis, studies have been extended to include the large family of methyl group-transferring enzymes. Methyl transfer is as ubiquitous in Nature as C–H activation, occurring at the levels of small molecule, protein, RNA and DNA modification. To initiate our studies, we chose an enzyme with well-established structural and kinetic characterizations, human catechol-*O*-methyltransferase (COMT) in which a methyl group is transferred from S-adenosylmethionine (AdoMet) to the oxygen of a catecholate.

Very early studies on this enzyme compared the magnitude of secondary KIEs, using AdoMet labeled with deuterium at the reactive methyl group, to a small molecule model reaction. A more inverse value for the KIE in the enzyme reaction led to the proposal of catalysis via transition state compression [8]. This was a unique and controversial finding that stimulated considerable activity among theoreticians. Following on our studies of C–H tunneling, that show control of active site compaction by bulky amino acid side chains, we pursued site-specific mutagenesis in COMT. The position targeted was Tyr68, a conserved residue that sits behind the sulfonium atom of AdoMet. From inspection of X-ray structures of COMT, key interactions involving Tyr68 include a close approach (within van der Waals contact) of its β carbon and the sulfonium center of cofactor, as well as an inferred hydrogen bonding interaction between its ring hydroxyl group and a nearby Glu. Kinetic studies were carried out using cofactor labeled with three tritiums at the activated methyl group, to increase the sensitivity of measurements beyond what is possible with deuterated cofactor [9]. The results indicate a trend in both k_{cat} and k_{cat}/K_M as Tyr68 is converted to either Phe or Ala; substitution of WT (108V) by a variant that maps to documented disease states (V108M) gives similar results. A plot of k_{cat}/K_M for reaction of the preformed COMT-AdoMet complex with dopamine (Fig. 2) shows a remarkable linear correlation between the size of the second order rate constant and the magnitude of the secondary KIE. One explanation for the effect, offered at the time of publication, was a transition state compaction in the native protein that is substantially lost on reduction in the size of the Tyr68 behind the methyl group donor. These studies have recently been extended, with the goal of establishing a molecular basis for the correlation shown in Fig. 2. A combination of experimental binding isotope effect measurements, high-level GPU-based QM/MM calculations that allow up to 30% of COMT to be analyzed using *ab initio* methods [10] and classical MD simulations point toward an important role for ground state compaction in COMT [11].

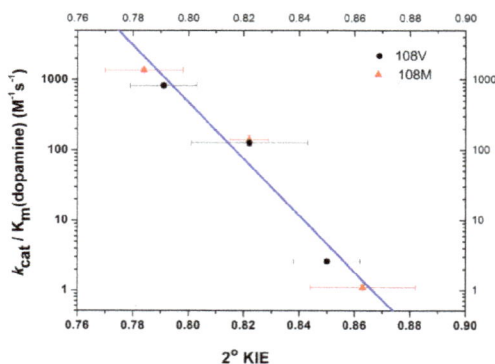

Fig. 2.   Linear relationship between k_{cat}/K_M(dopamine) and the magnitude of the 2° KIE in COMT [9].

Perspectives

The studies outlined above implicate a general role for active site compaction in enzyme catalysis. This can be achieved by a sampling of protein conformational substates (H-tunneling) or through a combination of structural and dynamical effects (methyl transfer). The methyl group transfer study extends the concept of active site compaction beyond the realm of systems that are dominated by quantum mechanical tunneling. For the future, it will be extremely important to pursue studies of a third enzymatic group transfer reaction. Phosphoryl transfer is a very attractive target, given its central role in biology. This will require developing a set of tools that can assess compaction and its sensitivity to alteration in protein structure/dynamics. The presence of active site compaction implicates a role for protein structure that goes beyond the "inner" active site region. Novel protein engineering approaches, that allow investigation of the role of overall protein size in catalysis, may be of considerable interest.

Acknowledgments

This work has been supported by the NIH (GM025765) and NSF (MCB0446395).

References

1. R. Wolfenden, *Chem. Rev.* **106**, 3379 (2006); D. R. Edwards, D. C. Lohman, R. Wolfenden, *J. Amer. Chem. Soc.* **134**, 525 (2012).
2. S. J. Benkovic, G. G. Hammes, S. Hammes-Schiffer, *Biochemistry* **47**, 3317 (2008).
3. S. Neidle, R. K. Allemann, N. S. Scrutton (eds.), *Quantum Tunnelling in Enzyme-Catalysed Reactions*, RSC Biomolecular Sciences, Cambridge, UK (2009).
4. J. P. Klinman, A. Kohen, *Annu. Rev. Biochem.* **82**, 471 (2013).
5. R. A. Marcus, *J. Chem. Phys.* **24**, 966 (1956).
6. M. P. Meyer, D. R. Tomchick, J. P. Klinman, *Proc. Natl. Acad. Sci. USA* **105**, 1146 (2008).
7. S. Hu, S. C. Sharma, A. D. Scouros, A. V. Soudackov, C. A. M. Carr *et al.*, *J. Amer. Chem. Soc.* **136**, 8157 (2014).
8. M. F. Hegazi, R. T. Borchardt, R. L. Schowen, *J. Amer. Chem. Soc.* **101**, 4359 (1979).
9. J. Y. Zhang, J. P. Klinman, *J. Amer. Chem. Soc.* **133**, 17134 (2011).
10. H. J. Kulik, J. Zhang, J. P. Klinman, T. J. Martinez, in preparation.
11. J. Zhang, H. J. Kulik, T. J. Martinez, J. P. Klinman, in preparation.

COMPUTATIONAL ENZYME DESIGN AND METHODS TO PREDICT THE ROLE OF REMOTE MUTATIONS

KENDALL N. HOUK

Department of Chemistry and Biochemistry, University of California
Los Angeles, CA 90095-1569, USA

My view of the present state of research on exploring enzyme families and enzyme catalysis

The computational design of functional proteins will boost the practical applications of biocatalysis. For nearly ten years, my group has been involved in the development of methods for the computational design of new enzymes. With David Baker, Stephen Mayo, Donald Hilvert, and others we have applied quantum mechanics, molecular dynamics, and combined bioinformatics/force-field approaches like Rosetta to predict protein sequences that will fold into functional enzymes.

The consequences of a successful approach of this type will be enormous. The scientific impact of success will be the demonstration that we understand how enzyme catalysts function at an atomistic level and have mastered the complexities of catalysis, enabling us to replace Nature's random evolutionary screening by prediction based on first-principles physics. More practically, catalysts, the mainstay of industrial synthesis, and even the proteins that enable and control metabolism, will be designed by computational prediction and then easily synthesized by bacteria programmed by DNA made automatically on DNA synthesizers. Indeed, groups around the world have successfully designed and redesigned functional proteins, but these accomplishments resulted from heroic efforts, not routine protocols.

The reality is that our success in designing enzymes is erratic at best, and usually only a small percentage of designs actually function properly. Methods to incorporate required functional groups into stable protein scaffolds have been devised, but directed evolution — not in general computationally predicted — often must be used to increase activity to useful levels.

In order to improve computational design methods, we have undertaken the computational investigation of the results of directed evolution. Our goal is to learn how to predict these mutations in advance of experiment, and to improve the methods for enzyme design.

My recent research contributions to exploring enzyme families and enzyme catalysis

General work in the computational enzyme design field has been reviewed [1]. The general inside-out scheme we devised with David Baker is shown below in Fig. 1. Computational enzyme design is a classic multi-scale problem, using high-accuracy quantum mechanics to position catalytic groups precisely around the transition states of the reaction and more approximate quantum, force-field and informatics methods for treatment of protein motions and solvation.

First, quantum mechanical calculations are used to design the active site to catalyze reactions of entirely new mechanisms or unnatural substrates by providing stabilizing interactions that lower the free energy of the transition state. We use motifs that Nature has evolved for different types of reactions, but quantum mechanics are used to determine optimal positioning of catalytic groups for the transition state of the reaction of interest. We have shown, in collaboration with Hilvert, that evolved active sites (*e.g.*, of esterases) can be predicted by quantum mechanical optimizations [2]. The computationally predicted three-dimensional arrangement of catalytic groups to stabilize a given transition state is called a theoretical enzyme

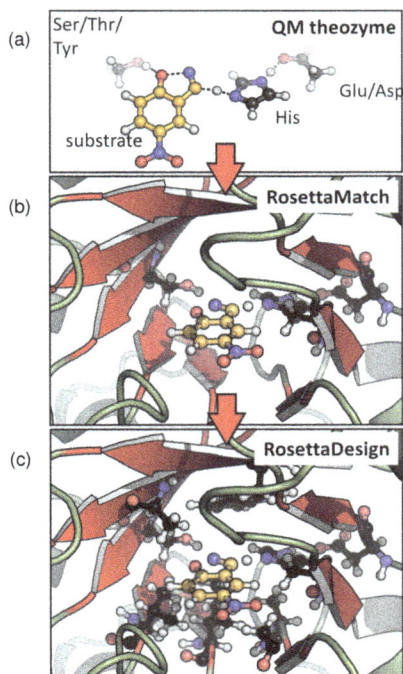

Fig. 1. The Inside-Out Protocol for enzyme design. The panels represent (a) the quantum mechanical prediction of a theozyme (here for the Kemp elimination), (b) the incorporation of a theozyme into a stable protein, and (c). the redesign of the non-catalytic residues in the active site.

or "theozyme" [1]. This theozyme is then incorporated into a stable protein of known structure in place of residues in the active site of the natural protein using the program from Baker's group, RosettaMatch (known as Matcher in the current implementation). Remaining side-chains that pack the active site are redesigned with RosettaDesign (known as EnzDes in the current implementation). Our group has also developed SABER [3] to search crystal structures of proteins in the PDB and to locate enzymes that already have the necessary arrangements of functional groups.

The designs produced from Rosetta or SABER are then evaluated according to Rosetta folding criteria and by comparison to theozyme geometries. The best are then subjected to molecular dynamics simulations in the presence of substrates, counterions, and water [4]. When MD causes the active site to deviate substantially from that predicted by QM to be required for catalysis, then that design is discarded.

This procedure has been successful for the design of Kemp eliminases, retro-aldolases, and stereospecific Diels–Alderases, plus some others. Variations on this approach have been successful in our collaborations with Mayo [5]. The role of molecular dynamics in designs has been demonstrated for many cases [6].

Our recent studies have turned to perfecting the post-Rosetta or post-SABER redesign process. We wish to computationally predict what now is only achieved experimentally by directed evolution [7]. To learn how to do this, we are evaluating the pathways of successful directed evolution experiments on several successful commercial enzymes. Our current efforts involve the theoretical study of enzymes evolved by our colleagues Yi Tang and Todd Yeates at UCLA [8] and by Gjalt Huisman's group at Codexis [9], to produce enzymes that can be used for the commercial production of drugs like sitagliptin, simvastatin, or montelukast.

One example of this involves a protein, LovD, that naturally accepts an alcohol to produce the drug lovastatin [8, 9]. LovD uses an acyl carrier protein to deliver the acyl group. Tang and Codexis evolved LovD to transfer a modified acyl group from a much smaller carrier. Figure 2 summarizes the results of our work on this evolutionary path [10], involving many long-timescale molecular dynamics (MD) runs on the special-purpose computer, ANTON [11].

Panel A shows an arrangement of functional groups in the catalytic triad that is ineffective in catalysis, while B shows an arrangement that is good for catalysis. The MD studies showed that remote mutations gradually produce catalytically potent arrangements of the catalytic triad.

Outlook to future developments of research on exploring enzyme families and enzyme catalysis

Armed with information on how remote mutations influence pre-organization of active sites, we are devising computational protocols to build such information into the design process. The first step in this is to use iterative Rosetta/MD [12] to monitor how remote mutations influence critical catalytic distances.

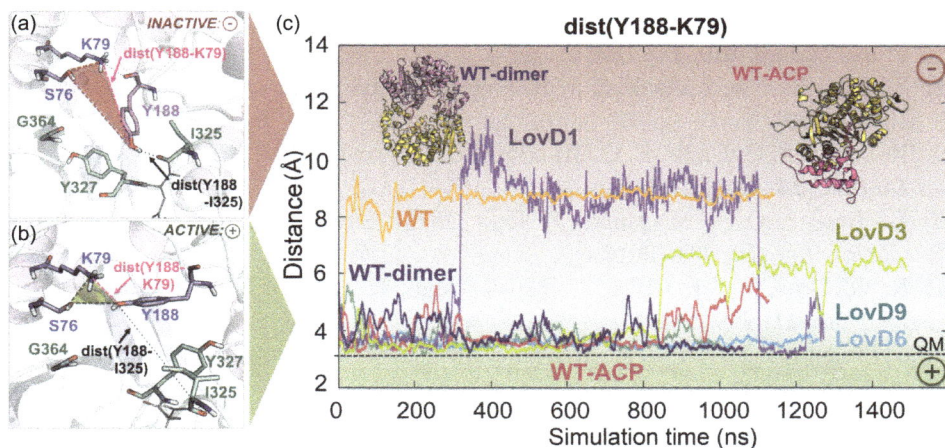

Fig. 2. Catalytically inactive (a) and active (b) arrangements of the Ser76–Lys79–Tyr188 triad in evolved LovD variants and protein-protein complexes revealed by MD simulations. (c) The catalytic Tyr188–Lys79 distance along the MD trajectories. Catalytic and non-catalytic regimes have been depicted in green and red, respectively. The less active mutants involve distances mostly in the red zone — signaling inactivity.

The fields of enzyme design and directed evolution are advancing rapidly. Simultaneously, computational methods — quantum mechanical and force-field-based, are being made more accurate and efficient, and the inexorable increases in computer power make larger and better calculations feasible. The importance of designed enzymes as catalysts will expand rapidly in research, commerce and medicine.

Acknowledgments

Our research has been supported by the Defense Advanced Research Projects Agency, the National Institute of General Medical Sciences, National Institutes of Health, and Maersk Oil Research and Technology Center. Computer services from XSEDE and PSC (ANTON) were provided by the National Science Foundation.

References

1. G. Kiss, N. Celebi-Olcum, R. Moretti, D. Baker, K. N. Houk, *Angew. Chem. Int. Ed.* **52**, 5700 (2013).
2. A. J. T. Smith, R. Muller, M. D. Toscano, P. Kast, H. Hellinga *et al.*, *J. Amer. Chem. Soc.* **130**, 15361 (2008).
3. G. R. Nosrati, K. N. Houk, *Prot. Sci.* **21**, 697 (2012).
4. G. Kiss, D. Rothlisberger, D. Baker, K. N. Houk, *Prot. Sci.* **19**, 1760 (2010).
5. H. K. Privett, G. Kiss, T. M. Lee, R. Blomberg, R. A. Chica *et al.*, *Proc. Natl. Acad. Sci.* **109**, 3790 (2012).
6. G. Kiss, V. S. Pande, K. N. Houk, *Meth. Enzymol.* **523**, 145 (2013).

7. U. T. Bornscheuer, G. W. Huisman, R. J. Kazlauskas, S. Lutz, J. C. Moore *et al.*, *Nature* **485**, 185 (2012).

8. X. Gao, X. Xie, I. Ashkov, J. R. Sawaya, J. Laidman *et al.*, *Chem. Biol.* **16**, 1064 (2009).

9. R. J. Fox, S. C. Davis, E. C. Mundorff, L. M. Newman, V. Gabrilovic *et al.*, *Nat. Biotech.* **25**, 338 (2007).

10. G. Jiménez-Osés, S. Osuna, X. Gao, M. R. Sawaya, L. Gilson *et al.*, *Nat. Chem. Biol.* **10**, 431 (2014).

11. R. O. Dror, R. M. Dirks, J. P. Grossman, H. Xu, D. E. Shaw, *Annu. Rev. Biophys.* **41**, 429 (2012).

12. S. Lindert, J. Meiler, J. A. McCammon, *J. Chem. Th. Comput.* **9**, 3843 (2013).

DISCOVERING NOVEL ENZYMES, METABOLITES AND PATHWAYS

JOHN A. GERLT

Institute for Genomic Biology, University of Illinois, Urbana-Champaign
1206 West Gregory Drive, Urbana, IL 61801, USA

My view of the present state of research on "exploring enzyme families and enzyme catalysis"

Homologous enzymes can catalyze the same overall reaction (isofunctional enzyme family), the same chemical reaction using different substrates (specificity diverse enzyme family), or different overall reactions that share a conserved partial reaction (mechanistically diverse enzyme superfamily). Before genome sequencing was routine, the members of enzyme (super)families were discovered by experimental characterization. However, now the members of enzyme (super)families are discovered without regard to either *in vitro* activity or *in vivo* metabolic/physiological function. As of August 2013, the nonredundant UniProtKB/TrEMBL database contained 42,821,879 sequences; this number is increasing rapidly. However, a conservative estimate is that $\geq 50\%$ of these sequences have uncertain, unknown, or incorrectly annotated functions. Indeed, as of August 2013, only 540,958 sequences had human-curated functions in the SwissProt database. Without high-throughput strategies and tools to decipher the functions of enzymes discovered in genome projects, the biomedical, industrial, and systems biology potential of these projects cannot be realized.

My recent research contributions to "exploring enzyme families and enzyme catalysis"

The Enzyme Function Initiative (EFI) [1], a large-scale collaborative project supported by the National Institute of Health, is addressing the challenge of identifying the *in vitro* activities and *in vivo* metabolic/physiological functions of enzymes discovered in genome projects. The EFI brings together expertise in bioinformatics, protein production and structure determination, homology modeling, *in silico* ligand docking [2, 3], experimental enzymology, genetics, transcriptomics, and metabolomics to devise tools and strategies that can be used to predict and verify novel enzymatic activities and metabolic pathways encoded by microbial genomes. The recognition of the importance of the human microbiome in human health justifies the EFI's focus on microbial enzymes, although many of the strategies and tools can be used to discover the functions of uncharacterized eukaryotic enzymes.

As an example of this approach [4], the EFI recently "deorphanized" an enzyme from the pelagic bacterium *Pelagibaca bermudensis* HTCC2601 for which the New York Structural Genomics Research Consortium (a high-throughput structural genomics project) had determined a high-resolution X-ray structure (PDB Code 2PMQ). The protein, now designated HpbD, is a member of the mechanistically diverse enolase superfamily [5].

The gene encoding HpbD (*hpb*D) is located in gene cluster that could encode a metabolic pathway (Fig. 1).

Pelagibaca bermudensis HTCC2601

Fig. 1.   Genome neighborhood context of the gene encoding 2PMQ (*hpb*D).

Homology models were constructed for both a periplasmic solute binding protein component of an ABC transport system (HpbJ) and a dioxygenase (HpbB1); *in silico* docking of virtual ligand libraries to all three proteins ("pathway docking" [6]) was used to predict metabolites that bind to the active sites of HpbD, HpbJ, and HpbB1. Taken together, the results allowed the prediction that HpbD binds *trans*-4-hydroxy-L-proline betaine (tHyp-B) and catalyzes its 2-epimerization to *cis*-4-hydroxy-D-proline betaine (cHyp-B). Subsequent enzymatic activity measurements confirmed this prediction (k_{cat}, 330 ± 30 sec^{-1}; k_{cat}/K_M, 7800 M^{-1} sec^{-1}). A high-resolution X-ray structure confirmed the validity of the predicted pose of tHyp-B in the active site of HpbD. Thus, the 4-hydroxyproline 2-epimerase function can be assigned to HpbD and its orthologues in the databases (Fig. 2).

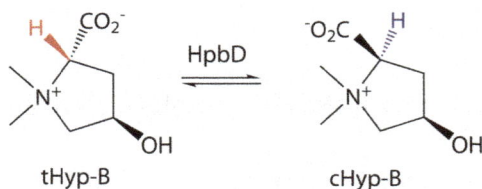

Fig. 2.   Reaction catalyzed by HpbD.

The family/superfamily memberships of the enzymes encoded by the neighboring genes allowed the prediction of a catabolic pathway for tHyp-B (Fig. 3): successive oxidative removal of both betaine methyl groups to yield cHyp which is oxidized to its imino acid (by a member of the D-amino acid oxidase family), dehydrated and "hydrolyzed" to α-ketoglutarate semialdehyde (by a member of the dihydrodipicolinate synthase superfamily), and, finally, oxidized to α-ketoglutarate

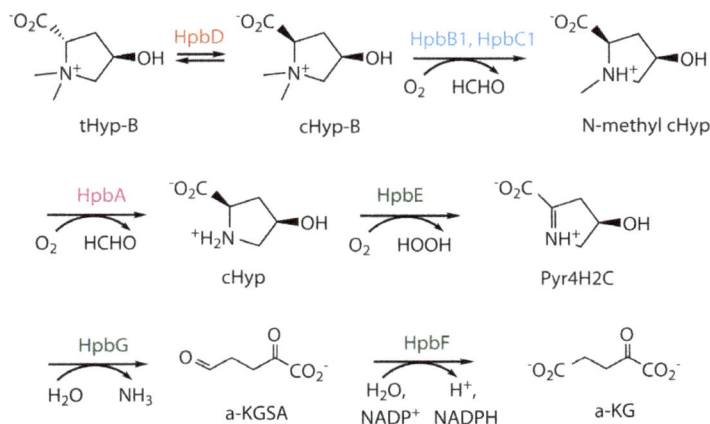

Fig. 3. Pathway for catabolism of tHyp-B.

(by a member of the aldehyde dehydrogenase family). This pathway was confirmed by metabolomics and genetics.

The *in vivo* physiological function of an orthologue of HpbD encoded by *Paracoccus denitrificans* PD1222 was determined by microbiology and transcriptomics (in contrast to *P. bermudensis*, *P. denitrificans* PD1222 can be grown in the laboratory and is genetically tractable). Betaines often are used by bacteria as osmolytes to reduce stress by the presence of high salt in the growth medium. The growth of *P. denitrificans* is inhibited by 0.5 M NaCl in glucose medium; this inhibition is alleviated by tHyp-B, establishing that tHyp-B is an osmolyte. The genes encoding transporters for tHyp-B are up-regulated by tHyp-B and cHyp-B, thereby allowing their accumulation (with 20 mM tHyp-B in the growth medium, the intracellular concentration of Hyp-B is 170 mM). However, the genes encoding HpbD as well as all of the downstream enzymes in the catabolic pathway are down-regulated by tHyp-B and cHyp-B; thus, the flux of tHyp-B through the catabolic pathway is regulated so that it is not catabolized when it is needed as an osmoprotectant. The "modest" value of the k_{cat}/K_M, 7800 $M^{-1}sec^{-1}$, for HpbD is physiologically relevant because of the multi-millimolar concentration of its substrate.

This example illustrates how multidisciplinary collaboration allows prediction and verification of *in vitro* activities and *in vivo* metabolic/physiological functions of uncharacterized enzymes discovered in genome projects. The EFI continues to develop new bioinformatic and computational tools to enable the biological community to adopt these strategies.

Outlook to future developments of research on "exploring enzyme families and enzyme catalysis"

The challenge of determining the *in vitro* activities and *in vivo* metabolic/physiological functions of uncharacterized enzymes discovered in genome projects will increase for the foreseeable future — the number of microbial species that are important for human health, bioremediation, and natural product isolation will continue to increase as new environmental niches are explored. Computational tools, including sequence similarity networks [7], homology modeling, and *in silico* ligand docking, will continue to be developed and refined to better meeting this challenge. These tools and the strategies with which they are applied must be disseminated to and used by the community.

Acknowledgment

The EFI is supported by the National Institutes of Health (U54GM093342).

References

1. J. A. Gerlt, K. N. Allen, S. C. Almo, R. N. Armstrong, P. C. Babbitt *et al.*, *Biochemistry* **50**, 9950 (2011).
2. J. C. Hermann, R. Marti-Arbona, A. A. Fedorov, E. Fedorov, S. C. Almo *et al.*, *Nature* **448**, 775 (2007).
3. L. Song, C. Kalyanaraman, A. A. Fedorov, E. V. Fedorov, M. E. Glasner *et al.*, *Nat. Chem. Biol.* **3**, 486 (2007).
4. S. Zhao, R. Kumar, A. Sakai, M. W. Vetting, B. M. Wood *et al.*, *Nature* **502**, 698 (2013).
5. J. A. Gerlt, P. C. Babbitt, M. P. Jacobson, S. C. Almo, *J. Biol. Chem.* **287**, 29 (2012).
6. C. Kalyanaraman, M. P. Jacobson, *Biochemistry* **49**, 4003 (2010).
7. H. J. Atkinson, J. H. Morris, T. E. Ferrin, P. C. Babbitt, *PLoS One* **4**, 34345 (2009).

PROGRAMMING NEW CHEMISTRY INTO THE GENETIC CODE OF CELLS AND ANIMALS

JASON W. CHIN

Centre for Chemical and Synthetic Biology, Division of Protein and Nucleic Acid Chemistry
Medical Research Council Laboratory of Molecular Biology, Francis Crick Avenue
Cambridge CB2 0QH, UK

My view of the present state of research on exploring enzyme families and enzyme catalysis

There has been rapid progress in developing approaches to incorporate unnatural amino acids into proteins. Early approaches used chemical aminoacylation of tRNAs in *in vitro* translation systems. Progress in this area has been facilitated by the development of RNA based aminoacylation systems that can be used *in vitro* as well as purified translation systems in which the ribosomes are isolated from cells but other translation components are made recombinantly. In contrast to cellular translation, *in vitro* translation systems commonly use stoichiometrically pre-aminoacylated tRNAs that are not- re-aminoacylated in the translation mix. It has been possible to select unnatural peptides and cyclic peptides bearing several unnatural monomers using *in vitro* translation and selection approaches and *in vitro* translation approaches have been key is clarifying the chemical scope of the ribosome for unnatural monomers.

Methods for incorporation unnatural amino acids into proteins *in vivo* promise scalable synthesis of modified proteins and provide the basis of powerful approaches for studying biological phenomena. Two main approaches for unnatural amino acid incorporation in cells have been described. One approach takes advantage of the permissivity of existing cellular aminoacyl-tRNA synthetases (or their active site mutants) to incorporate a structurally limited set of unnatural amino acids in response to sense codons. Typically these unnatural amino acids are incorporated sub-stoichiometrically creating a statistical mixture of proteins, though experiments in auxotrophic strains allow near quantitative incorporation of some amino acids in response to a sense codon in a target protein. These approaches have been useful for pulse-labeling the proteomes of cells. A distinct approach introduces orthogonal translational components and aims to genetically encode the quantitative installation of unnatural amino acids in response to a defined blank codon(s), most commonly the amber stop codon. An orthogonal ribosome has been created and evolved to decode new bank codons (including quadruplet codons), facilitating the incorporation of multiple distinct unnatural amino acids in a single polypeptide.

Orthogonal aminoacyl-tRNA synthetase/tRNA pairs (most-notably the pyrrolysyl-tRNA synthetase/tRNA pair) have been developed for the incorporation of diverse amino acids. Genetic code expansion has been developed in *E. coli*, yeast, mammalian cells and most recently in animals: *C. elegans* and *D. melanogaster*. The chemical scope of modifications has been augmented by combining genetic code expansion and chemoselective modification. The approaches developed have been used to obtain previously unattainable insights into the molecular basis of biological function, both by the synthesis of recombinant proteins for *in vitro* structure function studies and by developing approaches for controlling and imaging biological processes in cells and animals.

My recent research contributions to exploring enzyme families and enzyme catalysis

Reprogramming Cellular Translation: In the cell, DNA is copied to messenger RNA, and triplet codons (64) in the messenger RNA are decoded — in the process of translation — to synthesize polymers of the natural 20 amino acids. This process (DNA-RNA-protein) describes the central dogma of molecular biology and is conserved in terrestrial life. We are interested in re-writing the central dogma to create organisms that synthesize proteins containing unnatural amino acids and polymers composed of monomer building blocks beyond the 20 natural amino acids. We have invented and synthetically evolved new 'orthogonal' translational components (including ribosomes [1] and aminoacyl-tRNA synthetases) and demonstrated the assembly of these components into an 'orthogonal translation pathway' to address major challenges in re-writing the central dogma of biology in cells. The approaches we have developed address the outstanding problem of altering the function of conserved cellular hubs in cells that orchestrate essential functions (in this case the ribosome). We have shown that by creating orthogonal (and non-essential) versions of these hubs it is possible to evolve their molecular function. We have evolved the orthogonal ribosome to read the amber stop codon as a sense codon and to read quadruplet codons [2, 3]. Using these ribosomes that have additional decoding capacity in combination with mutually orthogonal synthetase/tRNA pairs we have demonstrated the first encoded incorporation of multiple useful unnatural amino acids into protein in cells [3]. By programming two amino acids that can react together into proteins we have demonstrated the genetically directed proximity acceleration of a cycloaddition reaction on a protein to cyclize a protein [3].

Developing the Pyrrolysyl tRNA synthetase tRNA pair for unnatural amino acid incorporation in cells and animals: We have extended the foundational technologies we have pioneered to develop new approaches to address unmet challenges in understanding biological processes with spatial, temporal and molecular precision [4]. Much of this work is underpinned by our development of the pyrrolysyl-tRNA synthetase/tRNA$_{CUA}$ pair for genetic code expansion. We have demonstrated that this pair, unlike the previous aminoacyl-tRNA synthetase/tRNA pairs that have been used for genetic code expansion, is orthogonal in *E. coli*, *S. cerevisiae*, mammalian cells, *C. elegans* and *D. melanogaster* [5–9] and this pair can be used to incorporate a range of aliphatic and aromatic amino acids. We have also demonstrated that this pair can be evolved in *E. coli* to specifically incorporate unnatural amino acids [5]. These advances have rapidly expanded the scope and accessibility of genetic code expansion because they allows aminoacyl-tRNA synthetases developed for the incorporation of new amino acids in *E. coli* to be transplanted and used for unnatural amino acid incorporation in eukaryotic cells and animals.

Insights into the molecular consequences of post-translational modifications via genetic code expansion: A catalog of post-translational modifications have been identified in proteins, and mass spectrometry based approaches are rapidly expanding the number of modifications identified. One way to provide insight into the roles of modifications in regulating protein structure and function is to make and study the modified protein. However, in many cases there is no way to make the modified proteins for structural and functional studies because the modifying enzymes are unknown or do not modify a given site specifically or quantitatively *in vitro*. We have developed approaches for genetically encoding post-translational modifications (including lysine acetylation [5] and mono- and di- methylation [4]), as well as the development of approaches that combine chemical synthesis and protein chemistry with genetic code expansion to allow the genetically directed installation of ubiquitination [10, 11], a post-translational modification that is challenging to install by either chemistry or genetic code expansion alone.

We have applied our approaches to installing post-translational modifications into recombinant proteins to obtain previously unattainable insight into biological regulation through interfacing our approaches with structural biology, enzymology and single molecule methods. Using these approaches we have defined the effects of a key acetylation on chromatin in regulating the wrapping of DNA around nucleosomes [4], the effect of cyclophilin acetylation in regulating cis-trans isomerization of HIV capsid and mediating cyclosporine binding (effects that may be important in immunosuppression and viral infection) [4], synthesized atypical ubiquitin linkages for the first time, determined the first structure of K6-linked ubiquitin linkages and profiled a variety of deubiquitinases on atypical ubiquitin linkages for the first time [4]. Our approaches for synthesizing post-translationally modified proteins have been widely adopted by many other groups and been used to provide additional insights into diverse biological processes.

Imaging biological processes in cells and organisms: Understanding how proteins function inside cells will be facilitated by approaches that allow proteins to be labeled and imaged with small molecule fluorophores. We have developed approaches for the rapid and site-specific labelling of proteins in cells with small molecule fluorophores. These approaches involve genetically encoding an unnatural amino acid at a defined site in a protein of interest and labeling this site with a fluorophore or other small molecule. Key to this strategy is the development and encoding of functional groups that (i) are bioorthogonal and (ii) can be labeled with probes in rapid reactions. We have demonstrated the encoding of strained alkenes and alkynes and their labeling with tetrazine based probes in fluorogenic inverse electron demand Diels-Alder reactions that produce nitrogen gas as the sole byproduct [12, 13]. This reaction (rate constants upto 10^5 $M^{-1}s^{-1}$) is 4-5 orders of magnitude faster than most previous bioorthogonal reactions reported, and this approach has allowed the labeling of proteins at genetically encoded unnatural amino acids in and on mammalian cells for the first time. The minimally perturbing imaging approaches we have developed allow imaging of proteins in cells, without large fluorescent tags and with very bright fluorophores. This is facilitating super-resolution imaging of proteins and other investigations into protein function.

Temporal dissection of signaling: Understanding the kinetics of elementary steps in signaling and the role of feedback and feedforward control in regulating cellular signals is an important challenge. We have demonstrated that by genetically encoding a photocaged lysine [7, 14] in place of a near universally conserved lysine in the active site of protein kinases (and introducing mutations that would make the wild-type kinase constitutively active) it is possible to create kinases in cells that can be activated by a pulse of light. This allows the activation of kinase signaling from any point in a cascade. This approach is allowing the dissection of the contribution of each step in a signaling cascade to signal propagation and adaptation [14].

Expanding the genetic code of animals: We have recently developed genetic code expansion approaches in whole animals (*C. elegans* and *D. melanogaster*) [8, 9] for the first time. One goal of this work is to develop and apply the tools we have created for manipulating and imaging processes in cell culture to biological processes including development, disease progression, and cognitive function best studies at the level of whole organisms.

Outlook to future developments of research on exploring enzyme families and enzyme catalysis

The ability to alter the functions of some of the most central and conserved biological systems and redirect these systems in cells to new ends provides a powerful paradigm for both the study of biology and the discovery and manufacture of new therapeutics and materials. The approaches we are developing to genetically encode unnatural amino acids into proteins in cells and animals are providing the foundation for a suite of new approaches to studying complex biological phenomena with molecular, spatial and temporal precision. By combining the chemist's ability to precisely tailor the properties of small molecules with redesigned biology that places these small molecules at precise positions in proteins in cells we can maximally leverage the unique advantages of chemistry and biology for synthesis and discovery.

Acknowledgments

Work in our laboratory is funded by the Medical Research Council, The European Research Council, The Human Frontiers of Science Programme, and The Louis-Jeantet Foundation.

References

1. O. Rackham, J. W. Chin, *Nat. Chem. Biol.* **1**, 159 (2005).
2. K. Wang, H. Neumann, S. Y. Peak-Chew, J. W. Chin, *Nat. Biotechnol.* **25**, 770 (2007).
3. H. Neumann, K. Wang, L. Davis, M. Garcia-Alai, J. W. Chin, *Nature* **464**, 441 (2010).
4. L. Davis, J. W. Chin, *Nat. Rev. Mol. Cell. Biol.* **13**, 168 (2012).
5. H. Neumann, S. Y. Peak-Chew, J. W. Chin, *Nat. Chem. Biol.* **4**, 232 (2008).
6. S. M. Hancock, R. Uprety, A. Deiters, J. W. Chin, *J. Amer. Chem. Soc.* **132**, 14819 (2010).
7. A. Gautier, D. P. Nguyen, H. Lusic, W. An, A. Deiters *et al.*, *J. Amer. Chem. Soc.* **132**, 4086 (2010).
8. S. Greiss, J. W. Chin, *J. Amer. Chem. Soc.* **133**, 14196 (2011).
9. A. Bianco, F. M. Townsley, S. Greiss, K. Lang, J. W. Chin, *Nat. Chem. Biol.* **8**, 748 (2012).
10. S. Virdee, P. B. Kapadnis, T. Elliott, K. Lang, J. Madrzak *et al.*, *J. Amer. Chem. Soc.* **133**, 10708 (2011).
11. S. Virdee, Y. Ye, D. P. Nguyen, D. Komander, J. W. Chin, *Nat. Chem. Biol.* **6**, 750 (2010).
12. K. Lang, L. Davis, J. Torres-Kolbus, C. Chou, A. Deiters *et al.*, *Nat. Chem.* **4**, 298 (2012).
13. K. Lang, L. Davis, S. Wallace, M. Mahesh, D. J. Cox *et al.*, *J. Amer. Chem. Soc.* **134**, 10317 (2012).
14. A. Gautier, A. Deiters, J. W. Chin, *J. Amer. Chem. Soc.* **133**, 2124 (2011).

EXPANDING THE ENZYME UNIVERSE THROUGH A MARRIAGE OF CHEMISTRY AND EVOLUTION

FRANCES H. ARNOLD

Division of Chemistry and Chemical Engineering, California Institute of Technology 210-41
1200 E. California Blvd., Pasadena, CA 91125, USA

Present state of research on expanding enzyme catalysis beyond Nature

For more than twenty years this laboratory has used directed evolution to modify enzymes. It is now widely accepted that directed evolution can change substrate specificity or reaction selectivity in desired ways, even if it sometimes remains difficult in practice. It is no longer surprising that enzymes adapt readily by accumulating beneficial mutations. And why should it be, since this is how Nature tailors them for myriad biological roles? More difficult to grasp is how Nature discovers new enzyme functions, particularly new catalytic activities. We know that the biological world's diverse catalytic repertoire is the product of evolution by natural selection, but we have little understanding of how Nature's tinkering generates new functions. Sometimes we are (un)lucky enough to catch them in the act — *e.g.*, the acquisition of antibiotic resistance or the ability to degrade man-made toxins. But for the vast majority of activities, the fossil record is nonexistent or too sparse to tell the molecular story. This leaves us without much guidance for evolving new enzymes in the laboratory. Consequently we are forced to contaminate our evolution experiments with knowledge — *e.g.*, computational design [1, 2] — in order to jumpstart the discovery process.

If every bad catalyst could become a good one by directed evolution, part of the problem would be solved. We would be able to take any catalytic antibody or computationally-designed enzyme (or bovine serum albumin) and convert it into a great catalyst with multiple rounds of random mutagenesis and screening. We have learned the hard way, however, that not every bad catalyst lies at the base of a tall fitness peak, at least one that can be scaled by a random uphill walk. We therefore want to know what features make a protein with a new catalytic activity the potential mother of a whole new enzyme family. And, what are good ways to find new enzyme activities in the first place?

Recent research contributions to creating new enzymes

My laboratory has been directing the evolution of a remarkable enzyme, a bacterial cytochrome P450, for at least a dozen years. This particular P450, from *Bacillus megaterium*, is quite specific — it catalyzes subterminal hydroxylation of fatty acids. As a family, however, the P450s are wonderfully diverse, catalyzing a wide range of reactions on an even wider range of substrates. Nature has demonstrated the 'evolvability' of this bit of protein that binds an iron-heme (along with its various electron-transfer partners), and we spent several happy years creating new versions of the bacterial enzyme that could mimic known functions of other (*e.g.*, human) P450 family members [3, 4]. We could also extend its function beyond what was known to be catalyzed by P450s, making, for example, versions that hydroxylate gaseous alkanes (propane and ethane), transformations thought to lie in the functional realm of the methane monooxygenase family [5].

The versatile iron-heme prosthetic group has been adopted by numerous transport, signaling, and catalytic proteins. Within just the P450 family the range of reactions catalyzed is impressive: hydroxylation, epoxidation, sulfoxidation, peroxidation, N- and O-dealkylation, and more. We found that we could access all of these reactivities, starting from wildtype or variants of our bacterial fatty acid monooxygenase and using directed evolution to increase the initially-low activities. We learned that even a single P450 is highly evolvable and lies at the base of many different fitness peaks. The P450 is also catalytically promiscuous — the bacterial enzyme or its close variants could catalyze low levels of many different reactions. Presumably natural selection made the same discovery many times, setting old P450s to new tasks in many new contexts.

More recently we decided to explore new reactions, not known in Nature. Reactions whose mechanisms share features of the P450 machinery but have not been discovered in Nature's many evolution experiments, because the context was not there. Understanding that natural selection can create new enzymes from promiscuous activities when presented with the opportunity to occupy a new niche (for example, a new substrate becomes available), we tested a collection of P450s for promiscuous activity in a few carbene and nitrene insertion reactions that are isoelectronic to the well-established formal 'oxene' transfer reaction of ferric-P450 enzymes with iodosylbenzene. The similarity is such that some of these reactions were explored in the 1980's by 'biomimetic' chemists, notably Breslow and Dawson [6] and others [7]. The poor turnover numbers they reported for P450s, however, discouraged further work, and no more was said for 30 years. But we knew from experience that when a P450 is a bad catalyst, it can often easily become a good one.

The wildtype P450 enzyme catalyzed just a few turnovers for cyclopropanation of styrene with EDA, fewer even than free hemin (Table 1) and other heme proteins [8]. However, it was the only catalyst to exhibit enantioselectivity, indicating the reaction took place in an active site that could exert some control on selectivity. We quickly identified variants in our collection, such as H2A10 and CIS (Table 1), that exhibited high selectivity for the *cis* diastereomer (opposite from that of hemin, which makes mostly the *trans* cyclopropane product), high activity, and high enantioselectivity.

The big breakthrough came, however, when we replaced the cysteine residue that ligates the heme iron with Ser [9]. Our goal was to have the reaction proceed *in vivo*, where it would have to rely on endogenous NADPH rather than the sodium dithionite used to reduce the heme Fe(III) to Fe(II) *in vitro*. This generated a very active cyclopropanation enzyme that functions extremely well *in vivo*: the 67,800 turnovers for the P411-CIS enzyme (Table 1) is, we believe, the highest activity ever reported for this reaction with any catalyst. The Cys-Ser mutation abolished all monooxygenase activity and caused the typical peak at 450 nm in the CO-difference spectrum to shift to 411 nm. Thus we call this new catalyst a P411. Much more active than P450-CIS, the P411-CIS has a crystal structure nearly identical to that of P450-CIS. We have started to diversify this enzyme by directed evolution to expand its substrate range and selectivity.

Table 1.   Activities of different P450 variants for styrene cyclopropanation [9]. Yields are based on EDA. TTN = total turnover number.

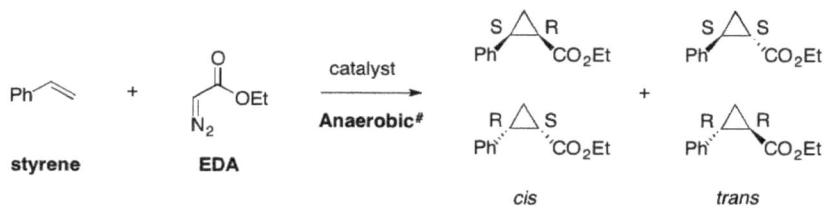

Catalyst	[EDA] (mM)	[P450] (μM)	% yield	TTN	cis:trans	%ee$_{cis}$	%ee$_{trans}$
Hemin	10	20	15	73	6:94	1	0
P450$_{BM3}$	10	20	1	5	37:63	27	2
9-10A TS F87V	10	20	1	7	35:65	41	8
H2A10	10	20	33	167	60:40	95	78
P450$_{BM3-heme}$-CIS	8.5	15	32	212	77:23	94	91
P411$_{BM3-heme}$-CIS	8.5	15	51	342	93:7	99	51
P450$_{BM3}$-CIS*	8.5	3.7	42	950	22:78	60	22
P411$_{BM3}$-CIS*	8.5	1.3	55	3700	76:24	96	25
P411$_{BM3}$-CIS*	170	1.8	72	67800	90:10	99	43

*Conducted with intact *E. coli* cells.
#Conditions for reactions with purified P450s: 1 equiv $Na_2S_2O_4$, Ar atmosphere, 0.1 M KPi pH 8.0.
Conditions for reactions with intact *E. coli* cells: 0.2 equiv glucose, Ar atmosphere, M9-N medium.

Olefin cyclopropanation is not (yet!) a biologically relevant transformation, because P450s do not encounter the reactive diazoesters in their native environments. They nonetheless have this promiscuous activity, which can be captured by evolution when the opportunity arises. Our work shows that this promiscuous activity can be enhanced significantly with just a few mutations, something that can happen readily in a protein evolving under selective pressure, either natural or forced.

Enzymes that catalyze the concerted oxidative amination of C-H bonds are also apparently absent from Nature's catalyst repertoire. Synthetic chemists, who are not limited to biologically accessible reagents and metals, have developed useful methods for C-H amination through a nitrenoid intermediate that also has no parallel in natural enzymes. Following up on studies performed in the 1980's [6], we investigated whether our P450 and P411 enzymes could catalyze intramolecular C-H amination of aryl sulfonylazides to form benzosultams (example shown in Table 2) [10]. Whereas wildtype P450 showed only weak activity, some P411 variants catalyzed several hundred TTN.

The purified enzymes as well as intact *E. coli* cells expressing the enzymes catalyze the amination reaction under anaerobic conditions (Table 2) [10]. P411-T268A and P411-CIS exhibited good activity and reasonable enantioselectivity

Table 2. P450 and P411 enzymes catalyze direct C-H amination. Comparison of total turnover numbers (TTN) and enantioselectivities of intact *E. coli* cells expressing P450 and P411 variants, with azide **1** at 0.1 mol% catalyst loading, giving sulfonamide **2** and benzosultam **3** [10].

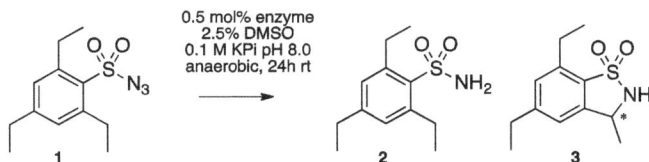

In vivo catalyst	[P450] or [P411] [(μM)]	Yield 3 [%]	TTN[a]	ee [%][b]
Empty vector	0	0	0	n.d.
P450$_{BM3}$	6.6	0.5	5.1	n.d
P450BM3-	5.8	7.8	26	84
P411BM3	4.3	6.7	29	16
P411BM3-	2.2	30	250	89
P411BM3-CIS	1.4	46	680	60
P411BM3-CIS-T438S	2.7	58	430	87

[a] TTN = total turnover number to benzosultam **3**.
[b] *(S−R)/(S+R).
n.d. = not determined.

(up to 89%); adding the T438S mutation to P411-CIS increased enantioselectivity (430 TTN, 86% *ee*). Optimization of expression conditions increased the productivity of whole-cell C-H amination, enabling conversions to **3** of up to 66% in small-scale reactions; higher yields have since been achieved at preparative scale.

Outlook for future enzymes

The tools are now in place to create enzymes that catalyze reactions not known in Nature. We have reported olefin cyclopropanation and C-H amination catalyzed by P450s and P411s, and there is clearly opportunity for more useful reactions based on this system. The highly evolvable P450 is an excellent starting point, both for discovering new enzyme-catalyzed reactions and for diversifying them through directed evolution. But there are likely many more such enzymes waiting for the right substrates to come along. A tasteful mix of chemical intuition, computational design where appropriate, and evolution (to circumvent our near complete ignorance of the details of the sequence-function code for catalysis) will generate whole new families of genetically-encoded catalysts, greatly expanding the catalyst repertoire for biosynthesis and for organic synthesis. What is more, we will be able to observe the creation of new biological functions and follow the mechanisms by which they arise and are diversified and optimized. The future will see Nature's chemical universe expand to include more of the clever chemistry of man.

Acknowledgments

I have described the work of highly talented graduate students and postdocs: Pedro Coelho, Eric Brustad, John McIntosh, Jane Wang and Chris Farwell. Support was received from the Jacobs Institute for Molecular Engineering for Medicine at Caltech, and the Gordon and Betty Moore Foundation through Grant GBMF2809 to the Caltech Programmable Molecular Technology Initiative.

References

1. F. Richter, R. Blomberg, S. D. Khare, G. Kiss, A. P. Kuzin *et al.*, *J. Amer. Chem. Soc.* **134**, 16197 (2012).
2. L. Giger, S. Caner, R. Obexer, P. Kast, D. Baker *et al.*, *Nat. Chem. Biol.* **19**, 494 (2013).
3. M. Landwehr, L. Hochrein, C. R. Otey, A. Kasrayan, J.-E. Bäckvall *et al.*, *J. Amer. Chem. Soc.* **128**, 6058 (2006).
4. A. M. Sawayama, M. M. Y. Chen, P. Kulanthaivel, M-S. Kuo, H. Hemmerle *et al.*, *Chem. Eur. J.* **15**, 11723 (2009).
5. R. Fasan, M. M. Chen, N. C. Crook, F. H. Arnold, *Angew. Chem.* **46**, 8414 (2007).
6. E. W. Svastis, J. H. Dawson, R. Breslow, S. H. Gellman, *J. Amer. Chem. Soc.* **107**, 6427 (1985).

7. J. R. Wolf, C. G. Hamaker, J.-P. Djukic, T. Kodadek, L. K. Woo, *J. Amer. Chem. Soc.* **117**, 9194 (1995).

8. P. S. Coelho, E. M. Brustad, A. Kannan, F. H. Arnold, *Science* **339**, 307 (2013).

9. P. S. Coelho, Z. J. Wang, M. E. Ener, S. A. Baril, A. Kannan *et al.*, *Nat. Chem. Biol.* **9**, 485 (2013).

10. J. A. McIntosh, P. S. Coelho, C. C. Farwell, Z. J. Wang, J. C. Lewis *et al.*, *Angew. Chem. Int. Ed.* **52**, 9309 (2013).

CONTROLLED RADICAL REACTIONS IN BIOLOGY AND THE IMPORTANCE OF METALLO-COFACTOR BIOSYNTHESIS

JOANNE STUBBE

Department of Chemistry, Massachusetts Institute of Technology
Cambridge, MA 02139, USA

My view of the present state of research on exploring enzyme families and enzyme catalysis

Radicals in general are considered to be inherently unstable and reactive and in Biology are almost always associated with destruction. The biological literature is rife with descriptions of reactive oxygen and nitrogen species generated as the inadvertent consequences of our environment and as byproducts of metabolism that if left unrestrained, result in modified macromolecules including DNA and protein. Nature has evolved small molecules (VitC, VitE) and enzymes that are able to inactivate the "bad" radicals directly or repair the damage they inflict. Radicals are thus vilified. One of the most exciting areas in enzyme catalysis for the next decade, in my opinion, is to understand how Nature has been able to harness the reactivity of radicals to carry out very difficult chemical transformations with exquisite specificity. Good radicals in biology play a central role in primary metabolism from making the building blocks required for DNA biosynthesis, to the light mediated oxidation of water to O_2. With the availability of the thousands of genomes and the ability to analyze them using bioinformatics, there are predicted to be $> 50,000$ reactions that involve free radical chemistry [1]. Thus defining the underlying principles of radical based reactions and the mechanisms by which radicals are generated will be an exciting area for the next decade.

My recent research contributions to exploring enzyme families and enzyme catalysis

My lab has been investigating for more than 30 years, how "good" radicals in biology are generated, protected, and used to catalyze the conversion of nucleoside 5'-diphosphates to deoxynucleotides by ribonucleotide reductases (RNRs). RNRs have been classified based on their metallo-cofactor that oxidizes an active site cysteine to a transient thiyl radical which then initiates nucleotide reduction in all classes. The class I RNRs are subdivided into Ia that use a diferric tyrosyl radical cofactor [2], Ib that use a dimanganese-tyrosyl radical cofactor [3, 4] and Ic, that use a Mn(IV)Fe(III) cofactor [5]. The class II and III RNRs use adenosylcobalamin and

S-adenosylmethionine and a $[4Fe4S]^{1+}$ cluster, respectively, both of which generate a 5'-deoxyadenosyl radical that initiates catalysis through a thiyl radical [6]. The RNR-catalyzed reactions involve stable and transient protein radicals (Fig. 1) and nucleotide radical intermediates. One of these cofactors SAM-$[4Fe4S]^{1+}$ (Fig. 2) is now proposed to be involved in thousands of reactions including the conversion of a methylphosphonate to methane gas [7] and the methylation of an adenine in the A site of the ribosome causing resistance to antibiotics [8].

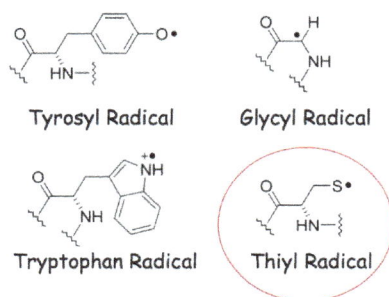

Fig. 1.   Good radicals in biology: both stable and transient protein radicals.

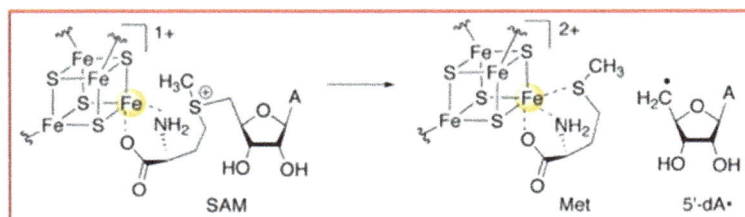

Fig. 2.   Good radicals in biology: the radical SAM superfamily of enzymes use SAM and $[4Fe4S]^{1+}$ to generate 5'-deoxyadenosyl radical which initates a broad range of reactions in both prokaryotes and eukaryotes.

New tools have been essential in analysis of how the metallo-cofactors initiate these radical reactions directly on the substrate or on the protein. These include rapid freeze quench high field EPR and ENDOR spectroscopies, transient absorption spectroscopies and the incorporation of isotopically labeled amino acids and unnatural amino acids site-specifically into proteins [9]. Daily, the structures of chemically and kinetically competent radicals are being revealed. The requirement for metallo-cofactors to initiate these reactions has now made clear the importance of understanding the biosynthetic and repair pathways that control the metallation process [10]. For example, the biosynthesis of the dimanganese-tyrosyl radical cofactor in one class Ib RNR requires a "bad" radical, $O_2^{\bullet-}$ as the oxidant [11]. Lessons learned from studies of RNR and other well studied enzymes such as lysine amino mutase [12], have set the stage for many chemical generalizations that have and will continue to emerge.

Outlook to future developments of research on exploring enzyme families and enzyme catalysis

The good news is that there are many novel mechanisms yet to be unraveled. The bad news is that many of the proteins require metallo-cofactors that are poorly incorporated in recombinant hosts during expression. When studying metal-based reactions, homogeneity of reaction centers is essential given the low sensitivity of the biophysical methods available. While we have a number of tools to look for, examine, and characterize biologically generated radicals, more sensitive tools that can be used *in vitro* and *in vivo* need to be invented. Identifying and understanding the factors that assist and control metal and oxidant delivery, requires further investigation. Despite the amazing progress that has been made in FeS cluster biosynthesis in the last decade [13], for example, the source of iron and small molecules involved in sensing iron, still remain unknown. Metal homeostasis and redox balance are intimately linked in the cell and are essential to understand. Finally, as educators, we need to do a better job incorporating radical-dependent reactions including electron transfer and proton coupled electron transfer mechanisms and the role of metals in biology into our introductory biochemistry courses.

Acknowledgment

I am grateful to the NIH in the USA for supporting my research throughout my entire career.

References

1. H. J. Sofia, G. Chen, B. G. Hetzler, J. F. Reyes-Spindoloa, N. E. Millerm, *Nucl. Acids Res.* **29**, 1097 (2001).
2. J. A. Cotruvo, J. Stubbe, *Annu. Rev. Biochem.* **80**, 733 (2011).
3. J. A. Cotruvo Jr., J. Stubbe, *Biochemistry* **49**, 1297 (2010).
4. N. Cox, H. Ogata, P. Stolle, E. Reijerse, G. Auling *et al.*, *J. Amer. Chem. Soc.* **132**, 11197 (2010).
5. W. Jiang, D. Yun, L. Saleh, E. W. Barr, G. Xing *et al.*, *Science* **316**, 1188 (2007).
6. J. Stubbe, W. A. van der Donk, *Chem. Rev.* **98**, 705 (1998).
7. S. S. Kamat, H. J. Williams, L. J. Dangott, M. Chakrabarti, F. M. Raushel, *Nature* **497**, 132 (2013).
8. T. L. Grove, J. S. Benner, M. I. Radle, J. H. Ahlum, B. J. Landgraf *et al.*, *Science* **332**, 604 (2011).
9. E. Minnihan, D. G. Nocera, J. Stubbe, *Acc. Chem. Res.* **46**, 2524 (2013).
10. J. A. Cotruvo, J. Stubbe, *Metallomics* **4**, 1020 (2012).

11. J. A. Cotruvo, T. A. Stich, R. D. Britt, J. Stubbe, *J. Amer. Chem. Soc.* **135**, 4027 (2013).

12. P. A. Frey, A. D. Hegeman, G. H. Reed, *Chem. Rev.* **106**, 3302 (2006).

13. R. Lill, *Nature* **460**, 831 (2009).

SESSION 2: EXPLORING ENZYME FAMILIES AND ENZYME CATALYSIS

CHAIR: DONALD HILVERT

AUDITORS: J.-M. FRÈRE[1], W. VERSÉES[2]

(1) Université de Liège, Centre for Protein Engineering, Institute of Chemistry,
B6 Sart-Tilman, 4000 Liège, Belgium
(2) Structural Biology Brussels, Vrije Universiteit Brussel and Structural Biology
Research Center, VIB, Pleinlaan 2, 1050 Brussel, Belgium

Discussion among panel members

Don Hilvert to Judith Klinman: I think the idea of compaction in catalysis is quite compelling. In fact, this is what we saw in the evolution of our enzymes. All the groups came together, yielding a very precise alignment. However, why did you conclude that this was a ground state phenomenon?

Judith Klinman: I will talk about methyl transfer first. If you use the labelled methyl group to look at the degree of compaction, basically you are looking at the increase in force constant at the methyl group when it is bound versus free. And you can look at it kinetically, to see what effect mutants have, particularly a residue that sits right behind the sulphur that is transferring the methyl group. In our first study, we showed that as you decrease the size of the group, the isotope effect approaches unity, which is what you see in solution, and the rate went down 10^3 fold. We concluded that the transition state — and this was very important — was more crowded in the wild-type enzyme, and when we mutated that essential tyrosine to either Phe or Ala, things moved apart and the transition state was more expanded. That would fit with the idea of the transition state being critical. But we went one step further: we simply looked at the binding of the S-AdoMet (S-adenosylmethionine), once again with tritium, three tritiums, in the methyl group to increase sensitivity. A lot of the effect that we had seen in the wild type kinetically was present in the ground state. We did not expect to see that, but it suggests that in the actual binding of this large, extended cofactor (bringing us back to ideas that Jencks developed years ago) the methyl group is presumably forced into a configuration in the ground state where it is quite restricted, either in its bond rotation — carbon-carbon bond rotation — or hyperconjugation. We don't have the full physical explanation. But we then performed these extended *ab initio* calculations with Todd Martinez (Stanford) using his GPU methods. Looking at it a little differently, he and Heather Kulik (Martinez lab) saw that the converged inter-nuclear distance between the methyl group and the inserted substrate depended on a huge number of atoms in the *ab initio* region. This is why I showed the picture with the scaffold together with all the soft squishy stuff in the middle. It is almost as if there are so many binding interactions that, in the ground state ternary complex you can force things closer than their van der Waals distances. You must pay the price somewhere: it is likely in the remote binding interactions but you need a lot of atoms to do that. That is a surprising result, but I am not all that confounded because in the tunneling situation you also need to get things very close. But there you do so transiently, by dynamical sampling. That's the model. Here, you are seeing all this compaction in the computation of the energetic minimum for the ground state ternary complex. So we need to really ask the following question: how, with *de novo* enzymes, can we reproduce the extended interactions that are going to bring the reactants so close? We know we need a lot of functional groups for catalysis, we need to go one step further, and we need to get the interacting atoms really close.

John Gerlt: Can I just follow up, Judith? In other words, did S-AdoMet bind more tightly to the mutant?

Judith Klinman: No, it does not. (*Note added in proof*: At the time of the conference, we had measured K_d values for the mutants using single tryptophan variants of the methyltransferase, seeing a very small trend, with the mutants binding the S-AdoMet slightly less tightly. We have now measured K_d for the binding of S-AdoMet to the WT methyltransferase and see an overall ca. 10x weaker binding in the direction K_d (Tyr) < Phe < Ala. That is the cofactor begins to bind *less* tightly to the mutants.) The whole idea is that you label the methyl group with three tritiums, and you look at the effect of binding on the discrimination for or against tritium in the bound S-AdoMet with the wild type or the mutant. And you can measure the discrimination between protium and tritium. That tells you what is happening in terms of the force constants at the C-H bonds within the transfered methyl group. And we see that the change in force constant is greater with the wild type. When you make a mutant of a residue behind the sulphur atom, a residue involved in the transfer of the methyl group, the discrepancy between what is happening in solution and in the enzymes starts to disappear. So you actually need that conserved tyrosine pushing on the sulphur. You know there were hints in the literature earlier that there were very strange binding effects in proteins. I remember that Vernon Anderson published some data years ago and everyone said that it must be wrong (and I must confess I was one of those persons). But in this instance, the unusual effect that we are seeing appears to be in the formation of the enzyme complex.

John Gerlt: It is not manifested in the binding constant?

Judith Klinman: The binding constants are not that different for S-AdoMet in the mutants. (See note added in proof above.)

<u>Kurt Wüthrich to Judith Klinman</u>: Did you replace this key tyrosine with phenylalanine? And if yes, what happened?

<u>Judith Klinman</u>: Yes, that was one of our mutants, tyrosine to phenylalanine. The tyrosine is probably important because you have a hydroxyl group hydrogen bonding to a glutamate fairly far away. That keeps the orientation fixed so that the β-methylene can be in the correct position to create this compaction. If we generate phenylalanine, we get some elongation between the methyl donor and acceptor atoms in the computation and diminution in the binding isotope effect. The big difference arises when we put alanine into that position. Now we get an internuclear distance that is not very different from what would be seen in solution. There does not seem to be much compaction left. There is still some with phenylalanine, but not as much as with the constrained tyrosine. Phe is funny because it is of course much more hydrophobic than tyrosine.

<u>Don Hilvert to Judith Klinman</u>: Along that line, concerning binding, in the phenomenal 500- or 700-fold isotope effect that you see on C-H activation, can you actually rule out the possibility that the labelled molecule, the deuterated molecule, does not bind in a non-productive way?

<u>Judith Klinman</u>: I cannot tell you anything about the actual binding. The structure that I showed you is the enzyme without the bound substrate. But why would deuterium bind non-productively where protium does not? This is what is so nice about these isotopic labels. All we are changing is the reduced mass of the bond. Honestly, I haven't thought about that as a possibility because it is not a real substitution. It is just putting in an isotope. What is so phenomenal is that the isotope effect, which is normally 80 anyway with the wild type enzyme — a very high value — goes into a regime where, in our mind, the donor and acceptor are just too far away. Within a full tunneling model, it is like the deuterium is facing a wall. The wavelength is just too short to go over that distance and the enzyme has become more rigid and cannot recover the wild-type distance. I would love to have some ideas on how to test that further. We are doing many things like X-rays at room temperature to see if we can look at changes in dynamics and H-D exchange in the double mutant. We have high-pressure effects that suggest it is actually more rigid than the wild type. All these things have to be tested now.

<u>Jason Chin to Don Hilvert</u>: I have a question about the experiment where you do computation and directed evolution. Is there an experiment in the literature where, starting from the same protein scaffold, someone has done directed evolution on the parent protein? Has someone done computational design and then directed evolution and compared both processes directly? I have seen a lot of papers where you do the computation and you do directed evolution: the computation clearly does something, the directed evolution does something more. But I would be curious to

see a comparison starting from the parent. Is that experiment in the literature? So maybe you can comment on both what you think would be the value of such an experiment and whether it has been done.

Don Hilvert: That precise experiment has not been done. David Baker's lab carried out a control experiment on retroaldolases that required a reactive lysine and Kemp eliminases using a scaffold, basically introducing mutations to mimic the composition of the active site that was generated computationally but without any computational information. And indeed, in those libraries he found active enzymes but at a much lower frequency. The activities were very similar, so I predict that if you use these hits as a starting point for evolution, you would get pretty far. For this type of simple model reaction, that may be a viable approach. If you would like to do something more difficult, my guess is that it will be hopeless because, in our experience, in screening or selection experiments where you start with no activity, you're in the noise at the beginning and you end up in the noise.

Kurt Wüthrich: to Jason Chin: These modified proteins give unlimited possibilities for the use of NMR spectroscopy, but the production yields must be sufficient. Can you make milligrams of these proteins with single amino acid residue modifications?

Jason Chin: We can routinely make milligrams of proteins, certainly in *E. coli*. We have, for example, solved a number of X-ray structures of proteins bearing post-translational modifications — acetylated proteins, ubiquitinylated proteins — using approaches that we have developed. The limitation, at least for us, is not in this area. The limitation is in finding questions well enough molecularly defined that you know that the outcome is going to be interesting. We could solve hundreds of X-ray structures, as could Tom and other people who have technologies for modifying proteins, but the number in which the questions can be reduced to something molecularly interesting is less clear. For the purposes of NMR, yield is a problem. You would like to put in things like traceless isotopes. There is, in principle, a conceptual problem because you would like an isotopically labelled amino acid identical to a natural amino acid, but you can't get a synthetase to recognize an isotopically labelled version of the amino acid. You have to put in a protected version, and deprotect it after the synthesis, so that you end up with the natural version. We have done that in ways that give you a completely native protein having a single isotope label at a single position in the protein.

Kurt Wüthrich: Do you still get good yields after these additional steps?

Jason Chin: Yes. For example, we have done experiments on a lysine side chain to which we have attached a para-nitrophenol group via a carbamate. The amino acid goes into the protein and we believe that, during the purification process, the nitro group is reduced and a spontaneous fragmentation process yields the lysine back. So we have been able to isotopically label a single lysine in a protein, take the NMR spectrum, and see a single peak, for effectively any protein you want to express in *E. coli*. For example, you could label an active site lysine and screen for drugs that modify the chemical shift in that one position. This could be achieved without having to deconvolute, or know the assignments of, the spectrum. It's not something we're done, but if anyone wants to collaborate on that, we can make the protein.

Don Hilvert: Along the same lines, I read a paper recently by Nediljko Budisa in Berlin who claimed that many of these modified aminoacyl-tRNA synthetases are extremely inefficient compared to the natural enzymes. Competition *in vivo* with natural processes or normal suppression can be quite severe. Yields of protein are low and also not very homogeneous. Could you comment?

Jason Chin: There are experiments in the literature where people have done *in vitro* pyrophosphate exchange, the first step of the aminoacylation process in the synthetase enzyme. Some of these experiments are quite interesting because they suggest that even though the exchange is well characterized, *in vivo* when you add, let's say, one millimolar amino acid to the cell and characterize the protein by mass spectrometry — which tells you the identity of the amino acid in the protein, and MS-MS which tells the position in the polypeptide where that amino acid is incorporated — those experiments show very clearly that what ends up in the protein is correct and you are able to quantify the yield. So that's all fine. The issue is that, *in vitro*, when you purify the synthetase enzyme and then ask how well do those unnatural amino acids bind to the enzyme active site, in some cases that number is quite high. And in some cases, at least in Nediljko Budisa's hands, the claim is that in *that* step of the reaction, the natural substrates are competitive. But that does not make this true for the overall catalytic process. Some of those could be dead end complexes and that would be consistent with the observation that what you are selecting on is the overall process and the ability to put the amino acids into proteins. So it depends on what question you are interested in asking. If you're asking can we make proteins containing these amino acids and make milligrams per litre, purify them, and then ask questions or do experiments in cells, such as the ones I focused on here today, I think the answer is clearly yes. If you're interested in what the exact enzymatic properties of the enzymes are, and I point out again that those *in vitro* measurements are only one step of a multistep process and there's really no controls that even say that those enzymes are behaving in the way that they are in cells or in fact that they're folded correctly *in vitro*, I think that's interesting characterisation. But I'm not really sure what we learn beyond that.

Ken Houk to John Gerlt: I want to ask more about the pathway docking procedure that you used. It was a quite impressive way to explore the function of an enzyme. The background of my question has to do with the fact that, in general, docking has a rather bad name because of its inaccuracy in calculating the binding energies of different kinds of molecules. Often, many docking programs will give micromolar to millimolar binding with a whole variety of substrates that might fit in the pocket. And the second issue is that, when you are looking for function, the enzyme is evolved to stabilize the transition state, not the substrate. I wonder if you can elaborate on this pathway docking procedure, which proves to be so good?

John Gerlt: Let me try to answer as Matt Jacobson would, since I am here representing a large-scale project. Docking is notorious for being not particularly good at predicting specific substrates. It was used initially in the pharmaceutical industry to try to identify potential lead candidates. That is not as demanding as, for example, screening against the KEGG database to ask which metabolites bind best and could be substrates for an enzyme of unknown function. When you dock into a single active site, cases in which the best predicted hit is the substrate is not very high. So the question is how could you improve that? This might be done by actually docking to a series of binding sites that constitute a metabolic pathway. We are presently considering catabolic bacterial systems where there are transport systems to bring the catabolite compound in, which is subsequently catabolised. Many transport systems have solute binding proteins in the periplasm and those binding sites can also be used for docking. They also have the advantage that if you produce the proteins, you can quantitate binding; you do not have to quantitate catalysis. And this can be done in a relatively easy way by looking for thermostabilisation by which you can screen a large number of compounds. But the idea is that in fact, if you dock to multiple enzymes in a metabolic sequence, perhaps no docking to any one enzyme would give the right answer. But if you look at trends for the docking hits, chemoinformatic methods can analyse those (and it is part of our project to do so), enhancing the ability to decide whether the most likely candidate is a substrate for the metabolic pathway. That is how we produced a trans-hydroxyproline betaïne, the substrate for that racemase. Again, none of this is perfect yet. If it were, the problem of predicting functions of unknown proteins would be easy. But we are certainly using that docking to two, three, four, five, six successive active sites with the idea that for the substrates, elements of specificity will be conserved through the pathway, allowing a better choice of compounds to test. The point of binding versus catalysis is very well taken. Matt Jacobson recently published a rapid report in Biochemistry, I think, about distinguishing between false-positives, which bind but are not catalytically active, and compounds that are real substrates. It was a QM/MM (Quantum Mechanics/Molecular Mechanics) approach sorting out why some compounds predicted to bind were not very good substrates, presumably due to electrostatic effects in the interaction of the various possible substrate candidates with the active site. I don't believe this has been totally sorted out. This docking doesn't work every time but that is why we're trying to improve it.

Judith Klinman: I just want to get back to the QM issue and how many atoms are going to be needed to really assess what is going on. In the catechol methyl-transferase, Todd and Heather needed to extend the number of atoms in the QM region way beyond what is usually done to get a true picture of what is happening in the binding. I would like to put this out as a major and important new direction to be tested in other enzymes. This is just one system, yet the theoretician Walter Thiel has been examining another class of enzymes, once again extending the QM region in a stepwise fashion to see how each new area changes final conclusions. If we are going to use QM, what are the criteria for concluding whether observations are related to function? Methodology will increasingly include many more atoms than the standard 20.

Ken Houk: I am guilty of using quantum mechanics. What you describe is a great problem in that QM, highly accurate for small systems, is combined with force field or MM methods to try to deal with the protein, and in particular to include dynamics so that one can calculate not just one but millions of different conformations and average them. While these techniques have been known for some time, they still are not very successful, unfortunately. My lab uses QM to predict the ideal arrangement of functional groups with respect to each other. In fact, for the substrate, precise geometrical arrangements in the transition state are extremely important for catalysis, as mentioned by Judith Klinman. Deviating by a few tenths of an Å is sufficient to lower the rate of catalysis substantially. Todd Martinez is a pioneer at using GPUs for quantum mechanics. Actually, this is much more common for molecular dynamics. We use GPUs for molecular dynamics too. This takes advantage of the gaming industry's huge audience for the latest machines having very inexpensive processors. This proved to be quite useful for molecular dynamics, but more difficult to apply to QM. We need to apply more accurate methods like QM to larger systems. As this will be more common in the future, one can hope to increase accuracy. Nevertheless, the problems we are discussing are huge, both in terms of the number of atoms involved and molecular dynamics, namely the necessity to include millions of conformations. The prospect of computation replacing the experimental approach is neither immediate nor even likely in the next decades.

Discussion among all attendees

Judith Klinman: I would like to use one of my own slides (Fig. 1) in generic fashion to address the question of how enzymes work. Combining experiment and theory is essential at this point. We constantly go back and forth, testing one against the other. I want to use an example on this slide to facilitate discussion. Here, an internal region of the protein, coloured on the figure, is surrounded by a scaffold which may be pretty rigid. But inside, approximately 600 atoms, a huge number, are interacting with each other. It is the entire ensemble that is essential for catalysis. This instance involves compaction between the methyl group donor and the

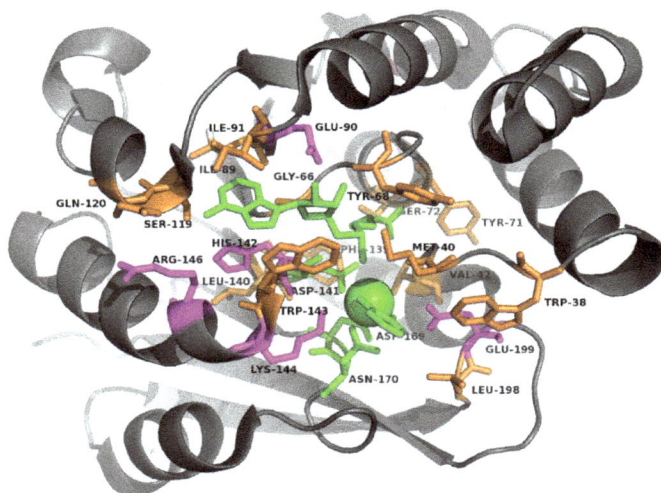

Fig. 1.   We can begin to test structural features capable of generating a short inter-nuclear distance between the methyl donor and acceptor using COMT as a model scaffold. Computations by Kulik and Martinez indicate the need for an extended QM region (shown by the atoms in color below), to obtain an energetically converged ground state structure for the ternary complex of COMT.

acceptor. So now you can start to do *in silico* experiments. You can take away a single side chain, one at a time, perhaps in combinatorial fashion later, and calculate how close the donor and the acceptor can get. Depending on what is seen, we can then go back to the laboratory to test these out. This kind of interactive prediction testing, from the computer to the lab and back, is the way we have to go.

Another comment concerning compaction: a young man came up to me in the coffee break, and he said: if you look at Morse curves, how in the world do you get things so close? We know from physics and physical chemistry that the energy to get something shorter than the van der Waals distance is very large, so that's another question that can be asked. What are all these remote interactions and compensating effects in proteins that can be translated into very specific reactive configurations? I'd like to put that out as a really important question.

Gebhard Schertler: I would like to present two slides very far from this topic but nonetheless related. We work on arrestin, a protein binding to G protein-coupled receptor. We have mutated every amino acid into alanines and tried to stabilize the interaction. We want to make a stronger binder to arrestin. On this graph, the red things bind less well and the blue things bind better. On the next figure, we include a structure. The better blue binders are very spread out, as are the sub-optimal reds in the interface. Intuition suggests changing the interface to get different binders, but these are too far away. What is the explanation? The blue patches are all far from the binding site. This is similar to what Judith Klinman just said: that things far away influence what is happening. In this case, it is most likely that we have two states of the protein. Disruption of the blue core with mutants favors an activated state of arrestin: an example of activation from afar. In directed evolution, where do the residues change? Are they changing in the first or outer shell? I predict that a lot of later evolution will happen in the outer shell. When we compare $\beta 1$ and $\beta 2$ adrenergic receptors, the innermost shell contacting the ligand itself is more or less identical. There are no different amino acids. In the outer shell, things are different. There, even small molecules like adrenaline and noradrenaline can bind differently. All these things have to do with the dynamics of the molecule. Events in the outer shell matter because they change protein dynamics.

Frances Arnold: You are right. Experiments show that when you evolve an enzyme for higher activity, allowing mutations to be randomly distributed across the protein, many improvements are found far from the active sites. Though this has been observed hundreds of times, the exact mechanisms remain to be found. In this respect, I think Ken Houk presented some very interesting data today.

Don Hilvert: Following on Frances Arnold's comment, active site mutations often have greatest impact on activity both favorable and unfavorable. This is to be expected given the starting assumption that interface residues have the largest effect. But it is true that proteins are attractive evolutionary objects because you can shape them through distant interactions and transmit that information to a binding pocket.

Richard Lerner: In evolving antibody binding, you often start with a pretty poor binder. Under selective pressure, it mutates by somatic random process. When you go from micromolar to nanomolar dissociation constants, you see both active site mutations and outer-shell mutations. Almost every somatically mutated antibody has outer-shell mutations. Schultz published a rather long paper about this in Science, saying that the general concept is not very complicated and that it is entropically very good to make the active site a better fit. I do not think that anybody has put it as elegantly as Judith Klinman in terms of compaction but you cannot find a somatically mutated antibody that doesn't make outershell mutations.

I have a more general question, not so much for the people in the room but for those who want to get a Gestalt of what we all do. Most people here are chemists including myself. Chemists like to make things and therefore, if there were another similar meeting of people who want to discover things, they might say to us: why do you bother to even make things for the good of mankind because there is so much which remains to be discovered? And the answer is: because we want to learn the principles. But an interesting question would be: how many important molecules, large or small, with say the importance and potency of steroids and prostaglandins, do the people in this room believe remain to be discovered? Or have we discovered 95% of the important ones? Let's just deal with small, very potent molecules. What percentage would you believe we have discovered (to Don Hilvert)?

Don Hilvert: People in the pharmaceutical industry always say that their hit rates are much higher if they search natural products libraries than if they start with small molecules from combinatorial libraries. But I do not think they represent probably more than 50% of the hits.

Richard Lerner: We've discovered 50% of the potent molecules, then?

Frances Arnold: When you ask what has been discovered and what remains to be discovered, are you talking about structures or function? Most people care about discovering functions.

Richard Lerner: I would say functions, Frances, quickly followed by structure. For this purpose, let us just quickly call them functions.

Frances Arnold: So then the context of the question is important: it depends on what you are asking the system to do.

Richard Lerner: OK. I am asking for function, for example anti-inflammatories.

Frances Arnold: But this is a non-natural function because it is something a human is interested in but not necessarily the biological system that made it. And the same is true for proteins, so many proteins to be discovered that Nature has not made because she could care less.

Chris Walsh: With regard to how many natural products remain undiscovered, there are people who do calculations based on the discovery rates of antibiotics of clinical relevance for human bacterial infections. One in ten *Streptomyces* make a molecule called streptothricin.The rate of success is 1 in 1000 for streptomycin, 1 in 10^4 for erythromycin, and 1 in 10^6 for daptomycin. And so the calculation has been made to find the next important molecule, you might have to screen 1 in 10^8. When the pharmaceutical industry stopped screening, they were testing about 3×10^6 molecules per year. So, under classical conditions of screening and looking for new cultures, they were never going to come up with the next new molecule. On the other hand, if you ask how many bacteria are there in the soil, you can count 10^{18}–10^{19}. By that calculation, 98% of the important natural products in terms of antibacterials could still be discovered because we have not screened more than 2% of the soil bacteria that have a history of producing molecules. That's just one area of therapeutics. What I say is based on an assumption about how rare truly interesting molecules have been, and whether that's an extrapolatable line.

Richard Lerner: Is somebody going to discover a new steroid in man?

Brian Roth: Yes.

Chris Walsh: If I were looking for interesting new molecules, I would not look in humans. I might look at the microbiome of the gut.

Brian Roth: My lab does lots of screening and we really love to screen natural products so if anyone here has any, please send them to us, the more interesting the better. And let me say that the approach that we take is to screen hundreds of molecular targets in a parallel fashion. I think you're right: if you take the pharmaceutical approach, which is one target at a time, the hit rate is very low. But if you have a large, genome-wide coverage of a particular target class, particularly the so-called orphans, then every time we do a screen, we find really interesting things. Our problem is having the manpower to analyse the results sufficiently rapidly to disseminate them.

Chris Walsh: That is a good and unusual problem to have.

Brian Roth: It is, unfortunately, true. I know there are just so many orphan targets that, on the basis of knockout data, have really interesting effects. At least in mice, there are really interesting steroid-like molecules that remain to be discovered. I can expand on that tomorrow. The other point relates to directed molecular evolution. I think we are the only laboratory that really has done this with GPCRs to any great extent. We were evolving a GPCR to bind an unnatural ligand, and be activated by it, to use as a tool eventually. This turned out to be extraordinarily valuable for many laboratories. By using saturation mutagenesis, we could screen millions of mutants within a couple of days, and found all of the interesting ones were actually in the binding pocket. So the uninteresting ones that caused non-specific activation were outside the binding pocket and had basically allosteric effects. But those that changed the pharmacology of the receptor (which were the ones we are interested in) were, as far as we could tell, right in the binding pocket and oriented towards the ligand. The situation may be a little different in other cases, depending on the particular reason for which you are trying to design the protein and whether these outer-sphere mutations that have an entropic effect are going to be effective. But we found that if you are trying to evolve a protein that binds an unnatural ligand, then, at least in our experience, those were the best.

Richard Lerner: How many enzymes are there in the universe?

Brian Roth: I don't know.

Richard Lerner: Maybe 10^4 more?

Chris Walsh to Judith Klinman: A question for Judith Klinman. I guess I have not thought about compaction very often in terms of catalysis. My question relates to the maturation of molecules like the green and red fluorescent proteins. On the one hand, triad of residues — serine, tyrosine, glycine in a row — get converted auto-catalytically or by auto-modification to the fluorophore in GFP or red fluorescent protein. But I understand that auto-modification only happens from a folded nascent precursor protein and not from the denatured protein. So is that a case of compaction? Presumably not, and a lot of calculations have been done on that auto-modification. So tell me how that is different.

Judith Klinman: Proteins fold and as they fold, they generate active sites that have a very large number of functional groups, which could also be part of the green fluorescent protein maturation. Actually, in my view, there is no counterpart in small molecule chemistry because entropically it is simply prevented: you can't get that many functional groups in one place at one time. So that is step 1. That is how proteins have evolved and of course the energy is paid at the time that the protein molecule is synthesized through GTP hydrolysis. But that is not the end of the story. That is a static ground state structure. And then the question is what else do we need to get these enormous rate accelerations. Having all the functional groups present does not seem sufficient. Now we have to go system by system, but the model that has emerged — at least in my lab from the hydrogen tunneling studies — is that these proteins are actually moving all the time. They are sampling many different ground states, and it is only those ground states that give you all the precise interactions, all the precise hydrogen bonds, electrostatics, etc., that allow catalysis to proceed so quickly. It's that level of dynamic motion. Of course, we think of proteins now as existing not as a single structure but as a family of conformational sub-states. Any perturbation you make will affect catalysis, whether or not it is a remote mutation. There is no question that putting a lot of functional groups around the bound substrate gives you a large rate acceleration. I think you could probably get a 10^8-, 10^9-, 10^{10}-fold rate acceleration. I do not think you are ever going to get to a 10^{25} fold rate acceleration from simply putting these active site residues together. There I think what you need is this distribution of conformational sub-states. That being said, in the methyl transfer reaction we appear to be seeing a lot of compaction in the ground state, at least at the ternary complex, but that is just one aspect of catalysis. I do not think that there is any way out of this kind of thinking right now. The problem is how do we show it computationally and experimentally and bring this all together? That is a challenge for going forward.

John Gerlt: In the complex, what fraction of the protein and substrate molecule is compacted and what fraction is relaxed? You are talking basically about a difference between the entropy of bringing things together and then, presumably, additional energy — enthalpy — is used to bring about this compacted state.

Judith Klinman: Before we did the methyl transfer experiment, I would have said that it was all in the dynamics, in some fraction of the total protein, and I could not have given you that number. That made it difficult to test the ideas, as well. It's just a transiently formed sub-state. But in the case of the methyl transfer, it looks as if there is enough going on in the ground state so that you can get it (the compacted state) as the dominant enzyme form. So then the question arises regarding charge effects: in C-H activation, the charge properties are very different from S-AdoMet-dependent reactions where you have the sulphur S^+ and nucleophilic substrate. The electrostatics are going to contribute to this phenomenon. I can't give you a number. All we can do currently is look for changes in the conformational distribution and correlate them with catalytic efficiency. This is being done more and more using biophysical tools.

Marina Rodnina to Judith Klinman: I also have a question for Judith. How common are these tunneling effects in enzymes? In other words, can it be that we are describing enzyme function in an entirely wrong way if we are not taking them into account?

Judith Klinman: We have been looking at tunneling for about 25 years. I would say that there is evidence that virtually every C-H activation has a strong tunneling component. It is much harder to say that in proton transfers from oxygen to oxygen, since you don't have the same tools available. But I would say it is very prevalent. I think the reason that my group started to think about dynamics is that once you invoke hydrogen tunneling, the movement within the chemical step is just wave function overlap and then all the motions that give rise to the chemistry have to lie within the heavy atoms of the protein. My guess is that this is probably true for all catalysis. So yes, there is something wrong with the idea of transition state stabilisation. It's just looking for all the catalysis in the enhanced binding at the transition state, and not giving us the physical principles that allow these enormous rate accelerations.

Ray Stevens: A question for Ken Houk. First, an observation: you made a comment that docking is not very reliable for enzymes, and sometimes we hear kinases have anywhere between 5 and 15% success rates. Whereas with GPCRs, not talking specifically about enzymes *per se*, the hit rate is closer to 60 to 80%. You get much better hit rates with docking than with enzymes. And then Richard Lerner made the observation about enzymes, where you see some mutations on the inside (in catalytic antibody enzymes) but you also find a lot of mutations on the outside. Brian Roth made the observation that most of the mutations are occurring directly in the binding site. So I am curious, from a computational perspective: how do you use this knowledge concerning the hit rates and the position of the mutations? Does it have something to do with dynamics? How do you use this and how do you connect those dots?

Ken Houk: We can't use that information directly in a computation. A computation is built on the physics of interactions, so we would like to use quantum mechanics since we think that is the right physics, but it is too time-consuming computationally to be used for all systems. There is another complication, which is that the sites of mutation often aren't directly related to catalysis. You can measure increased catalysis, but, for example, in experiments that I was trying to help explain, Codexis is trying to develop a commercially viable catalyst to be used in somewhat unnatural conditions — that is, mixtures of water and DMSO for example. The mutations that occur are contributing to the lifetime of the enzyme and to other factors that are different from catalysis. That makes it difficult to mix in directly with computation. When we do a computation, we are trying to essentially measure the free energy difference between the bound substrate and the transition state in the enzyme. I didn't describe any calculations like that today, but that would be the ultimate in using computations to explain catalysis. To do that correctly, unfortunately, it is not possible to use a lot of empirical information. What we need is to have much more accurate force fields if we are going to use classical mechanics or faster quantum mechanics. In both fields, progress is constantly being made but we still have to wait for those developments to occur.

Jan Steyaert: I have a hypothesis rather than a question. If conformational sampling is indeed so important for catalysis, I would think that conformational antibodies that recognize these enzymes would act as allosteric modulators. Is there evidence for that?

Richard Lerner: Jim Wells has published a series of very nice papers where he actually selected antibodies that are allosteric modulators.

Jan Steyaert: But those were large conformational changes. This is more subtle, no?

Richard Lerner: 700 Å^2 of protein surface. That is pretty large, yes.

Tom Muir: I have a question for the protein evolutionists. We've had some beautiful examples of using in-laboratory evolution to achieve new functions. In one example, Jason Chin is really looking for a synthetase with new specificity in terms of an unnatural amino acid that you can bind to the synthetase and activate. But it is a slightly different situation with a catalyst. You obviously want to have activity toward a specific reaction, but you also would like to have some promiscuity with respect to substrate scope. I wonder what the state of the art is and the current thinking about how one balances these two effects. To me it seems that there are interacting effects: the enzyme is going to evolve toward the particular substrate that you happened to use, not necessarily toward a whole panel of related substrates. What is your thinking on that?

Frances Arnold: This is the sort of question I spend a lot of time thinking about. When chemists say, "I love lipases because they will work on anything," I answer, "What they do (hydrolysis) isn't very interesting, but they do have a broad substrate range and reasonably good selectivities." The real trade-off is usually between having a broad substrate scope and having the regioselectivity or enantioselectivity that you really want on each substrate. Right now, it's just trial and error. It is very hard to know *a priori* whether you'll have broad scope or not. Another group of people would say that doesn't matter, because all you need is a quick screen and a couple of mutations to fine-tune the catalyst for any substrate or selectivity that you want. If you buy that argument, then you really don't need to have super broad range, but you have to be able to access the substrate scope that you need.

Don Hilvert: If you want high activity, you have to have very high specificity as well. Many computational designs are attractive in this regard because you get a foot in the door: you have a promiscuous catalyst that you can then potentially evolve in many different directions.

Kurt Wüthrich to John Gerlt: John, you now have this marvellous machinery set up, and the PDB includes at least 5300 domains of unknown function. How long will it take to annotate just this sample of structures with your machinery? You should not work in empty space, here you have material to test your machinery.

John Gerlt: We are developing the machinery to allow such de-orphanization to occur. We spend a lot of time worrying about the proper way to evaluate the success of the Enzyme Function Initiative. Is it evaluated by how many PDB structures we can de-orphanize ourselves, or are we developing strategies and procedures to do that? I would say that we are not necessarily a machine to crank through every structure and de-orphanize it, but what we are trying to do is to develop tools and strategies so that other people can do this. I would say that a problem with many PDB structures is that they were presumably determined in an unliganded state. One of the proteins I talked about today, PDB 2PMQ, was actually crystallized in what can be called a dockable state. After the fact, the structure was determined with the ligand bound and found to be virtually superimposable on the unliganded structure. I would argue that in cases like that the success rate for docking should be high. And we would not necessarily be interested in all of those. We alone are not developing a machine that is going to be totally responsible for doing this. I don't wish to be evasive; our success will be developing pathway docking. Because docking is not a hundred percent accurate in each case, is there a way to do multiple docking experiments and successive catalytic events that enhance that rate? This is not a challenge that one large-scale project is going to solve.

Kurt Wüthrich: Should I translate this by saying that you are developing a car but don't want to run the risk of driving it on the highway?

John Gerlt: No, not at all, but it is undergoing design continuously.

Chris Walsh: Maybe you have a car but you do not have any gas to put in it.

Wilfred van der Donk to Jason Chin: This question is for Jason Chin. We are doing some incorporation of unnatural amino acids using the methodologies that you and others have developed. In our very limited experience thus far, success seems to be very variable. In one position, for the same amino acid, the same tRNA, and the same aminoacyl-tRNA synthetase, incorporation is great. In the same peptide at another position, we do not seem to get much suppression efficiency. So are there any rules? Is it understood why that is and is it something that you see as well? Are we just doing things wrong?

Jason Chin: There is a series of papers by Miller in JMB, probably in the seventies, that describe codon context effects in amber suppression, for example. We have done things like develop ribosomes that direct more efficient incorporation at amber stop codons. And there are other strategies that address this, which seem to even out some of those effects that you see if you do a simple experiment with a synthetase and tRNA. The simplest thing to do is to change the codon context, if you can, around the amber stop codon. There are rules based on this paper by Miller that will tell you what to put directly after the amber stop. Of course in some cases that may result in an amino acid change and in other cases it may just be silent. But that is the simplest thing to do. There may also be contextual effects related to where you are in the polypeptide sequence. You know there is this issue with translational ramping in the N-terminal region of proteins versus further on the polypeptide sequence, as some people have reported. My lab is trying to put together a database of our internal experiments and open that up to the community, sharing what the incorporation efficiency is at a particular site, which experiments worked well or which experiments worked not so well, to empirically determine as a community what the rules might be. I think that would be a resource for people. But I think the simplest thing for you is to try this codon context.

Henk Lekkerkerker: As a total outsider, I am impressed by these wonderful developments in genome mining, computation, specialist gene lines, etc. In 2007, President Bush predicted that by 2017 the U.S. would be using 35 billion U.S. gallons of bio-ethanol. It's generally accepted that only 15 billion of that would come from Midwest corn. That leaves about 20 billion gallons to come from cellulose feedstock. When you talk to people from DuPont, they complain they can get good enzymes (cellulases) only from fungal, not bacterial, sources. They need doses about 25 times as high as for the corn conversion and it is slow. But with these eminent people here, I expect that with the combination of genome mining, calculations, and other approaches, we can solve the problem of DuPont by producing 20 billion gallons of bio-ethanol. We produce about one billion now.

Frances Arnold: It is not the enzymes' fault. If you compare cellulases from many different organisms, they are all pretty much the same in terms of their catalytic capabilities. There are no huge factors of difference. So the only opportunity for increasing intrinsic activity is to raise the temperature of the process or to pre-treat the material so that you make the cellulose available. All those things take money.

The reason not much bio-ethanol is being made is that the price of oil is too low and the price of corn and other feedstocks is too high. Nobody can make money out of it. It is certainly not the enzymes' fault.

Henk Lekkerkerker: I will tell it to these people, and they will be happy, but I understand that they are building factories for which the technology is not yet fully developed. Nevertheless, to meet this aim of 20 billion gallons, they are going ahead. I agree that with Shell Gas etc. that the price of oil is low. So it is temperature and not enzymes.

Frances Arnold: Certainly there is some room for improving the enzymes because they are being asked to work under non-natural conditions. But the claims that you can dramatically increase the intrinsic activity of these enzymes are completely unfounded. That's not going to happen. You are going to improve the process context in which they are going to work, and you are going to reduce the cost of production, which Novozymes and Genencor have been very successful in doing, and they are keeping the rest of the profit in their pocket.

Bernard Henrissat: Building on what Frances has just said, the problem is not so much the enzymes. It is that the substrate has evolved for millions of years to resist enzymatic degradation. You know a tree must survive for an entire lifetime, sometimes hundreds of years, without being destroyed. There has been a lot of invention on the plant side, on the biomass side, to resist degradation.

Frances Arnold: The biomass is very hard to degrade. It has evolved to be hard to degrade.

Henk Lekkerkerker: I agree it is difficult to make a solution of cellulose and it either has to be highly acidic or highly basic, but the pre-treatment problem seems to be solved.

Bernard Henrissat: It is not only cellulose in there.

Henk Lekkerkerker: Agreed.

Lode Wyns: It is a pity that JoAnne Stubbe is not here. She sent in a very nice text, with a lot of interesting manganese bioorganic and inorganic chemistry. In going through the text, I noticed that she was the only one who mentioned education and the problem that education is lacking on chemical elements that Nature doesn't utilize. Looking at myself in the mirror, I say "My own knowledge of those elements is weak." At one time, we tried to make antibodies against vanadate in order to make kinases, in the same way as they were making antibody enzymes against phosphates to cleave esters. But I experienced that my knowledge, and my inorganic chemistry colleagues' knowledge, was not sufficient to make the right happen. I guess, when we talk about new enzymes, that there is a whole world of inorganic elements that we don't deal with. In that respect, education is a major problem. That is what JoAnne wrote. I think that she was the only one to mention education in this context.

Don Hilvert: That is a very good point, but there are inorganic chemists who are trying to design enzymes. Thomas Ward at the University of Basel, for example, has done very nice work combining the intrinsic reactivity of metal ions with the selectivity that can be offered by protein binding pockets. There's a lot more to do, obviously: the periodic table is very large.

Chi-Huey Wong: Going back to the long distance effects, I have a question for Judith Klinman and perhaps the others. There are so many examples now of long distance effects on active site binding and catalysis. So the question is: how far are we from being able to predict the long distance effect on binding and catalysis, and what do we have to do next to comprehend this?

Judith Klinman: This is the area that makes enzymology a real challenge right now. We can do it on a system-by-system basis. I can give you one example where we have analysed tunneling as a function of a single residue at a dimer interface 25 Å away, and can actually show a network that controls the active site compaction. Would I have predicted that in advance? Probably not. The only way we identified it was by comparing a psychrophile to a thermophile. You know, you have families of homologous enzymes that have to work at very high temperature or low temperature. In that case we got lucky. By comparing the two, we could actually identify the distal residues in the network. But in general, as I indicated with this the methyl transfer example, we need to go back and forth between *in silico* computation and experimentation. We don't have rules yet and I wish I could say more. We don't even have the required temporal resolution. There has been a great deal of controversy: enzymes turn over with rates of milliseconds, so for a long time everyone was focussing on millisecond motions to control catalysis. In fact, probably most of the millisecond motions are an accommodation as the enzyme moves through its catalytic trajectory because the enzyme changes its shape as you go from the substrate to an intermediate to the product. Those millisecond motions may just be an accommodation to having different structures in the active site. Most of the motions are going to be in that sub-millisecond to nanosecond, even picosecond, regime. It's tool development that we've discussed here. We have to design the appropriate tools to get that temporal resolution. We need spatial and temporal resolution. These are big problems, and I would like to get back to education because I think the way we teach students enzymology is simply not helping the field at all. It is not just bioinorganic or bioorganic chemistry. We really need another approach. Hopefully, the next generation will have the background as well as the tools. We need more people working on the problem.

Ken Houk: In response to Chi-Huey Wong's question, the way we are approaching this is not really predictive but it is computational screening. Related to what Judith just said, the only approach so far is to do mutations and then do molecular dynamics to understand the change in the enzyme average structure, and to determine whether that improves the active site arrangement or makes it worse. This is unfortunately a very time-consuming procedure, but it is the only way we really can understand how long-distance mutations affect the activity. Unfortunately, sometimes there are loop-motions and so forth that are slow enough so it takes microseconds to milliseconds dynamics to try to account for that.

Don Hilvert: How large do the changes have to be to quantify reliably?

Ken Houk: I guess we know anecdotally that a change of several tenths of an Å from ideal is sufficient to reduce activities substantially. We're looking to design in smaller differences than that.

Frances Arnold: I have been asked over the years by multiple journal editors to speculate on effects of mutations that we discovered by evolution. I always refuse, saying that I want to write a paper of lasting value. That usually works. The reason is that I know that I can tell ten different stories and they would all be acceptable but I would not know which one is true. So, now that you are making strides and are answering some of these really difficult questions, I wonder how do you know the story, the explanation that you provide for the effects of a mutation is true? Judith (Klinman), this question is also for you.

Ken Houk: If I can follow-up and respond to that. I agree that currently it is faster to do random evolution than it is to do computational predictions, but once we are able to do reliable predictions, then we will know that we understand.. I wouldn't claim we can do that yet.

Judith Klinman: I just want to comment that so many mutations, especially those that are remote from the active site, do not change the X-ray structure at all. That means that you have to look elsewhere for an explanation as to what is going on. When you see structural changes then, I think it becomes really uninterpretable, but that's where the next round of experiments and computation picks up. When we have profound effects on activity, and the solved structure looks identical, we need to look beyond that to understand what is going on.

Kurt Wüthrich: The more this discussion goes on, the more I feel like asking whether we should hope to soon get a general explanation for enzyme mechanisms, or should we rather aim at investigating the mechanisms of one, two or maybe five enzymes during the next ten years? What is your opinion? What can we hope for realistically?

Ken Houk: Enzymes take advantage of everything to maximize activity. I don't think we will get a single answer to that question, other than the very general answer about changing the free energy of activation no matter what the effect. One can hope that, by studying more systems, we will be able to develop some general rules about how remote mutations influence activity, but I do not know what those rules are going to be.

Judith Klinman: I think that there are some fundamental reactions that occur in biology. One of them is phosphoryl transfer, for example, and we have already talked about C-H activation and methyl transfer today. There are some very fundamental classes of reactions essential for life. If I were to choose, I would focus on these very general classes of bond activation and cleavage, to see if we can in fact develop a core set of principles that will allow us to go forward.

Frances Arnold: But that does relatively little to help the design problem, right? If you understand the mechanism, it does not necessarily mean that you know how to design it or even how to predict the effects of mutations.

Judith Klinman: Well, I would like to get back to the example of methyl transfer because we defined a core region we think is essential for compaction. Maybe compaction is a design principle. Maybe by going back and forth between *in silico* and experimental results, we can validate that and see how compaction is generated. And then that is a design principle.

Frances Arnold: But I see that everyone has his pet explanation, and tends to answer the problem in terms of that explanation without really pursuing alternatives. And, equally, addressing them side-by-side. Probably computation is the best and only way that can be done, but do people really do that? First of all, we also need a tremendous amount of data because prediction is critical, and then we will need to predict many times because any individual prediction is an anecdote. Any paper that says that a prediction worked without showing a lot of data could just be showing the prediction that worked on that particular day, while the results of the other nine students whose predictions did not work were not published. We really don't know what is going on behind the scenes here.

Judith Klinman: I agree with you. I think the field has come to be dominated by computation. We need experiments as well. When new ideas emerge, this occurs because you make an observation which can't be explained by previously existing theories. It is not, let's throw this out or let's throw that out. It happens because you have a dominant paradigm and then new data emerge that cannot be explained by that dominant paradigm. Then you move on to a new way of thinking. I think that often starts from experiments, not from computation, and then you need this interplay between both. Or maybe it can also come from computation. Since I am an experimentalist, I'm obviously biased, but I think it's the interplay of observations that no longer fit the dominant paradigm.

<u>John Gerlt</u>: I ask Judith what she regards as the rate-limiting step in bringing together computation and experimentalists. I think that I hear you say that computationalists are not sufficiently engaged, or is there a lack of interaction with computationalists? I am not quite sure what the issue is. I don't mean to be argumentative. The point I want to make is that this requires collaboration.

To bring up again what Kurt (Wüthrich) said about the Enzyme Function Initiatives being a machine, I could also regard the DARPA project on Computational Enzyme Design (I do not know how it is currently supported) — with design, theoretical analysis, directed evolution, and structure analysis of proteins — as a machine that produced impressive results. I am hoping that our machine will produce impressive results. I am in complete agreement that one needs to somehow figure out how one can bring together experimentalists and computationalists to analyse and solve this problem. Is the problem achieving this a lack of coordination or interest?

<u>Judith Klinman</u>: Part of it is due to a lack of resources. We need the money to go forward. This is a really important area of research. Another part of it, in my experience over the years, has been language difficulties. Physical chemists have their own language. These experiments need to be explained to the theoreticians so that they are also fully on board. Therefore I would think that workshops to promote that kind of communication could help. But generally the way science works is on a one-on-one basis: you have a question that you cannot answer and then you look around for someone who has a skill that you do not have, such as computation, and then you move forward. It often takes a long time to lay the groundwork so that you fully understand what each person is doing both experimentally and computationally.

<u>John Gerlt</u>: I would say that these machines take some time and thought to put together, but when they have been thought out and put together they in fact can move forward in a significant way. In the United States, there is a "RO1 mentality" for doing science, and bringing together RO1-supported independent scientists to collaborate on a problem is not necessarily the most effective way to solve the problem. Establishing machines, if you can do it, becomes more productive.

<u>Judith Klinman</u>: I think we're getting into philosophy. There is big science, and there is that individual investigator following his or her nose and seeing where it takes them, but, yes, with resources you would be able to do that.

<u>Gebhard Schertler</u>: There are very good problems for big science. Biologists in the genome project have started to really use that. Looking at new organisms and mechanisms, or finding functions, is a good opportunity for big science, as long as we can design a screening process. There are also very good reasons for small science and idea-driven science, and I think we need both to work together. Biologists

certainly have to learn (for example from the physicists) to organize to get the resources, and that often might need a change away from individual contributions that we very much value in biology research.

But let us go back to the idea-driven thing and the interesting discussion. We talked about time-scale and catalysis and how fast the chemical bond is actually made or broken, and it is quite interesting to think about this. I obviously only know the answer for my pet molecule. And in rhodopsin, this decision is made in 200 femtoseconds. The primary effect that determines the quantum yield in the enzyme is determined in 200 femtoseconds. Now I ask you whether that time scale is not valid for all enzymes. Chemistry operates in a time scale of making and breaking bonds, and this must be in a timeframe of femtoseconds to picoseconds but not longer. You (Judith Klinman) mentioned microseconds and I think this is not correct. But what then happens in biology, and that is what we observe a lot, is that the chemistry has to be absorbed by the system. Some people have spoken about a protein quake for example: when you have thrown a ball into an enzyme, you somehow have to dissipate the energy. This is also the case when you put for example a photon in, and I do not know if that happens in catalysis so much. But we also have sub-states, and we have so many intermediate states when we activate rhodopsin because of timing. So at that point chemistry has happened, but the protein has to time a process in such a way that it is adequate for a signalling process, which is in this case a protein-protein interaction. So in biology, we deal with very different time-scales. In signalling we need hundreds of microseconds to milliseconds, which allows the diffusion to happen and the partners to find each other. But in catalysis, when you make or break bonds, I think everything is decided on 200 to 500 femtoseconds, and we need to support the tools, which can address these time-scales.

Judith Klinman: I agree with you completely. The chemistry is happening on femtosecond time scales. The motions that I and others are talking about are within the protein. It is the sampling of the different conformers that sets up an environment that allows this passage via the activated complex or the trajectory that can lead to product. The actual making and breaking of the bonds is very fast. So the microsecond time scale does not refer to the chemical step, it refers to the class of motions that sets up the environment to allow the bond to be cleaved. You can think of it as an environmental reorganisation; enzymes have evolved as the genius of environmental reorganisation. You use the same thinking for reactions in solution, but what is happening on the enzyme is quite distinct.

Ken Houk: What you say is entirely correct. In transition state theory or in trajectory studies of reactions this happens in a hundred femtoseconds or less. But even for reactions that take hours the actual bond making and breaking still only takes a few hundred femtoseconds. But, as Judith says, it is the probability of that one event happening that can vary, and that probability is influenced by the environment and the dynamic motions that are much slower either in solution or in an enzyme.

Hiroaki Suga: I am always excited by the combination of *in silico* simulations and *in vitro* evolution to get catalysis. But right now one needs to start from a certain protein scaffold. So, how do you pick the protein scaffold and how many proteins scaffolds are available at this moment? How many scaffolds are required to achieve the variety of different catalyses?

Don Hilvert: In the original experiments about 200 scaffolds were hand-picked because they were abundant, monomeric and thermostable. Subsequently, it has been possible to really look at every protein in the PDB. I think the success rate with those other proteins though tends to be much lower, because what you find is that when you introduce ten mutations at the active site of such a scaffold you more often than not screw things up. This means that the protein aggregates and precipitates or can't be produced. In the Baker lab they have also introduced additional ways of compensating for these destabilization mutations that can then rescue failed designs. With regard to which scaffold you choose, the largest fraction of successful enzyme designs has been done in TIM barrels, which is perhaps not so surprising since these are very abundant in enzyme families in Nature as well. It is very easy to bring catalytic functionality together convergently around the molecule. I guess scaffolds like that are going to be the best in the future.

Don Hilvert: I think with that we should bring the session to an end and I would like to thank everybody for the vigorous discussion.

Session 3
Microbiomes and Carbohydrate Chemistry

3b5q

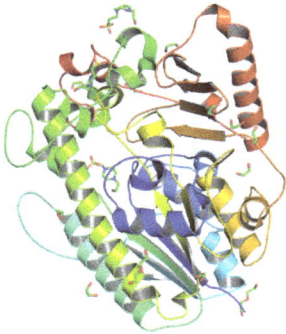

3b7f

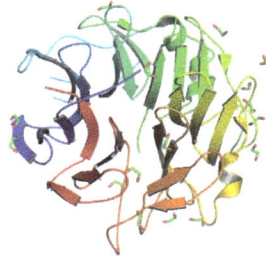

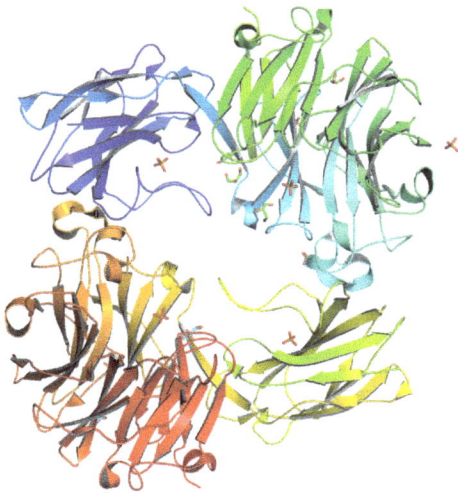

4fj6

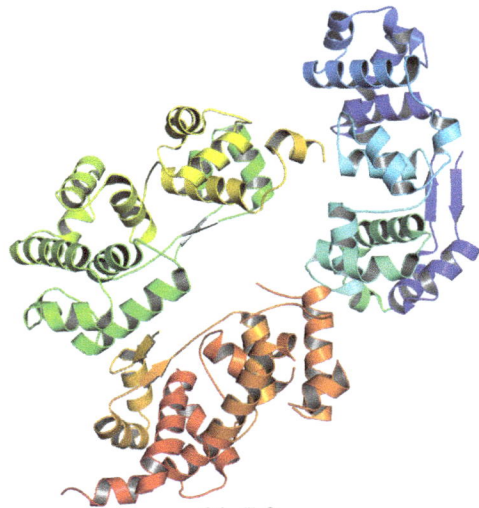

2h56

Structures determined by the Joint Center of Structure Genomics of gut microbiome resident bacterial proteins responsible for the degradation of carbohydrates, sulfated sugars, and glycoproteins, including putative glycosyl hydrolases from *Ralstonia eutropha* (PDB 3b7f) and *Parabacteroides distasonis* (PDB 4fj6), *Bacillus halodurans* glycosidase (PDB 2h56), and *Bacteroides thetaiotaomicron* sulfatase (PDB 3b5q).

STRUCTURAL BASIS FOR HOST/COMMENSAL-MICROBE INTERACTIONS IN THE HUMAN DISTAL GUT MICROBIOME

IAN A. WILSON

Department of Integrative Structural and Computational Biology, The Scripps Research Institute, 10550 North Torrey Pines Road, La Jolla, CA 92037, USA

My view of the present state of human microbiome research

The continued deluge of sequences from genome sequencing efforts are providing "omics" and related fields with unheralded opportunities to explore the richness and diversity of life forms and the fundamental processes that allow organisms to evolve and function in their own particular niches and environments. Metagenomics has uncovered a vast diversity of microorganisms that colonize the human body, which is estimated to contain over 10 times more microbial cells than human cells. The NIH Human Microbiome Project (HMP) was launched in 2008 with the goal of exploring these microorganisms and to investigate their role in human health and disease [1–3]. However, microbiome projects to date been have dominated by efforts to sequence the bacterial genomes and elucidate the types and composition of the bacteria that inhabit specific environments, such as the human gut. These studies have enabled identification of a large number of novel protein families that are specific to the gut microbial communities. Over the past decade or so, the field of structural genomics (SG) was founded to address questions on a genome scale [4–6]. As such, high-throughput structural biology (HTSB) was pioneered to develop tools and methodologies to tackle biomedical and biological questions on a scale not previously contemplated but critical if the fruits of the genomes efforts were to be realized at the protein structure and function level. These HTSB approaches have indeed enabled broader explorations into the rapidly expanding protein universe, as well as structural coverage of entire organelles, organisms, and collections that inhabit specific niches, such as the human microbiome.

The human body is colonized by vast communities of microorganisms that thrive in specific niches such as the human gut, skin, oral cavity, and vagina and engage in complex symbioses with the host. These largely commensal microorganisms interact extensively with their host through secreted and membrane-bound proteins and production of small molecule metabolites. However, identifying the precise function of the proteins involved in the microbiome is a major challenge. As a first step in investigating the basis of this complex symbiosis, efforts, such as the HMP, are cataloging and characterizing the diversity of the human microbiota. Substantial genetic diversity has already been discovered in the gut (300 to 1000 unique species, depending on the methods used) [1, 7] and the activities of their gene products supplement the host-encoded metabolic capabilities and, therefore, are extremely beneficial to the host. Despite the dynamic flux and the species variability between individuals, certain microbes are consistently detected in high numbers. Diversity is even greater on the gene level, where organism diversity is compounded by extensive polymorphism in bacterial species. Relative gene composition between species and the distribution of genes within species may thus be more important than species diversity itself [8].

Historically, microbes were typically recognized as infectious agents of disease. However, Koch's postulates may not hold for complex commensal systems. Human health as well as disease is strongly affected by the commensal/host interactions. While the human microbiome has been implicated in various disorders, only rarely has a single species been identified as a causative agent. In some cases, such as obesity, non-alcoholic fatty liver disease, ulcers and cancer, a specific group of bacteria within the microbiome may be involved [8, 9]. In other cases, such as inflammatory bowel disease (IBD) and allergies, aberrant inflammatory responses deriving from an imbalance of commensals may contribute to disease (reviewed in [10]). The role of microbes in these complex disorders suggests that many potential disease targets will be identified from study of these commensal microorganisms. The gut microbiota are, therefore, a largely untapped source of novel and potentially "druggable" targets with the potential to revolutionize therapeutic strategies for many important human diseases and conditions [11].

My recent research contributions on the distal gut human microbiome

The Joint Center for Structural Genomics (JCSG) which I direct, has established a robust and scalable high-throughput (HT) gene-to-structure pipeline that has already delivered over 1450 novel structures to the community, by either X-ray crystallography or NMR, on a wide range of targets from bacteria to human [5]. The JCSG is currently leveraging its HT platform to address challenging new frontiers in structural biology, including exploration of the human microbiome and other biomedically important areas, such as regulation of stem cells, T cells, and human nuclear receptors through interaction with specific biological partners (see www.JCSG.org). The JCSG is capitalizing on its extensive experience over the past 13 years to develop the best strategies to enhance chances of success, particularly in these large-scale projects. Integrative structural/systems biology methods and technologies developed in PSI-2 to analysis of the complete proteome of a single organism, such as our *Thermotoga maritima* project, and the subsequent modeling of its central metabolic network, have been invaluable for developing similar applications to the human gut microbiome [4, 12, 13]. These well-tested "engines" of SG are well suited for exploring metagenomic projects in collaboration with specialists in the field.

Since the start of PSI:Biology (July 2010), the JCSG has focused on microbial gut secreted proteins as its main biological theme project. To date ∼3000 human microbiome specific proteins have been processed and screened for potential bioactive activity. To date, over 300 unique protein structures from commensals from the human gut have been determined. Most of these are gut-secreted proteins that likely interact with the host and cover a variety of biological important host-microbe interactions, such as carbohydrate processing and assimilation, small molecule interactions, and cell signaling. However, as expected from this type of exploratory project, the specific function of most of these targets remains unknown. A significant portion of the genome of these human gut symbionts encodes carbohydrate-processing enzymes. Sequencing efforts are greatly outpacing the functional and structural characterization of these interesting novel classes of enzymes, but progress has been made on the larger families from this class of proteins.

SG efforts further aim to close this gap by enhancing our understanding of sequence and structure diversity and the inter-relationship between structure-function. Indeed, JCSG structures determined are increasing our knowledge of the human gut microbiome. For instance, the genome of the gram-negative anaerobe dominant symbiont of human distal intestinal tract, *Bacteroides thetaiotaomicron (Bt)*, was sequenced in 2004 and its 4816 genes have a strong representation of glycosyl hydrolases (259 or >5% of all genes), as well as 30 recognized proteins with carbohydrate binding modules (CBMs). At the time of sequencing, >40% of its genes were annotated as "hypothetical proteins", indicating that these numbers may be underestimated. Indeed, the JCSG has identified at least 40 additional (putative) glycosyl hydrolases and almost 100 additional CBMs. These enzymes play a crucial role in the metabolism of polysaccharides and are intimately linked to the host's metabolism. To date, the JCSG has solved 79 structures from *Bt* (out of 262), resulting in an additional ∼20% of *Bt* proteins that can now be annotated based on the insights from these structures.

This new set of structures has helped us identify new (likely) host signaling proteins, new classes of enzymes, new classes of bacterial fimbriae, and novel families of leucine-rich repeat (LRR) proteins with never-before seen arrangements of LRRs. LRRs are typically found in eukaryotic proteins and are usually involved in protein and ligand binding. The best known-human LRR proteins are the innate immunity receptors, TLRs and NLRs. All known eukaryotic LRR proteins adopt a very similar, torus-like shape. The JCSG structural characterization of LRRs stemmed from unexpectedly identification of bacterial LRR proteins when exploring human gut microbiome targets of unknown function. Prior to this, very few bacterial LRR proteins were reported and those that were, are predominantly virulence factors, such as *Listeria internalin* [14]. To date, we have determined 14 bacterial LRR structures, some of which are revealing unexpected novel structural features and variations compared to eukaryotic LRRs that could not have been predicted from the sequence alone. They show an amazing structural diversity, with almost every one displaying novel structural features.

In collaboration with Dr. Koji Nakayama (Nagasaki University, Japan) and Dr. Lucy Shapiro (Stanford University School of Medicine), the JCSG is exploring novel types of pilus found in oral pathogens and other gut bacteria. We have determined structures of 15 novel pilins, which define a large diverse protein superfamily in the animal gut. Our structures provide strong evidence for a novel mechanism of pilus assembly [15].

Together, this new knowledge is being incorporated into a metabolic network reconstruction, developed in the collaboration with Drs. Andrei Osterman (Sanford-Burnham Medical Research Institute, San Diego, CA), similar to that accomplished with our ground-breaking *T. maritima* studies that led to the first complete structural reconstruction of a metabolic network [13]. *B. thetaiotaomicron* is now our new model system for structural and function explorations into the human gut microbiome.

Beyond nutrient utilization and immune modulation, the microbiome also influences the host through its small-molecule metabolites. Our colleagues at the Genomics Institute of the Novartis Research Foundation (San Diego, CA) and others found a number of metabolites in serum and urine derived from human microbiome [16, 17]. For examples, these studies demonstrated that secondary bile acids are modulated by the gut microbiome. Beyond emulsifying dietary fats, bile acids influence gene expression through nuclear receptors like FXR, and are implicated in diseases, such as pancreatitis, diabetes, and are the progenitors of gallstones [18]. Bile-modifying pathways were found in only a limited subset of species [19], and the responsible genes (the *bai* operon of *Clostridia*) have been identified [20]. The JCSG has recently determined the structure of BaiA [21]. This study represents the first structure-function characterization of a human gut microbial enzyme involved in secondary bile acid synthesis and provides key insights into the mechanism of catalysis and uncovers structural features that could be utilized to design modulators of secondary bile acid levels in the human body. Another obvious impact of microbial metabolism on the host is the bacterial processing of complex carbohydrates. Human gut microbes contribute to host caloric uptake (estimated at 30%) [9] through breakdown of complex carbohydrates into usable sugars or conversion to fatty acids. Gut microbes also metabolize host surface glycans, effectively recycling shed epithelial cells [22–24]. These examples highlight the important role commensals can play in regulating host metabolism. Furthermore, microbial physiology impacts delivery of drugs, were oral delivery can fail due to metabolism by commensals that can inactivate a drug or metabolite compound, or convert it into harmful products, like carcinogens. The microbial enzymatic repertoire can, therefore, influence circulating metabolites and drugs and, thereby, influence the host in locations well outside of the gut itself [16].

Outlook for future developments on human microbiomes

Understanding how commensals have succeeded in occupying a particular niche without causing disease, while closely related strains or microbiome dysfunction cause disease provides opportunities for generating strategies not only to defeat pathogenic invasion, but also to harness the full potential of the beneficial effects of the microbiome for human health and well being. Mammalian cells detect microbes through pattern recognition receptors, which can bind microbial-specific components. Commensals modulate the host's innate immune system despite containing molecules, which bind and activate these receptors. The immune system removes pathogens, but there is mounting evidence that the immune response may be just as devoted to maintaining beneficial populations [7]. Recognition of commensal versus pathogen is largely defined by interactions between extracellular proteins of the host and microbe.

SG is therefore playing a significant role in understanding the complex symbiosis of the human microbiome. Many poorly characterized bacterial and host proteins represent potential therapeutic targets. Microbial enzymes producing bioactive metabolites [16] form pathways where structural information is essential in developing specific inhibitors to evaluate *in vivo* effects on the host. Likewise, extracellular microbial proteins function in cell adhesion, immunity and signaling. Beyond structures themselves, the SG high-throughput "engines" (gene to structure to function) can be applied in collaborations with biologists to better leverage the impact of these fascinating metagenomics projects.

Acknowledgments

Marc Elsliger is gratefully acknowledged for help with preparation of this manuscript, Dennis Wolan for helpful suggestions, and the NIH NIGMS Protein Structure Initiative, PSI:Biology (U54 GM094586) for support to the JCSG.

References

1. P. J. Turnbaugh, R. E. Ley, M. Hamady, C. M. Fraser-Liggett, R. Knight *et al.*, *Nature* **449**, 804 (2007).
2. C. M. Lewis Jr., A. Obregón-Tito, R. Y. Tito, M. W. Foster, P. G. Spicer *et al.*, *Trends Microbiol.* **20**, 1 (2012).
3. M. J. Friedrich, *JAMA* **300**, 777 (2008).
4. S. K. Burley, A. Joachimiak, G. T. Montelione, I. A. Wilson, *Structure* **16**, 5 (2008).
5. M. A. Elsliger, A. M. Deacon, A. Godzik, S. A. Lesley, J. Wooley *et al.*, *Acta Crystallogr. Sect. F Struct. Biol. Cryst. Commun.* **66**, 1137 (2010).
6. T. C. Terwilliger, D. Stuart, S. Yokoyama, *Annu. Rev. Biophys.* **38**, 371 (2009).
7. L. V. Hooper, *Nat. Rev. Microbiol.* **7**, 367 (2009).

8. P. J. Turnbaugh, M. Hamady, T. Yatsunenko, B. L. Cantarel, A. Duncan *et al.*, *Nature* **457**, 480 (2009).

9. F. Sommer, F. Backhed, *Nat. Rev. Microbiol.* **11**, 227 (2013).

10. E. A. Grice, H. H. Kong, G. Renaud, A. C. Young, G. G. Bouffard *et al.*, *Genome Res.* **18**, 1043 (2008).

11. W. Jia, H. Li, L. Zhao, J. K. Nicholson, *Nat. Rev. Drug Discov.* **7**, 123 (2008).

12. S. A. Lesley, P. Kuhn, A. Godzik, A. M. Deacon, I. Mathews *et al.*, *Proc. Natl. Acad. Sci. USA* **99**, 11664 (2002).

13. Y. Zhang, I. Thiele, D. Weekes, Z. Li, L. Jaroszewski *et al.*, *Science* **325**, 1544 (2009).

14. W. D. Schubert, C. Urbanke, T. Ziehm, V. Beier, M. P. Machner *et al.*, *Cell* **111**, 825 (2002).

15. Q. Xu, B. Christen, H. J. Chiu, L. Jaroszewski, H. E. Klock *et al.*, *Mol. Microbiol.* **83**, 712 (2012).

16. W. R. Wikoff, A. T. Anfora, J. Liu, P. G. Schultz, S. A. Lesley *et al.*, *Proc. Natl. Acad. Sci. USA* **106**, 3698 (2009).

17. M. Li, B. Wang, M. Zhang, M. Rantalainen, S. Wang *et al.*, *Proc. Natl. Acad. Sci. USA* **105**, 2117 (2008).

18. A. Moschetta, A. L. Bookout, D. J. Mangelsdorf, *Nat. Med.* **10**, 1352 (2004).

19. P. B. Eckburg, E. M. Bik, C. N. Bernstein, E. Purdom, L. Dethlefsen *et al.*, *Science* **308**, 1635 (2005).

20. J. M. Ridlon, D. J. Kang, P. B. Hylemon, *J. Lipid Res.* **47**, 241 (2006).

21. S. Bhowmik, D. H. Jones, H. P. Chiu, I. H. Park, H. J. Chiu *et al.*, *Proteins* **82**, 216 (2014).

22. L. Bry, P. G. Falk, T. Midtvedt, J. I. Gordon, *Science* **273**, 1380 (1996).

23. E. C. Martens, H. C. Chiang, J. I. Gordon, *Cell Host Microbe* **4**, 447 (2008).

24. J. L. Sonnenburg, J. Xu, D. D. Leip, C. H. Chen, B. P. Westover *et al.*, *Science* **307**, 1955 (2005).

CARBOHYDRATE CHEMISTRY AND BIOLOGY

CHI-HUEY WONG

The Genomics Research Center, Academia Sinica
Taipei, 115, Taiwan
Department of Chemistry, The Scripps Research Institute
La Jolla, CA 92130, USA

My view of the present state of research on carbohydrate chemistry

Carbohydrates on cell surface exist as glycoconjugates and are involved in protein folding and numerous biological recognition events such as differentiation, development, cell adhesion, pathogen-host interactions, cancer progression and many other intercellular interaction events. However, most of these biological functions are not well understood, mainly due to the lack of tools available for the study, and as a result the pace of the development of carbohydrate-based therapeutics and diagnostics is relatively slow. In the past 10 years, however, many new tools and methods have been developed and used to solve major problems and create new opportunities in carbohydrate chemistry and biology associated with diseases. As a result, the field of glycoscience has attracted a great deal of attention that stimulates more research activities.

My recent research contributions to carbohydrate chemistry

Our interests are to develop new methods and strategies to tackle the problems of carbohydrate-mediated biological recognition and associated diseases. The followings list our contributions to the subject.

1. Synthesis of oligosaccharides on large scales using glycosyltransferases coupled with regeneration of sugar nucleotides *in situ*. This strategy is considered to be one of the most practical methods available to date for the large-scale synthesis of oligosaccharides, making possible the clinical evaluation and manufacture of this class of biomolecules. The concept was first conceived in 1982 and then further improved, as illustrated in the synthesis of sialyl Le^X tetrasaccharide and Globo-H hexasaccharide [1–3].

2. Preparation of homogeneous glycoproteins with well-defined glycan structures. Development of such a methodology is important in order to understand the

Glycans on Cell Surface and Proteins: Functional Study and Opportunities

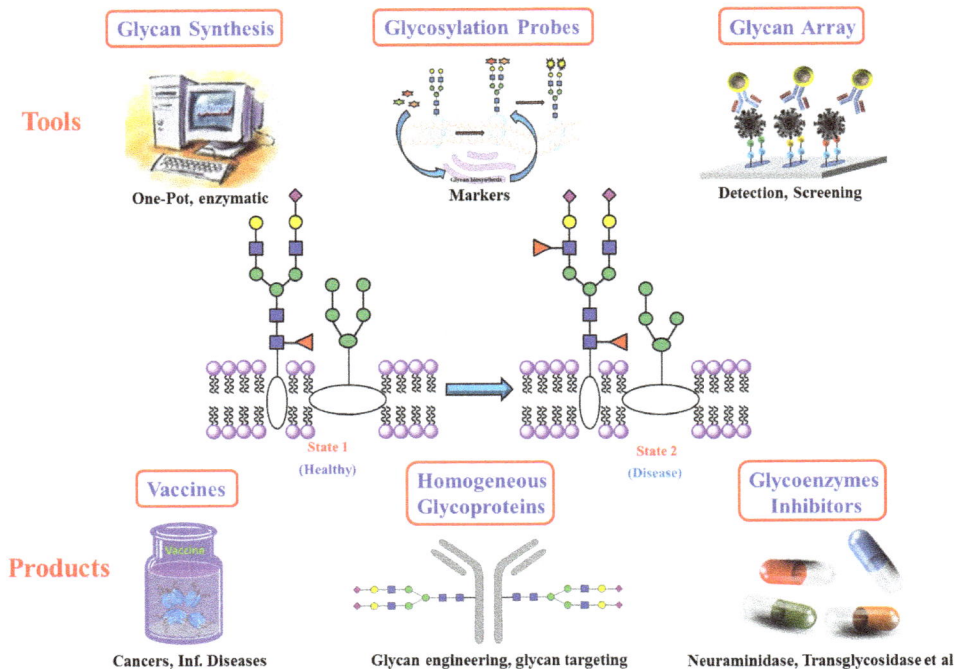

Fig. 1.   Identification of glycan markers associated with diseases and development of glycan arrays, vaccines and therapeutics.

effect of glycosylation on glycoprotein structure and function and to develop glycoprotein pharmaceuticals. Ribonuclease B was used as a model to develop methodologies for this purpose [4–6], where the glycan moiety was trimmed down with endoglycosidases to a homogeneous glycoform followed by addition of new glycans with glycosyltransfer enzymes. Using this strategy, homogeneous glycoproteins with well-defined glycan structures can be prepared from the original mixtures of glycoforms [6]. Alternatively, glycoproteins can be assembled through native chemical ligation or sugar-assisted glycopeptide ligation, followed by elongation of the glycan moiety as described above [6–7]. The methodology was also used to investigate the contribution of each monosaccharide in a glycan to the folding energetics and stability of the glycoprotein. This study has led to the discovery that the monosaccharide (O-GlcNAc or N-GlcNAc) attached to protein contributes most significantly to glycoprotein folding and stability [8–10], and the finding was used in the development of monoglycosylated hemagglutinins as influenza vaccines with broader protection activity, pointing to a new direction of universal vaccine design.

3. Automated oligosaccharide synthesis. This is a long-standing problem and a major challenge in the field. Development of automated methods for oligosaccharide synthesis will facilitate the discovery research in glycoscience, and toward this goal, we reported the first automated oligosaccharide synthesis method using the reactivity-based approach in a one-pot and programmable manner [11, 12]. We continue the effort to design more building blocks with well-defined reactivity to be used for the assembly of various glycans for glycan array development and study of protein-glycan interaction.

4. Development of glycan microarray for analysis of protein-carbohydrate interaction. The idea of glycan array development was conceived in 2002 and our laboratory is one of the three groups involved in the development [12, 13]. We started with the use of synthetic glycans covalently or non-covalently attached to microtiter plates, then to glass slides and to the surface of aluminum oxide coated glass slides [14]. We also developed new methods for determination of dissociation constant [12, 15] and study of multivalency, including heteroligand binding. This work has provided a new platform for the rapid and quantitative binding analysis of protein-carbohydrate interaction, allowing us to understand the specificity of carbohydrate binding proteins and to dissect their binding energy, including the analysis and detection of influenza and antibodies that recognize glycans [16, 17].

5. Identification of glycan markers and drug discovery. To follow the process of biological glycosylation and identify glycan markers associated with cancer progression, we have developed several molecular probes for use to study post-translational glycosylation and identify specific markers on cancer cells and cancer stem cells [18–20]. This work has led to a better understanding of EGFR dimerization and intracellular signal transduction related to glycosylation [21], and a study of fucosyltransferases 8 overexpression related to cancer invasion [22]. In addition, new glycoenzymes or glycan-binding proteins have been identified from the use of these probes as targets for drug discovery and development of carbohydrate-based vaccines against cancers and infectious diseases [23–25].

Outlook to future developments of research on carbohydrate chemistry

With more tools available, especially the tools for better synthesis and imaging, the field of glycoscience will be further advanced to facilitate our understanding of the biological glycosylation process in disease progression, and hopefully from the understanding, new therapeutic and diagnostic strategies will be developed.

References

1. Y. Ichikawa, Y. C. Lin, D. P. Dumas, G.-J. Shen, E. Garcia-Junceda *et al.*, *J. Amer. Chem. Soc.* **114**, 9283 (1992).
2. K. M. Koeller, C.-H. Wong, *Nature* **409**, 232 (2001).
3. T.-I. Tsai, H.-Y. Lee, S.-H. Chang, C.-H. Wang, Y.-C. Tu *et al.*, *J. Amer. Chem. Soc.* **135**, 14831 (2013).
4. K. Witte, P. Sears, R. Martin, C.-H. Wong, *J. Amer. Chem. Soc.* **119**, 2114 (1997).
5. P. Sears, C.-H. Wong, *Science* **291**, 2344 (2001).
6. R. Schmaltz, S. R. Hanson, C.-H. Wong, *Chem. Rev.* **111**, 4259 (2011).
7. A. Brik, Y.-Y. Yang, S. Ficht, C.-H. Wong, *J. Amer. Chem. Soc.* **128**, 5626 (2006).
8. E. E. Simanek, D.-H. Huang, O. Seitz, C.-H. Wong, *J. Amer. Chem. Soc.* **45**, 11567 (1998).
9. S. R. Hanson, E. K. Culyba, T. L. Hsu, C.-H. Wong, J. W. Kelly *et al.*, *Proc. Natl. Acad. Sci. USA* **106**, 3131 (2009).
10. W. Chen, S. Enck, J. Price, D. Powers, E. Powers *et al.*, *J. Amer. Chem. Soc.* **135**, 9877 (2013).
11. Z. Zhang, I. R. Ollmann, X.-S. Ye, R. Wischnat T. Baasov *et al.*, *J. Amer. Chem. Soc.* **121**, 734 (1999).
12. C.-H. Hsu, S.-C. Hung, C.-Y. Wu, C.-H. Wong, *Angew. Chem. Int. Ed.* **50**, 11872 (2011).
13. F. Fazio, M. C. Bryan, O. Blix, J. C. Paulson, C.-H. Wong, *J. Amer. Chem. Soc.* **124**, 14397 (2002).
14. S.-H. Chang, J.-L. Han, S. Y. Tseng, H.-Y. Lee, C.-W. Lin *et al.*, *J. Amer. Chem. Soc.* **132**, 13371 (2010).
15. P. H. Liang, S.-K. Wang, C.-H. Wong, *J. Amer. Chem. Soc.* **129**, 11177 (2007).
16. C.-C. Wang, J.-R. Chen, Y.-C. Tseng, C.-H. Hsu, Y.-F. Hung *et al.*, *Proc. Natl. Acad. Sci. USA* **106**, 18137 (2009).
17. H.-Y. Liao, C.-H. Hsu, S.-C. Wang, C.-H. Liang, H.-Y. Yen *et al.*, *J. Amer. Chem. Soc.* **132**, 14849 (2010).
18. M. Sawa, T. L Hsu, T. Ito, M. Sugiyama, S. R. Hanson *et al.*, *Proc. Natl. Acad. Sci. USA* **103**, 12371 (2006).
19. T.-L. Hsu, S. R. Hanson, K. Kishikawa, S.-K. Wang, M. Sawa *et al.*, *Proc. Natl. Acad. Sci. USA* **104**, 2614 (2007).
20. C.-S. Tsai, H.-Y. Yen,, M.-I. Lin, T.-I. Tsai, S.-Y. Wang *et al.*, *Proc. Natl. Acad. Sci. USA* **110**, 2466 (2013).
21. Y.-C. Liu, H.-Y. Yen, C.-Y. Chen, C.-H. Chen, P.-F. Cheng *et al.*, *Proc. Natl. Acad. Sci. USA* **108**, 11332 (2011).
22. C.-Y. Chen, Y.-H. Jan, Y.-H. Juan, C.-J. Yang, M.-S. Huang *et al.*, *Proc. Natl. Acad. Sci. USA* **110**, 630 (2013).

23. K.-C. Chu, C.-T. Ren, C.-P Lu, C.-H. Hsu, T.-H. Sun *et al.*, *Angew. Chem. Int. Ed.* **50**, 9391 (2011).

24. Y.-L. Huang, J.-T. Hung, S. K. C. Cheung, H.-Y. Lee, K.-C. Chu *et al.*, *Proc. Natl. Acad. Sci. USA* **110**, 2517 (2013).

25. H.-Y. Chuang, C.-T. Ren, C.-A. Chao, C.-Y. Wu, S. Shivatare *et al.*, *J. Amer. Chem. Soc.* **135**, 11140 (2013).

CHEMICAL BIOLOGICAL PROTEOMICS OF BACTERIAL PROTEIN FUNCTIONALITIES IN THE HUMAN DISTAL GUT MICROBIOME

DENNIS W. WOLAN

Department of Molecular and Experimental Medicine, The Scripps Research Institute
10550 North Torrey Pines Road, La Jolla, CA 92037, USA

My view of the present state of human microbiome research

The distal gut microbiome represents a vast collection of commensal bacteria that are essential to human metabolism, immune development and homeostasis, epithelial cell angiogenesis, and protection from pathogenic bacteria infiltration [1]. Alterations in gut bacteria populations directly govern and dictate the onset of microbiome-related diseases, such as obesity [2], malnutrition [3], inflammatory bowel diseases (IBD) (*i.e.*, Crohn's disease and ulcerative colitis) [4], diabetes [5], circulatory diseases [6], and colorectal cancer [7]. Recent *in vivo* manipulations of the bacteria that colonize the distal gut in mouse models have clearly shown the enormous effect that these bacteria have on the host. For example, genetically compromised IBD mice have chronic intestinal inflammation that can be alleviated by administration of general antibiotics [8]. Conversely, IBD mice raised in a germ-free environment exhibit no inflammation-related problems until colonized with microbiota from a healthy donor [9]. These results, and many others, support the development of targeted therapies as innovative approaches to treat gut microbiome-related diseases.

Great strides in microbiome research have been made over the last decade and are almost entirely attributable to the pioneering development of culture-independent pyrosequencing technologies. These methods generate vast amounts of metagenomic 16S rRNA and deep DNA sequencing information and have provided valuable insights into the ecological diversity and composition of bacteria that colonize the human intestinal tract [10]. Based on the outcome of these studies and the concerted efforts of the Human Microbiome Project consortium, we now know that the taxonomic composition is highly divergent among individuals, populations, and geography and that a universal "core" microbiome of bacterial components is not likely to exist [11]. Furthermore, we have not yet determined what contributing factors help shape the microbial composition of the distal gut. Strong evidence exists for environmental influences governing the commensal ecology, as differences in diet (*e.g.*, vegetarian vs. high animal protein) have significant dissimilarities in the highly abundant bacterial species [12, 13]. Conversely, host genetics may play a role as conservation among mother-daughter pairs have more similar microbiota compositions than unrelated individuals [14]. The ability to correlate microbiota alterations to external factors and host genetics will greatly improve our ability to address the causative or casual links between the gut microbiome and diseases.

We need to improve our comprehensive understanding of the role of the human microbiome and we can readily accomplish this task by the application of novel chemical and biological technologies. Considerable limitations in RNA and DNA sequencing analyses restrict the extent to which they can be used to develop a deeper and more fundamental understanding of the biological and biomedical impact that microbiomes have on human health and well being. Despite providing fascinating insights into the species diversity and composition of the colonizing bacteria, these metagenomic studies can only speculate on the bacterial protein composition and its interaction with the host as well as only qualitatively assess variation among samples without consideration of other resident microbiota organisms, including fungi, viruses, and single-cell eukaryotes. Furthermore, the application of different sets of universal primers results in significant variability in the predicted bacterial composition of an individual sample [15]. As such, new methods, chemical biology tools, and interdisciplinary technologies must be developed to further evolve our biological understanding of host:microbiome interactions and to harness and control the enormous impact that enteric bacteria have on human health and disease.

My recent research contributions on the distal gut human microbiome

We seek to advance the field of human distal gut microbiome research from its original basis in genomics into proteomics and quantitative biology through identification and characterization of bacterial proteins and enzymatic activities that are expressed within healthy and diseased distal gut microbiomes. Unlike metagenomic approaches, our chemical biological studies are not focused on population differences among commensal bacteria species as the root cause of microbiome-associated diseases. We hypothesize that regardless of the microbial composition within the distal gut, conserved bacterial enzymatic and functional activities are altered in the microbiome and spur inflammation, tumorigenesis, and other host-driven responses to aberrant microbiota. Importantly, our results over time will also shed light on whether conservation in function is preserved in the face of diversity of the composition of the microbiota within the microbiomes of different healthy individuals and populations.

Negotiating the complexity of the microbial protein content is an enormous challenge and requires methods that directly and rapidly sample biological systems, yet permit the systematic analysis and quantitation of individual classes of proteins. To accomplish our goals, we design and use reactive chemical probes, termed activity-based protein probes (ABPPs), that irreversibly label a variety of bacterial protein families and permit the isolation of these tagged proteins from complex proteomes

(Fig. 1). ABPPs are tremendously versatile and contain two essential elements that enable target proteins to be labeled and isolated from complex proteomes: 1) a reactive group that covalently modifies active sites of proteins with shared catalytic scaffolds and 2) an identification tag, such as biotin for isolation and/or visualization of labeled proteins (Fig. 1(a)). These types of small molecule probes have been employed with great success in the identification of dysregulated proteins in a variety of human ailments, such as cancerous tumors [16] and parasitic infections [17] (Fig. 1(a)). An advantage for incorporating ABPP into the proteomic analysis of the gut microbiome is the ability to engineer either promiscuity or specificity into the chemical scaffolds for either general mining of the proteome or for explicit searches for protein functionality, respectively.

A preliminary goal of ours is to create a universal proteomic methodology that can be applied rapidly and thoroughly to a variety of gut microbiomes. As such, we have begun to develop our own covalent modifying chloromethyl ketone (CMK) probes that covalently adhere to reactive serine and cysteine residues located within enzyme active sites (Fig. 1(b)). Our molecules exploit the use of "click" chemistry so that a range of synthetic substructures, including visualization labels (e.g., fluorophores) or enrichment tags (e.g., biotin) can be introduced after labeling of the microbiome proteome (Fig. 1).

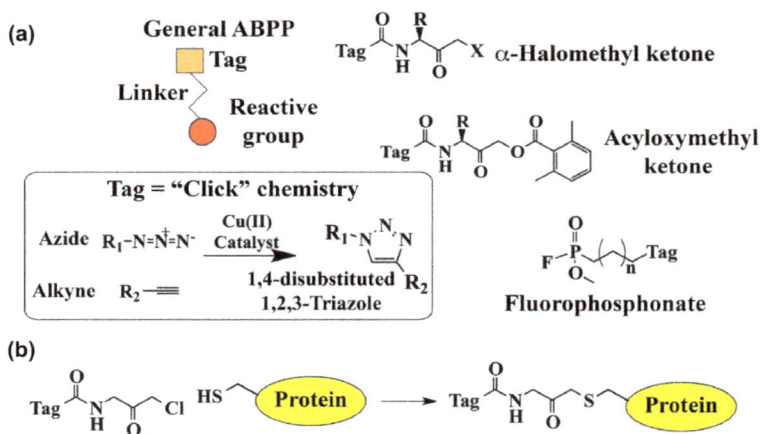

Fig. 1.   Examples of general ABPP and "click" chemistry (a) and reaction of a chloromethyl ketone probe with surface-exposed cysteine residue (b).

This chemical-based enrichment technique is used in conjunction with liquid chromatography coupled to tandem mass spectrometry (LC-MS/MS) for bacterial protein identification and quantitation (Fig. 2). Our development and application of the powerful combination of small molecule probes and LC-MS/MS proteomics allows us to elucidate aberrant enzymatic activities and functionalities associated with diseased microbiomes as well as provide: (1) a novel chemical biological technique to complement metagenomic studies; (2) a foundation on which to map the comprehensive host:microbiome proteomic network; (3) a set of valuable chemical tools directed towards understanding the microbiome biology; and (4) an initial list of potential therapeutic targets to combat or protect against microbiome-related diseases.

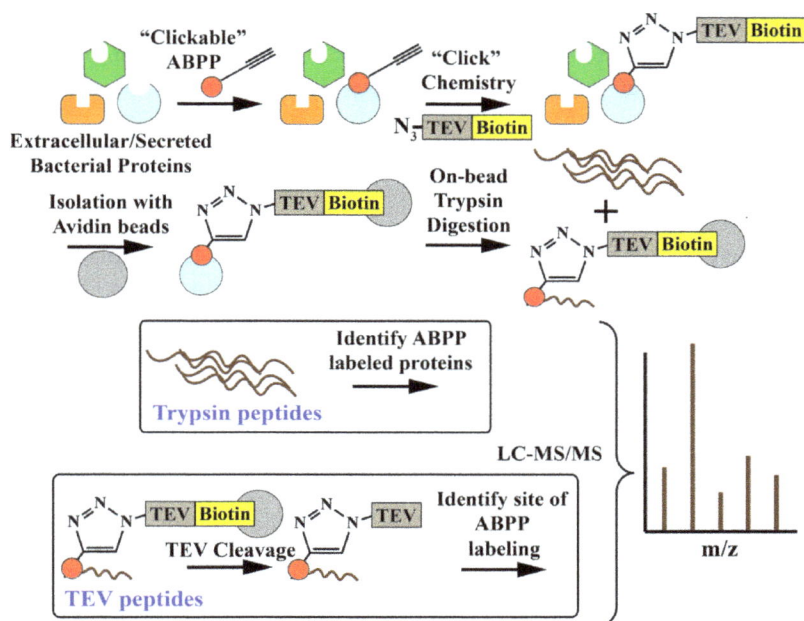

Fig. 2. ABPP proteome labeling, "click" chemistry, and enrichment.

We have recently employed our novel small molecule CMK-based ABPPs in the identification of commensal bacterial proteins from a "normal" human subject. For example, an LC-MS/MS dataset yielded approximately 500 different bacterial proteins that are primarily involved in lipid, carbohydrate, and protein metabolism as assessed by KEGG pathway classification and an additional 15% of the labeled proteins are listed as hypothetical or putative. Importantly, out of the estimated 400 bacterial species that are responsible for production of the ABPP-labeled proteins, metagenomic approaches have not reported the majority of these species as members of the commensal gut microbiota. These results strongly support our chemical-based methodologies as a complementary approach to metagenomics. We are currently expanding our proteomic approaches to establish a baseline enzymatic landscape of proteins among "normal" human gut microbiomes with a variety of ABPP. We will then interrogate microbiomes from patients with a variety of gut-related diseases in order to elucidate compromised bacterial protein functions.

Outlook for future developments on human microbiomes

One of the most significant questions is if altered microbiota drive propagation of host disease phenotypes. Several lines of evidence suggest bacterially produced metabolites, including short-chain fatty acids directly control the Treg cells of the host's adaptive immune response [18]. These recent studies are beginning to address the immense potential influence the microbiota have on our health. Moreover, substantial evidence has revealed that manipulation of the microbiome by external intervention, such as antibiotic treatment of IBD, results in ablation of disease. Unfortunately, such drugs jeopardize the composition of the gut microbial flora and encourage colonization of the now vulnerable intestinal tract by pathogenic bacteria [19]. These findings strongly suggest that new chemical and biological approaches are urgently needed to specifically alter microbial enzymatic function and alleviate host physiological responses to gut bacteria dysbiosis. Most importantly, they illustrate that such therapeutic development will require specific enzymatic targets, as nondescript off-target effects may result in drastic changes in the population and/or composition of commensal gut flora and have deleterious effects on host symbiosis.

In summary, we need to develop and apply novel and innovative complementary methodologies and approaches to provide higher spatial and temporal resolution to the many environmental and genetic variables that contribute to understanding of microbiome homeostasis and enhance our ability to manipulate gut bacteria to combat a variety of human ailments. These advancements will only be possible with paradigm shifts in the approaches and techniques applied to the *in vitro*, cell-based, and animal model studies associated with microbiome research.

Acknowledgment

We gratefully acknowledge The Scripps Research Institute for funding.

References

1. J. I. Gordon, *Science* **336**, 1251 (2012).
2. P. J. Turnbaugh, R. E. Ley, M. A. Mahowald, V. Magrini, E. R. Mardis *et al.*, *Nature* **444**, 1027 (2006).
3. A. L. Kau, P. P. Ahern, N. W. Griffin, A. L. Goodman, J. I. Gordon, *Nature* **474**, 327 (2011).
4. J. R. Marchesi, E. Holmes, F. Khan, S. Kochhar, P. Scanlan *et al.*, *J. Proteome Res.* **6**, 546 (2007).
5. A. Giongo, K. A. Gano, D. B. Crabb, N. Mukherjee, L. L. Novelo *et al.*, *ISME J.* **5**, 82 (2011).
6. E. Holmes, R. L. Loo, J. Stamler, M. Bictash, I. K. Yap *et al.*, *Nature* **453**, 396 (2008).
7. J. R. Marchesi, B. E. Dutilh, N. Hall, W. H. Peters, R. Roelofs *et al.*, *PLoS ONE* **6**, e20477 (2011).

8. S. S. Kang, S. M. Bloom, L. A. Norian, M. J. Geske, R. A. Flavell *et al.*, *PLoS Med.* **5**, e41 (2008).

9. W. S. Garrett, G. M. Lord, S. Punit, G. Lugo-Villarino, S. K. Mazmanian *et al.*, *Cell* **131**, 33 (2007).

10. L. V. Hooper, J.I. Gordon, *Science* **292**, 1115 (2001).

11. C. A. Lozupone, J. I. Stombaugh, J. I. Gordon, J. K. Jansson, R. Knight, *Nature* **489**, 220 (2012).

12. C. De Filippo, D. Cavalieri, M. Di Paola, M. Ramazzotti, J. B. Poullet *et al.*, *Proc. Natl. Acad. Sci. USA* **107**, 14691 (2010).

13. T. Yatsunenko, F. E. Rey, M. J. Manary, I. Trehan, M. G. Dominguez-Bello *et al.*, *Nature* **486**, 222 (2012).

14. P. J. Turnbaugh, M. Hamady, T. Yatsunenko, B. L. Cantarel, A. Duncan *et al.*, *Nature* **457**, 480 (2009).

15. D. A. W. Soergel, N. Dey, R. Knight, S. E. Brenner, *ISME J.* **6**, 1440 (2012).

16. N. Jessani, M. Humphrey, W. H. McDonald, S. Niessen, K. Masuda *et al.*, *Proc. Natl. Acad. Sci. USA* **101**, 13756 (2004).

17. D. C. Greenbaum, A. Baruch, M. Grainger, Z. Bozdech, K. F. Medzihradszky *et al.*, *Science* **298**, 2002 (2002).

18. P. M. Smith, M. R. Howitt, N. Panikov, M. Michaud, C. A. Gallini *et al.*, *Science* **341**, 569 (2013).

19. C. Jernberg, S. Löfmark, C. Edlund, J. K. Jansson, *ISME J.* **1**, 56 (2007).

AUTOMATED OLIGOSACCHARIDE SYNTHESIS: FROM INSIGHTS INTO FUNDAMENTAL GLYCOBIOLOGY TO VACCINES AND DIAGNOSTICS

PETER H. SEEBERGER

Max-Planck Institute for Colloids and Interfaces, Department for Biomolecular Systems, and Freie Universität Berlin, Arnimallee 22, 14195 Berlin, Germany

My view of the present state of research on carbohydrate chemistry

Access to pure, defined carbohydrates is key to a detailed understanding of the role of glycans in biological systems. Glycans are of particular importance in immunology and the interactions of the microbiome with the human body. Given the fact that complex glycans cover bacterial as well as human cells it is not surprising that carbohydrates play an important role for attachment of bacteria but also in immunomodulation. In the immunological context, glycan have both beneficial as well as detrimental effects and a detailed understanding of structure–function relationships is needed.

While biological studies involving nucleic acids (genetics) and proteins (proteomics) have made huge strides since the 1970s due to automated sequencing and synthesis methods, glycobiology has suffered from the lack of good general methods for both sequencing and synthesis. Matters are more complicated because the term "carbohydrates" really relates to several rather different structural classes of molecules. For this reason, the synthesis of glycans has also proven greatly more challenging that that of the other two biopolymers. Glycans are banched and each glycosidic linkage constitutes a stereogenic center, thus requiring region — as well as stereocontrol — two aspects that do not enter into the challenge of oligopeptide and oligonucleotide synthesis. All four of the major classes of glycans, glycoproteins, glycolipids, glycosaminoglycans and glycoslyphosphatidylinsotial (GPI) anchors), are glycoconjugates — a combination of an oligosaccharide and another biomolecule, and complicates their synthesis drastically. At the same time isolation of glycoconjugates is also extremely difficult and no amplification processes analogous to PCR exist.

Carbohydrate chemistry has been practiced for more than one century now and increasingly more complex molecules have become accessible as chemical techniques have improved. Until today carbohydrate assembly has, for the most part, remained an art as few highly specialized laboratories have taken months and often years to procure a glycan molecule of interest. While the synthetic molecules have contributed decisively to our current understanding of structure-activity relationships in glycans, the speed (or better lack thereof) in glycan assembly has been a serious obstacle for glycobiology. Glycobiology needs rapid and reliable access to pure and defined glycans that can be incorporated into tools such as glycan arrays. In the case of oligonucleotides and peptides automated solid phase synthesis has produced the desired molecules for many years.

My recent research contributions to carbohydrate chemistry

Automated Carbohydrate Synthesis: Based on the precedence of DNA and peptide assembly, the Seeberger laboratory developed, by solving a host of chemical problems, the first automated oligosaccharide synthesizer [1]. After more than ten years of development and a complete overhaul of the synthetic strategy [2], a commercial synthesizer (Glyconeer 3.0 by GlycoUniverse) is now available to provide **access to many complex carbohydrates in *days* rather than *years*.** This platform has been used to access the major classes of glycans up to 30-mers [3] including glycosaminoglycans [4], GPI-anchors as well as *O*- and *N*-linked glycans. The glycans prepared by the fully automated synthesis system are ready for printing onto surfaces and particles or for conjugation to carrier proteins for vaccine development. Using the methods we developed, non-specialists can access defined sugars for biological or medical applications.

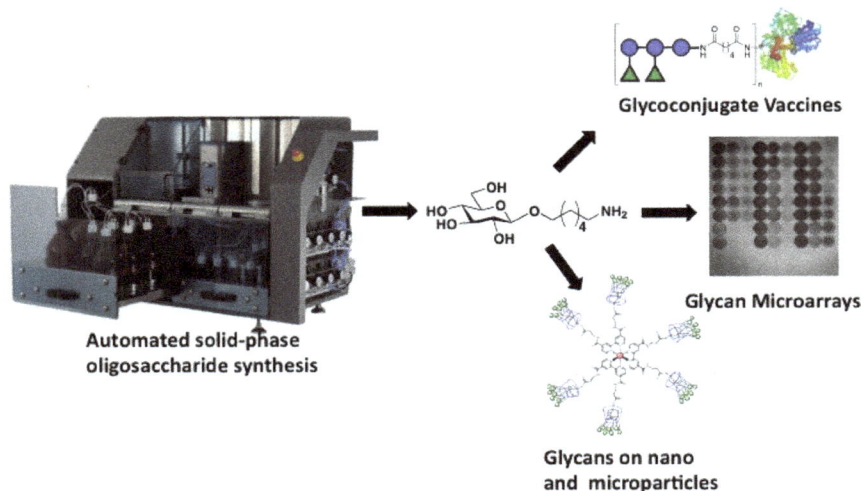

Fig. 1.   The automated solid-phase oligosaccharide synthesis platform Glyconeer 3.0 as basis for chemical glycobiology.

Synthetic Tools for Glycobiology: Rapid access to usable quantities of defined oligosaccharides has enabled us to create synthetic tools that have been commonplace in genomics and proteomics research. The Seeberger group pioneered the use of carbohydrate microarrays and printed the very first glycan arrays using robotic printing [5] to determine the ligands for carbohydrate-binding proteins and to define oligosaccharide vaccine antigens. We screened blood for anti-carbohydrate antibodies that correlate with disease patterns. In this context we were able to demonstrate partial resistance to malaria infection based on anti-GPI toxin antibodies [6]. Blood tests for other infectious agents such as parasites threatening pregnant women and for autoimmune diseases are currently under development. Carbohydrate-functionalized, highly fluorescent nanoparticles were prepared and utilized to image carbohydrate-based targeting *in vivo* [7].

Investigations Into the Role of GPI Anchors in Biology and Medicine. The chemical synthesis of (GPI)-anchor glycans was key to the identification of the malaria toxin [8]. A general approach to synthetic GPI anchors provided the basis for detailed biological studies into the role these complex molecules play in biology [9]. In addition to the first synthetic GPI-anchored prion protein we prepared a host of other GPI glycolipids. Synthetically derived GPIs served to establish the parameters for calculations concerning the structure of GPI [10]. Biophysical studies using grazing incidence X-ray diffraction (GIXD) shed light on the structure of GPI-glycolipids in the lipid bilayer of the cell membrane [11]. Investigations into the role of glycolipid signaling in the inflammatory cascade, nerve growth and other biologically relevant are currently underway.

Defining Glycosaminoglycan-Protein Interactions Responsible for Cellular Signaling. Proteoglycans are major components of the extracellular matrix that surround all mammalian cells. Core proteins anchor glycosaminoglycan (GAG) polysaccharides in the outside of the lipid bilayer. GAGs mediate blood coagulation, virus entry, and angiogenesis as well as many other important biological events by binding to proteins such as growth factors, chemokines, and cytokines. To define structure-activity relationships of GAGs, the Seeberger laboratory introduced a modular synthesis of this class of complex carbohydrates [12] that was the basis for automated assembly [4]. Microarrays of defined heparin oligosaccharides [13] served as valuable molecular tools to rapidly identify heparin-protein interactions and have been used widely to establish GAG binding motifs of importance in signaling cascades in biological systems.

Synthetic Carbohydrate Vaccines. The presence of specific cell-surface polysaccharides on parasites, bacteria, viruses and cancer has been the basis for three marketed vaccines against meningococci, pneumococci and *Haemophylis influenza* type b (Hib) that are saving millions of lives each year. Since many pathogens cannot be cultured and their cell-surface carbohydrates cannot be isolated in pure form synthetic access to the desired antigens is an attractive alternative. We have utilized synthetic chemistry to access defined oligosaccharides as a basis for vaccine design [14]. Using our synthetic capabilities (*vide supra*) we have prepared vaccine candidates against several parasitic and bacterial diseases. Vaccine candidates against *malaria* [8], Group A *Streptococcus* [15], *C. difficile* [16], and Neisseria meningitides [17] as well as *Streptococcus pneumoniae* are currently at different stages of development.

Diagnostic Approaches. Monoclonal antibodies raised against synthetic antigens can recognize bacteria very sensitively and specifically. Antibodies against the antigen on *Bacillus anthracis* are now used in a marketed anthrax detection kit [18] while a test for plague based on an antibody is currently under development [19]. Glycan arrays carrying synthetic structures are used to diagnose different infectious diseases as well as unwanted immune reactions to drug substances.

Outlook to future developments of research on carbohydrate chemistry

Carbohydrate chemistry will be the deciding technology to further progress in the field of glycobiology. In the absence of amplification methods analogous to PCR access to pure carbohydrates is restricted to synthesis. The products of these syntheses will be important standards for glycan sequencing efforts, they will be used to create ever more comprehensive glycan arrays and other tools. Our fundamental understanding in the creation of glycosidic linkages is still far from adequate, even after 100 years of research. A better understanding of mechanisms and factors that influence the outcome of glycosylations will eventually allow us to predict the outcome of glycosylations reliable and not to have to rely on trial and error. In addition to mammalian glycobiology, glycans that play a role in bacterial communication and for plant biology will come into focus. In addition to communication, access to ever more complex glycans will allow us to ask fundamental questions of structure. The role of glycans in plant structure is paramount and by understanding the forces involved in building complex yet stable molecular assemblies will give us a basis to think about creating architectures based on first principles using simply glycans.

The classes of GPI-anchors and GAGs are areas that will benefit from synthetic access. An atomic understanding of GPI-anchored proteins will allow us to understand how such proteins interact in the cell membrane and get involved in signaling process between cells and into cells. The role of GAGs as key mediators of signals via the extracellular matrix is in principle clear. However, what exact sequences or patterns or sequences of GAGs are responsible for interactions with a host of proteins is still unclear. The next few years will see an explosive growth in information regarding glycan-protein interactions that will become available. With an improved understanding of how GAGs and GPIs are involved in signaling mechanisms, new biology and new modes of therapeutic or prevention will become available. Particularly, relatively weak, temporary arrangements of biomolecules in membranes will be of interest that include carbohydrate-carbohydrate interactions and interactions of GPI-anchored proteins. In time glycans will take their place next to oligonucleotides and polypeptides as key signaling molecules.

Acknowledgments

Generous funding of work in my laboratory has been provided by the Max-Planck Society, the ERC (Advanced Grant), the EU (FP7 program), the BMBF and the Körber Foundation.

References

1. O. J. Plante, E. R. Palmacci, P. H. Seeberger, *Science* **291**, 1523 (2001).
2. B. Castagner, L. Kröck, D. Esposito, C.-C. Wang, P. Bindschädler *et al.*, *Chem. Sci.* **3**, 1617 (2012).
3. O. Calin, S. Eller, P. H. Seeberger, *Angew. Chem. Int. Ed.* **52**, 5862 (2013).
4. S. Eller, M. Collot, J. Yin, H.-S. Hahm, P. H. Seeberger, *Angew. Chem. Int. Ed.* **52**, 5858 (2013).
5. D. M. Ratner, E. W. Adams, J. Su, B. R. O'Keefe, M. Mrksich *et al.*, *ChemBioChem.* **5**, 379 (2004).
6. F. Kamena, M. Tamborrini, X. Liu, Y.-U. Kwon, F. Thompson *et al.*, *Nat. Chem. Biol.* **4**, 238 (2008).
7. R. Kikkeri, B. Lepenies, A. Adibekian, P. Laurino, P. H. Seeberger, *J. Amer. Chem. Soc.* **131**, 2110 (2009).
8. L. Schofield, M. C. Hewitt, K. Evans, M. A. Siomos, P. H. Seeberger, *Nature* **418**, 785 (2002).
9. Y.-H. Tsai, X. Liu, P. H. Seeberger, *Angew. Chem. Int. Ed.* **51**, 11438 (2012).
10. M. Santer, M. Wehle, I. Vilotijevic, R. Lipowsky, P. H. Seeberger *et al.*, *J. Amer. Chem. Soc.* **134**, 18964 (2012).
11. C. Stefaniu, I. Vilotijevic, D. Varón Silva, M. Santer, G. Brezesinski *et al.*, *Angew. Chem. Int. Ed.* **51**, 12874 (2012).
12. H. A. Orgueira, A. Bartolozzi, P. H. Schell, R. Litjens, E. R. Palmacci *et al.*, *Chem. Eur. J.* **9**, 140 (2003).
13. J. L. de Paz, C. Noti, P. H. Seeberger, *J. Amer. Chem. Soc.* **128**, 2766 (2006).
14. P. H. Seeberger, D. B. Werz, *Nature* **446**, 1046 (2007).
15. A. Kabanova, I. Margarit, F. Berti, M. R. Romano, G. Grandi *et al.*, *Vaccine* **29**, 104 (2010).
16. C. E. Martin, F. Broecker, M. A. Oberli, J. Komor, J. Mattner *et al.*, *J. Amer. Chem. Soc.* **135**, 9713 (2013).
17. Y. Yang, S. Oishi, C. E. Martin, P. H. Seeberger, *J. Amer. Chem. Soc.* **135**, 6262 (2013).
18. M. Tamborrini, D. B. Werz, J. Frey, G. Pluschke, P. H. Seeberger, *Angew. Chem. Int. Ed.* **45**, 6581 (2006).
19. C. Anish, X. Guo, A. Wahlbrink, P. H. Seeberger, *Angew. Chem. Int. Ed.* **52**, 9524 (2013).

CARBOHYDRATE-ACTIVE ENZYMES IN MICROBIOMES

BERNARD HENRISSAT

Centre National de la Recherche Scientifique and Aix-Marseille University
Campus de Luminy, 13288 Marseille, France

My view of the present state of research on microbiomes and carbohydrate chemistry

Despite their apparently unexciting chemical composition, carbohydrates can form an astronomical number of combinations through the stereochemical diversity of the hydroxyl groups that they carry, through the multiple possibilities to assemble monosaccharides to each other, and through the many non-carbohydrate substituents that can decorate the resulting oligo- and polysaccharides. Laine has calculated that there exists more than 10^{12} possible isomers for a reducing hexasaccharide [1]. Although not all possible isomers are found in Nature, Nature has long exploited the stereochemical and physical diversity of carbohydrates for a large variety of biological roles, from carbon reserve (starch, glycogen, laminarin, etc.), to structure (cellulose, hemicelluloses, pectins, peptidoglycan, chitin, mannans, glycosaminoglycans, carrageenans, agarose, porphyrans, fucans, etc.), or as the mediators of intra- and intercellular recognition within one organism or between organisms [2]. The complex glycan structures borne by glycoproteins or the multitude of bacterial exopolysaccharides mediate the recognition between a host and commensals, symbionts or pathogens. The complexity of carbohydrate recognition reaches probably its maximum in the distal intestine where hundreds of different bacterial strains digest the colossal diversity of our food and epithelial glycans, while harboring a multitude of exopolysaccharides that participate to the maturation of the immune system [3].

Nature's utilization of carbohydrates seem only limited by the ability of enzymes to assemble and disassemble a glycan without interfering with another. The number of protein folds being limited, Nature has created carbohydrate-active enzymes (CAZymes) from a small number of primordial scaffolds. This creative process has taken place all along the evolution of living organisms and has enabled the emergence of a multitude of complex glycans which became substrates for breakdown enzymes. For instance, cellulose was a relatively minor macromolecule when life was essentially marine. The successful emergence of terrestrial plants was made possible by the exceptional mechanical properties of cellulose. From then on, cellulose became the most abundant macromolecule of photosynthetic origin on Earth, thereby creating an immense new ecological niche for organisms that would be able to digest plant cell walls. Various microbes and fungi jumped on the new opportunities offered by land plant biomass by "recycling" a number of pre-existing enzymes that progressively adapted to the new substrate. This process took place independently several times from different protein scaffolds, explaining why nowadays cellulases are found in more than 10 different families.

The evolutionary processes by which novel substrate specificities are acquired from a common ancestor leave traces that can be detected in the sequence of contemporary proteins [4, 5]. Thus, unexpectedly, what is the usual drawback of carbohydrates (their chemical resemblance) is at the origin of their success in the postgenomic era: whilst the fine specificity of DNAses, RNAses, proteases and esterases are practically impossible to derive from their sole sequences, the hierarchical classification system that we have proposed for CAZymes [6–8] allows, already in several cases, the prediction of the carbohydrate substrate based on the assignment to a family [9]. In principle, this provides the opportunity to examine the glycobiological profile of an organism based on its genome sequence. However, the acquisition of novel specificities from older CAZymes often groups together in the same family enzymes that act on different substrates and this results in significant problems during the functional annotation of a number of CAZyme-related genes. Today our knowledge of functional variations within CAZyme families is still insufficient to accurately and reliably predict the substrate specificity of all CAZymes identified during genomic or metagenomic sequencing. The organization of the sequence information in clans, families and subfamilies can serve as a guide to select putative CAZymes for functional studies by identifying the poorly explored regions of each family, i.e., where homology-based specificity predictions are unreliable. Our classification of CAZymes therefore can contribute simultaneously to the rational identification of targets for functional studies and to the utilization of this information to decipher the Glycobiology organisms based on genomic or metagenomic sequence. This virtuous circle will progressively contribute to the emergence of Glycogenomics, and will be instrumental to establish a Systems Biology that will capture the peculiarities of carbohydrates.

My recent research contributions to microbiomes and carbohydrate chemistry

The spectacular progress of modern DNA sequencing technologies together with the launch of large multinational efforts allowed (i) to isolate and sequence the genome of a multitude of gut bacteria and (ii) to generate metagenomic data of increasing quality and depth [10–12].

During the last 5 years, I have explored the potential of CAZymes in the functional profiling of genomic and metagenomic data derived from the digestive microbiota of humans and animals. The main results based on CAZyme family assignment and the resulting broad substrate categorization have led to several observations. First, an early metagenomic study of cow rumen has revealed that rumen

CAZymes appear to contain a number of appended carbohydrate-binding modules much smaller than CAZymes from environmental bacteria [13]. This is attributable to the confinement and highly substrate concentration found in the digestive ecosystems.

Second, we noted that the CAZyme profile of obese individuals was less diverse than that of control subjects [14]. Next, through the analysis of fecal metagenomic data derived from 33 mammalian species and 18 humans, we showed that diet shapes the CAZyme profile in animals and humans [15]. Finally, we recently analyzed the enormous genomic and metagenomic datasets generated by the HMP consortium [12], which sampled not only gut, but also bucal, urogenital, nasal and skin body sites [9]. Examination of 493 bacterial reference genomes isolated from different body sites, showed that carbohydrate degradation capabilities appear more similar within a bacterial taxonomical family than among bacteria colonizing the same habitat. Yet, the analysis of 520 metagenomic samples from the five major body sites showed that even when the community composition varies the CAZyme profiles are very similar within a body site, suggesting that the observed functional profiles have adapted to the local carbohydrate composition [9].

All in all, and despite the current limitations due to their incompletely studied sequence-to-function relationships, CAZyme profiles appears to correlate with both taxonomy and habitat (lifestyle) of individual organisms (genomes) or communities thereof (metagenomes). Irrespective of the expectable improvements of sequencing technologies, future progress will depend on our capacity to predict with increased accuracy the fine substrate specificity of CAZymes identified in genomes and metagenomes.

Outlook to future developments of research on microbiomes and carbohydrate chemistry

In principle, the clan-family-subfamily classification of CAZyme sequences has the potential to determine the glycobiological profile of an organism simply based on its genome sequence. However, the multiple specificities that are often grouped together in a given sequence-based CAZyme family is an obstacle to the simple and straightforward functional annotation of CAZyme-related genes. Whilst this problem can be overcome by the definition of smaller subfamilies [16, 17], our knowledge of functional variations within CAZyme families and subfamilies is still insufficient to accurately and reliably predict the substrate specificity of all CAZymes identified during genomic or metagenomic sequencing. The organization of the sequence information in clans, families and subfamilies should serve as a guide to select and study candidate CAZymes for functional studies by identifying the poorly explored regions of each family, *i.e.*, where homology-based specificity predictions are unreliable. Coupled to the discovery of novel families by the glycobiological community, the hierarchical classification of CAZymes therefore can contribute simultaneously to the rational identification of targets for functional studies and to the utilization of this information to decipher the Glycobiology of organisms based on genomic or metagenomic sequence. This virtuous circle will progressively improve and the precise specificity of most CAZymes that interact closely with their substrate will be predictable. If coordinated, this effort will contribute to the emergence of Glycogenomics, and will be instrumental to establish a Systems Biology able to capture the peculiarities of carbohydrates.

Applications could be massive as carbohydrate polymers constitute abundant renewable materials for use by a variety of industries and for the production of transportation biofuels. Complex carbohydrates also constitute a major fraction of human and animal diet and could be used to manipulate, stabilize or restore the composition of the microbial communities along the digestive tract. Due to its wide variety and limited overlap with the human metabolic pathways, the human microbiome constitutes a potential source of therapeutic drug targets [18]. The breakdown or modification of orally administrated drugs by the microbiota also attracts attention from a health perspective [19]. Although the microbiota has been linked to a number of diseases, it is presently unclear whether the changes in the microbiota composition are the cause or the result of the disease [20]. The interplay between the microbiota and heath is still in its infancy, but certainly constitutes a mine of opportunities for the future.

Acknowledgments

Current research in my laboratory is funded by the following institutions: Centre National de la Recherche Scientifique (CNRS), European Research Council (ERC), European Commission (EC), Agence Nationale de la Recherche Scientifique (ANR), Fondation Infectiopole Sud, and University of Aix-Marseille.

References

1. R. A. Laine, *Glycobiology* **4**, 759 (1994).
2. A. Varki, N. Sharon, *Essentials of Glycobiology* 2nd edition, 1 (2009).
3. A. El Kaoutari, F. Armougom, J. I. Gordon, D. Raoult, B. Henrissat, *Nat. Rev. Microbiol.* **11**, 497 (2013).
4. C. Chothia, J. Gough, C. Vogel, S. A. Teichmann, *Science* **300**, 1701 (2003).
5. S. Kumar, K. Tamura, M. Nei, *Brief Bioinform.* **5**, 150 (2004).
6. B. Henrissat, *Biochem. J.* **280**, 309 (1991).
7. B. Henrissat, G. Davies, *Curr. Opin. Struct. Biol.* **7**, 637 (1997).
8. B. L. Cantarel, P. M. Coutinho, C. Rancurel, T. Bernard, V. Lombard, B. Henrissat, *Nucl. Acids Res.* **37**, D233 (2009).
9. B. L. Cantarel, V. Lombard, B. Henrissat, *PLoS One* **7**, e28742 (2012).
10. P. J. Turnbaugh, R. E. Ley, M. Hamady, C. M. Fraser-Liggett, R. Knight, J. I. Gordon, *Nature* **449**, 804 (2007).
11. J. Qin, R. Li, J. Raes, M. Arumugam, K. S. Burgdorf *et al.*, *Nature* **464**, 59 (2010).

12. The Human Microbiome Project Consortium, *Nature* **486**, 207 (2012).

13. J. M. Brulc, D. A. Antonopoulos, M. E. Miller, M. K. Wilson, A. C. Yannarell *et al.*, *Proc. Natl. Acad. Sci. USA* **106**, 1948 (2009).

14. P. J. Turnbaugh, M. Hamady, T. Yatsunenko, B. L. Cantarel, A. Duncan *et al.*, *Nature* **457**, 480 (2009).

15. B. D. Muegge, J. Kuczynski, D. Knights, J. C. Clemente, A. González *et al.*, *Science* **332**, 970 (2011).

16. M. R. Stam, E. G. Danchin, C. Rancurel, P. M. Coutinho, B. Henrissat, *Protein Eng. Des. Sel.* **19**, 555 (2006).

17. H. Aspeborg, P. M. Coutinho, Y. Wang, H. Brumer 3rd, B. Henrissat, *BMC Evol. Biol.* **12**, 186 (2012).

18. B. D. Wallace, M. R. Redinbo, *Curr. Opin. Chem. Biol.* **17**, 379 (2013).

19. I. D. Wilson, J. K. Nicholson, *Curr. Pharm. Des.* **15**, 1519 (2009).

20. L. Zhao, *Nat. Rev. Microbiol.* **11**, 639 (2013).

THE MICROBIOME(S): MICROBIOTA, FAMILIES, FUNCTIONS

ADAM GODZIK

Program in Bioinformatics and Systems Biology, Sanford-Burnham Medical Research Institute
10901 N. Torrey Pines Rd, La Jolla, CA 92037, USA

My view of the present state of research on microbiomes and carbohydrate chemistry

Complex microbial communities thrive in every environmental niche on Earth and recent breakthroughs in DNA sequencing provided us with unprecedented level of information about these communities, effectively rediscovering them both for science and popular imagination [1].

For over a century, microbiology was dominated by a "pure culture paradigm", where the ability to grow a microbe in a pure culture was seen as a prerequisite to any scientific study of its function and/or characteristics. While very successful in identifying microbial causes of many diseases, in cases where individual members are difficult or even impossible to study in isolation it slowed down and sometimes reversed studies of microbial communities. Existence of such communities, for instance multi-bacterial biofilms, was well known since the invention of the microscope — already in 1677 Anton van Leeuwenhoek reported seeing multiple "Animals in the scurf of a mans Teeth", a claim that was widely questions and almost led to his expulsion from the Royal Society of Lodon [2]. However, only technical advances in DNA sequencing that allowed direct sequencing of environmental samples (metagenomics) brought these observations to the forefront of modern science. Surveys of ocean [3] and human microbiomes [4] made metagenomics and microbiomes a subject of popular press [5]. Especially a realization that humans, and to a varying extend all multicellular organisms, are in fact super-organisms that depend on their microbial components for normal functioning captured public imagination. Human microbiome consists in fact of hundreds of smaller microbiomes, as every surface and even smallest region of our body hosts separate, ofter very different microbiomes [6]. An often repeated estimate suggest that in human body, microbes outnumber human cells by a 10:1 margin, and microbial genes outnumber human ones at least 100:1. When we include viruses, which outnumber microbes by a similar margin, we are indeed outnumbered in our own body.

Previous microbe-human interaction paradigm viewed microbes as agents of disease and lack of them as a symbol of health, it seems that the opposite is closer to the truth. For instance a first symptom of many diseases is diminishing the diversity and numeric count of our microbiome [7] and the absence of some bacterial species was found to be correlated to diseases [8]. The most obvious role of human microbiome, especially the gut microbiome, is providing humans with the ability to digest food particles that are normally not accessible to human enzymes. Human gut bacteria, such as *Bacteroides*, developed extensive networks of carbohydrate metabolism able to process complex polysaccharides of plant origin, as well of peptoglycans (protein-carbohydrate polymers) of animal origin. For rats, 30% of an caloric intake is provided by byproducts of microbial metabolism [9]. The second role is protecting the host, mostly by overcrowding the pathogenic bacteria, but also by priming host immune system. Microbiome also may play other, at this point little understood roles, such as contributing to the maturity of host nervous system [10]. Research on human microbiome is still in a very early phase, but it is clear that our microbes form dense networks of interactions with the host, us, on every level — from metabolites, to proteins to cell-cell interactions.

As much as human microbiome captured the headlines, as microbiomes go it is not particularly complex or diverse. Many environmental microbiomes display astonishing diversity with for instance at least 9 distinct phyla are usually abundant in soil microbiomes [11], compared to two major ones in the human gut microbiome. Bacterial communities are found in most exotic environments, from the Atacama desert [12], to inside of glacial lakes [13], isolated for thousands or in some cases for millions on years. Novel carbohydrate metabolism enzymes found in human gut microbiome are just a tip of the iceberg of fascinating discoveries we can expect from the barely touched or still to be explored regions of protein universe in other environments, driven by unusual and often unknown chemistry and we can expect many fascinating discoveries from the barely touched or still to be explored regions of protein universe.

My recent research contributions to microbiomes and carbohydrate chemistry

In my group we have been collecting and analyzing gene repertoires from different microbiomes and developing a family fingerprint approach to directly compare them for the functional content, an alternative to the phylogenetic profiling such as done by the 16S rRNA.

By comparing presence/absence of protein families in different functional groups, we can for instance see an functional complementarity between family repertoires in the human genome and in the human gut microbiome, where families lost in the human lineage since the last eukaryotic common ancestor (LECA) can be found in the human gut bacteria, thus making the human+microbiome superorganism having a similar metabolic capacity as LECA (Fig. 1) [15].

We also put a big effort into defining new protein families, specific to given types of microbiomes [16]. Since most protein families defined in public databases were identified from model, broadly distributed microbes, they rather obviously focus on broadly distributed families, which members have universally conserved functions. In well studied microbes, over 80% of all proteins can be classified into known families. At the same time, only 40–60% of predicted proteins coded by genes identified in metagenomic studies can be identified, suggesting that microbes forming microbiomes in many environments could have developed novel functions, tuned to the specific challenges of their environments. Therefore databases of protein families have to be expanded to include novel families found in newly surveyed environments. We have defined and deposited to Pfam over 400 protein families specific to the human gut environment and we are now using them to analyze and compare microbiomes associated with diseases and study geographical diversity of human microbiomes.

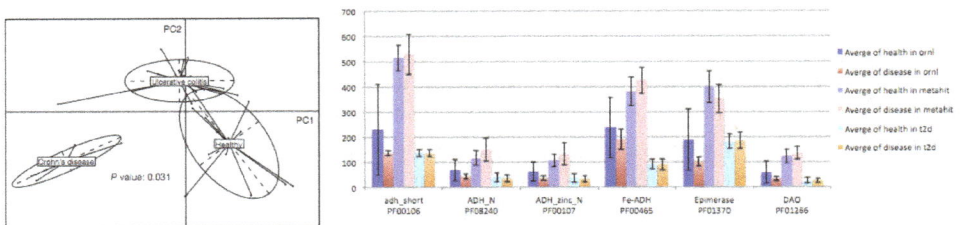

Fig. 1.　Comparison of bacterial composition of different microbiomes, here between human microbiomes associated with two forms of inflammatory bowel disease (Crohn's disease and ulcerative colitis): (left) principal component analysis of species distribution [14], (right) relative frequencies of proteins from specific families, here proteins involved in butyrate synthesis (Godzik lab, unpublished).

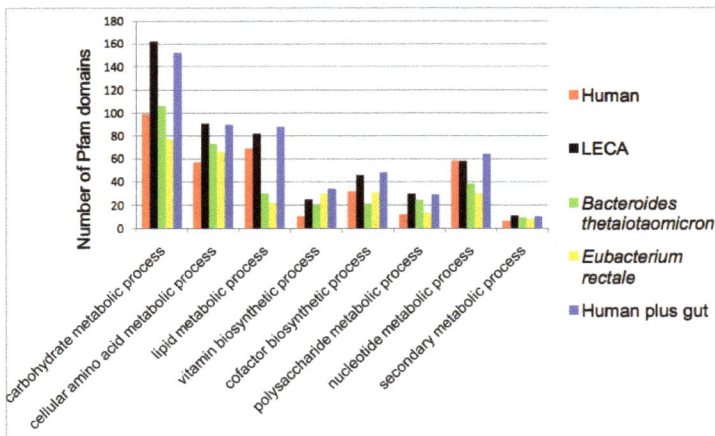

Fig. 2.　Human gut microbiome complements the domains lost in the human lineage since the last eukaryotic common ancestor (LECA). Reconstruction of ancestral domain repertoires suggests that LECA had an extensive repertoire of metabolic functions (black bars), many of which were lost in the branch leading to vertebrates, including humans (red bars). While individual human gut microbes (green and yellow bars) have limited repertoires of such domains, together with the host they almost achieve the same numbers as those in LECA.

Outlook to future developments of research on microbiomes and carbohydrate chemistry

We have just begun to survey the world of microbiomes and the amazing diversity of genes they contain. Even if a relatively small percent of these new genes code for enzymes, we still face billions of novel enzymes, performing potentially novel functions.

Acknowledgment

Research described here was funded by the NIH grant U54 GM094586 (JCSG).

References

1. National Research Council (U.S.). Committee on Metagenomics: Challenges and Functional Applications. and National Academies Press (U.S.), *The new science of metagenomics : revealing the secrets of our microbial planet*. 2007, Washington, DC: National Academies Press. xii, p. 158.
2. F. N. Egerton, *J. History of Biol.* **1**, 1 (1968).
3. J. C. Venter, K. Remington, J. F. Heidelberg, A. L. Halpern, D. Rusch *et al.*, *Science* **304**, 66 (2004).
4. S. R. Gill, M. Pop, R. T. Deboy, P. B. Eckburg, P. J. Turnbaugh *et al.*, *Science* **312**, 1355 (2006).
5. C. Zimmer, *New York Times* **2010**, (2010).
6. P. J. Turnbaugh, R. E. Ley, M. Hamady, C. M. Fraser-Liggett, R. Knight *et al.*, *Nature* **449**, 804 (2007).
7. D. C. Baumgart, W. J. Sandborn, *Lancet* **380**, 1590 (2012).
8. H. Sokol, B. Pigneur, L. Watterlot, O. Lakhdari, L. G. Bermúdez-Humarán, *et al.*, *Proc. Natl. Acad. Sci. USA* **105**, 16731 (2008).
9. D. L. Sewell, B. S. Wostmann, C. Gairola, M. I. Aleem, *Amer. J. Physiol.* **228**, 526 (1975).
10. C. L. Maynard, C. O. Elson, R. D. Hatton, C. T. Weaver, *Nature* **489**, 231 (2012).
11. V. Torsvik, L. Ovreas, *Curr. Opin. Microbiol.* **5**, 240 (2002).
12. A. de los Ríos, S. Valea, C. Ascaso, A. Davila, J. Kastovsky *et al.*, *Int. Microbiol.* **13**, 79 (2010).
13. Y. M. Shtarkman, Z. A. Koçer, R. Edgar, R. S. Veerapaneni, T. D'Elia *et al.*, *PLoS One* **8**, e67221 (2013).
14. J. Qin, R. Li, J. Raes, M. Arumugam, K. S. Burgdorf *et al.*, *Nature* **464**, 59 (2010).
15. C. M. Zmasek, A. Godzik, *Genome Biol.* **12**, R4 (2011).
16. K. Ellrott, L. Jaroszewski, W. Li, J. C. Wooley, A. Godzik *et al.*, *PLoS Comput. Biol.* **6**, e1000798 (2010).

N-LINKED PROTEIN GLYCOSYLATION

MARKUS AEBI

Institute of Microbiology, Department of Biology, ETH Zürich, CH-8093 Zürich, Switzerland

My view of the present state of research on microbiomes and carbohydrate chemistry: Protein glycosylation

For a long time, carbohydrates as protein modifications have received little attention and very little was known about their functions. This lack of knowledge was mainly due to the lack of analytical tools that specifically detect and identify complex carbohydrates on proteins. In addition, complex carbohydrates are the product of biosynthetic pathways in which the structure of the product is determined by the specificity of the enzymes (glycosyltransferases) involved and not by a linear template such as RNA in the case of proteins. A good example of the fact that carbohydrate modifications of proteins have been overlooked is the relatively recent discovery of the dynamic modification of a large number of cytoplasmic and nuclear proteins by N-acetylglucosamine on serine and threonine residues (O-GlcNAcylation) [1]. This protein modification is as frequent and as dynamic as protein phosphorylation but the lack of a simple and specific detection system has slowed down the functional analysis of this highly conserved process in eukaryotic cells. Similarly, protein glycosylation was for a long time considered to be eukaryote-specific but research in the last ten years revealed that such modifications are very frequent in prokaryotes and that protein glycosylation is highly divers [2]. Novel analytical tools such as MS-based methods and oligosaccharide-specific antisera have been instrumental in the identification of such novel protein glycosylation processes in prokaryotes.

Reverse genetics in model organisms is an excellent tool to study the function of proteins *in vivo* and this tool can also be applied to reveal the function of carbohydrate modifications. However, this requires a detailed analysis of the biosynthetic processes and the characterization of the different glycosyltransferases that are involved generation of protein-linked oligosaccharides. Therefore, genetic tools have only been applied recently in glycobiology but they revealed a large functional diversity of protein glycosylations. Powered by genetic studies of glycosylation processes in model organisms such as yeast and mouse, a large number of human congential disorders have been recently linked to deficiencies in protein glycosylation pathways [3].

My recent research contributions to microbiomes and carbohydrate chemistry: Protein glycosylation

The research of my group has focused on the process of N-linked protein glycosylation. Using the model organism *Saccharomyces cerevisiae* we have characterized the N-glycosylation pathway in the Endoplasmic Reticulum and characterized many of the steps in the assembly and the transfer of the N-linked glycan to protein [4]. We also studied a homologous process of N-linked protein glycosylation found in the bacterium *Campylobacter jejuni* and characterized it at a molecular level after functional transfer of the pathway into *Escherichia coli* [5]. In collaboration with the group of Kaspar Locher, ETH Zürich, the crystal structure of the bacterial oligosaccharyltransferase (OST) was solved [6]. This structure revealed the molecular basis for the recognition of the conserved N-X-S/T substrate motif and was the basis for the "twisted amide" model of the reaction mechanism.

It is interesting to note that N-linked protein glycosylation became the most frequent protein modification in eukaryotes: more than 20,000 different N-glycosylation sites are proposed to be glycosylated in the eukaryotic proteome [7]. A short recognition element (N-X-S/T) is a prerequisite for the observed evolutionary trends in N-glycan multiplicity in specific sets of proteins. Due to the fact that OST requires a flexible peptide for substrate recognition, the execution of N-linked protein glycosylation before the folding process of substrate proteins is a specific feature of eukaryotic N-linked protein glycosylation as compared to its prokaryotic counterparts. Additional subunits of the eukaryotic OST can slow down the folding process of substrate proteins [8]. In particular, Ost3/6p exhibit oxidoreductase activity and bind specific polypeptides, both non-covalently and via transient disulfide bonds and thereby maintain the substrate peptide in a "glyosylatable" stage. This shift from the prokaryotic order of events (translocation of the polypeptide – folding – glycosylation) to the eukaryotic order (translocation – glycosylation – folding) results in an increase of potential glycosylation sites. At the same time, the transfer of the highly hydrophilic oligosaccharide to a polypeptide alters its folding parameters significantly: N-glycosylation extends the folding landscape of polypeptides. However, the folding of polypeptides is not only affected by the altered biophysical properties due to attached oligosaccharide, the oligosaccharide itself is used as an anchoring point for the assembly of a glycan dependent chaperone and folding machinery, the calnexin/calreticulin complex [9]. In addition, the processing of the N-linked glycan is used to signal the folding status of the protein and defined carbohydrate structures are essential signal for the degradation of not properly folded proteins in the ER of eukaryotic cells [10]: N-linked protein glycosylation is a central factor in the folding of secretory and membrane proteins in the eukaryotic cell.

Once in place on properly processed proteins, the N-linked glycans can serve additional functions and the eukaryotic cell has devoted a whole organelle to the remodeling of protein-bound glycans: the Golgi apparatus. In a species- and cell-type specific process, N-linked glycans are trimmed and reconstructed by a series of different hydrolases and glycosyltransferase resulting in the well-known structural diversity of N-linked glycans. Using a novel, MS-based analytical method we were able to quantify site-specific glycan structures on proteins with multiple N-linked glycans (C-W. Lin, R. Gauss, S. Fleurkens, I. and M. Aebi, unpublished results), Our results revealed a site-specific processing of the N-glycan by the Golgi machinery, suggesting a kinetically controlled process that is influenced by the interaction of the oligosaccharide with the covalently bound polypeptide. It is the processing machinery in the Golgi, the structure of the glycoprotein and the location of the attachment site that determines the site-specific oligosaccharide structure. The site-specific structures of N-linked glycans can provide novel functionalities to N-glycoproteins.

Outlook to future developments of research on microbiomes and carbohydrate chemistry: Protein glycosylation

The structural complexity of carbohydrate modifications will remain a challenge for analytical approaches. New MS-based approaches that are fast and that give a detailed structural information are urgently needed. It will be of outmost importance that the analysis of glycan functions does not remain a domain of specialized glycobiologists and glycochemists. Once glyco-analytics can be performed in any decently equipped laboratory, we will be able to identify a very high diversity of functional roles of protein-bound glycans. It is evident that glycans are the main components of the extracellular matrix in all cells, pro- and eukaryotic. Cells interact with their environment via their glycan interphase and we will be able to decipher the underlaying "glycocode". The biophysical properties of carbohydrates, different from those of nucleic acids and polypeptides, are the basis for this "code" but it is evident that carbohydrate-binding proteins, lectins, are the main interpreters. A large number of lectins are encoded in all genomes that are sequenced so far, but for only a few of these proteins we know their biological function.

Acknowledgments

Work in my lab has been generously supported by the Swiss National Science Foundation, the EU (framework 7) and ETH Zürich.

References

1. G. W. Hart, C. Slawson, G. Ramirez-Correa, O. Lagerlof, *Annu. Rev. Biochem.* **80**, 825 (2011).
2. H. Nothaft, C. M. Szymanski, *Nat. Rev. Microbiol.* **8**, 765 (2010).
3. H. H. Freeze, V. Sharma, *Semin. Cell Dev. Biol.* **21**, 655 (2010).
4. P. Burda, M. Aebi, *Biochim. Biophys. Acta* **1426**, 239 (1999).

5. M. Wacker, D. Linton, P. G. Hitchen, M. Nita-Lazar, S. M. Haslam *et al.*, *Science* **298**, 1790 (2002).

6. C. Lizak, S. Gerber, S. Numao, M. Aebi, K. P. Locher, *Nature* **474**, 350 (2011).

7. D. F. Zielinska, F. Gnad, J. R. Wisniewski, M. Mann, *Cell* **141**, 897 (2010).

8. E. Mohorko, R. Glockshuber, M. Aebi, *J. Inherit Metab. Dis.* **34**, 869 (2011).

9. A. Helenius, M. Aebi, *Annu. Rev. Biochem.* **73**, 1019 (2004).

10. R. Gauss, K. Kanehara, P. Carvalho, D. T. Ng, M. Aebi, *Mol. Cell.* **42**, 782 (2011).

SESSION 3: MICROBIOMES AND CARBOHYDRATE CHEMISTRY

CHAIR: IAN A. WILSON

AUDITORS: N. CALLEWAERT[1], R. LORIS[2]

(1) Unit for Medical Biotechnology, Inflammation Research Center, VIB and Department of Biochemistry and Microbiology, Ghent University, Technologiepark 927, B-9052 Gent-Zwijnaarde, Belgium

(2) Structural Biology Brussels, Vrije Universiteit Brussel and Department of Structural Biology, VIB, Pleinlaan 2, B-1050 Brussels, Belgium

Discussion among panel members

Karen Nelson could not make it but was kindly replaced by *Adam Godzik*.

Ian Wilson: Let me first start by asking about the microbiome and this problem of associating it with disease and are we ever going to get to that situation where we will be able to do that?

Adam Godzik: I am sure we are. I think that there are a lot of interesting correlations. What we are still having problem with is whether they are really causes of these diseases or just bystanders. There are some very interesting results on mice and model systems, which suggest that this is the driving force, but this is still not proven.

Ian Wilson: What about the sample variation and the difficulty of actually assessing... this is a real problem, isn't it?

Adam Godzik: Yes, the diversity is an enormous problem but there are some signatures which survive these differences. There are specific combinations of species and specific combinations of pathways, which are overrepresented or missing despite all the variations.

Dennis Wolan: There are plenty of examples in the literature, particularly of the butyrate producing species, not even species but phyla, that are involved in disease states, lack thereof actually. Jeremy Nichols I think has been doing quite a lot of work, particularly in colorectal cancers, to show that particular metabolisms, like cresol (and cresol sulfate) are overproduced in colorectal cancer patients. So taking these kind of molecules and potentially bating for the proteins that do interact with them is a potential way of going after drug discovery targets.

Adam Godzik: Let me add something. One of the interesting observations people make is that very often disease states are associated with upsets of certain groups of proteins of pathways. So very often it seems that they are not causing the disease but, if they are missing, this propagates the disease state.

Bernard Henrissat: I would like to complete this (or try to contribute to this) by saying also, our view of what is going on is limited by the fact of what we sample. We sample what is coming out. We should be sampling, using animals for instance, what is going on all along the digestive tract. Because sampling of what is going on in the mid colon or at the beginning of this colon is very difficult, and it is rarely done on these cohorts of people.

Dennis Wolan: The resolution that we are at right now is quite minimal. We do know that the different types of bacterial species that reside for instance in the mucosal layer are highly different from those in the luminal content. The other issue is as far as animals are concerned; there is some debate as to how useful they are, particularly as the mucin layer composition is quite different in mouse as compared to human.

Bernard Henrissat: I very much agree with this when I mentioned animals, I was not thinking about mice. We just take a whole animal, such as little piglet or pig. And you can perform some experiments with an animal than you cannot perform in a human person.

Ian Wilson: Anyone else want to contribute to this topic?

Richard Lerner: What about mycoplasma? Bacteria live in a tube, mycoplasma live on and in the cells. There doesn't seem to be much attention on mycoplasma.

Adam Godzik: As I mentioned, in one of the unexplored areas, that all the issues about intracellular bacteria are just beginning to be appreciated. We just obtained very interesting results on a possible role of intracellular bacteria in colorectal cancer and it definitely seems tumors are infected by specific strains of intracellular bacteria. We don't know if they are cause or effect of colorectal cancer, but definitely mycoplasma, and many other bacteria of this type, are driving things. They are just not studied yet.

Judith Klinman: I was wondering if bacterial metabolites can serve a function in signaling in mammalian cells. Is anything known about that? (Like bacterial metabolites being used by mammalian enzymes)

Dennis Wolan: There are some examples, for instance in butyrate-producing *Clostridium* bacteria, butyrate does block NFκB activity. There are clear examples where metabolites that are produced by bacteria do have an effect on the physiological aspect of the human. A lot of it has to do with inflammation, as far as what is in the literature at least.

<u>Ian Wilson</u>: How are we going to get at all these different functions? Do you think we are going to get it through these chemical probes, through structures, or through bioinformatics approaches? How are we going to figure out all of the diversity of all functions that are going on in the microbiome?

<u>Bernard Henrissat</u>: I can take a try on this one. I think we need to integrate everything, we cannot leave something aside and say "This is not going to work out". We will need high-throughput assays for functions, we will need three-dimensional structures, we will need everything to acquire enough knowledge so we can go into prediction. We cannot leave anything aside — we will take it all.

<u>Dennis Wolan</u>: Along those lines, the Human Microbiome Project is expanding into trying to incorporate a wide variety of "-omics" kind of technologies: metabolomics, transcriptomics, genomics, proteomics, etc. to decipher the kind of differences that are observed in disease versus normal states. However, there is definitely something to be said about taking examples of particular proteins or working from a bottom-up kind of approach to understand the functional significance and how they can umbrella into other functions of bacterial proteins.

<u>Adam Godzik</u>: I agree that this problem is very complicated. There are now two broad areas/ways to approach it. One uses larger studies, where we hope that by comparing large cohorts of disease versus healthy people, you will notice the single biggest signal. Then identify the signal and feed it back into model systems. There are several studies like this on inflammation and several other diseases, where they start from observations in large human cohorts and then go to mouse models and try to see whether single identified features can be used to modulate the system. On the other hand, there are a lot of *ad hoc* random trials and with some of them, like probiotics for instance which are not regulated, something can be discovered but it also generates a lot of noise.

<u>Ian Wilson</u>: Bernard said something which I thought is quite provocative: that function is not determined by the fold. What do you exactly mean by that? You are probably talking about fine specificities because there are carbohydrates processing enzymes, signature sequences... Can you explain a little more?

<u>Bernard Henrissat</u>: I will take just one example that is been under our nose for so many years, that is lysozyme. If you did not know what α-lactalbumin is doing, you would probable predict it to be a lysozyme. By fold, by sequence similarity, by everything. It is only when you inspect the catalytic machinery that is not there that you may have a hint that it is no longer functioning as a lysozyme. But you cannot guess what the new function is. You could not guess that one ancestral fold has been has been recruited as the binding partner for a glycosyl transferase to alter its acceptor/donor specificity. This is what I mean there: the fold will not predict the function necessarily. Often it will do.

Ian Wilson: I get that. But I was really referring to the carbohydrate-processing enzymes. Are there so many differences between the folds? So the fold is already telling you it is a carbohydrate-processing enzyme. What you don't know is probably the substrate specificity. Is that what you are referring to?

Bernard Henrissat: It is right. It is "determining the specificity": knowing that you cleave a sugar is not sufficient. You would like to know which sugar. You would like to know whether you have an ensemble of enzymes, in a operon or a cluster, that target a complicated substrate, or the substrates that gets in the microbiota. It is very complex and needs the co-operation of many enzymes. Knowing that you have a sugar-cleaving enzyme is not fully satisfying. I would like to have more!

Ian Wilson: Are there any questions from the discussion panel members?

John Gerlt: I have a question about the carbohydrate enzymes. Are these carbohydrate-active hydrolases and transferases excreted and used to degrade polysaccharides so that we can use these as carbon sources? What is the diversity of catabolic pathways within the human microbiome species because you only talked about the hydrolases, but there are kinases, aldolases, dehydrogenases, for example.

Bernard Henrissat: I can only speak about the enzymes I know. There is a great variety also of glycosyl transferases that are elaborated by the bacteria of our microbiome, for instance. The reason for this variety is presently unclear but when we look at the number of different glycosyl transferases that are encoded in the microbiome, there is an enormous diversity of carbohydrate structures that are produced at the surface of the bacteria in our gut. One of the ideas I have on the role of such a variety is for protection against phages. If all our bacteria in the gut had the same carbohydrate structure at their surface, they would probably be very sensitive to phage attack — what you have in a dairy factory where you have a single species of *Lactobacillus* for instance. By having a variety of carbohydrate structures you ensure resilience of the microbiota.

John Gerlt: But I am so curious about the catabolism of carbohydrates by the gut bacteria. Is there much known about it? Is it a simple catabolism to maintain the human microbiome? Is it producing the hydrolases so that we then can utilize a wider variety of dietary oligo- and polysaccharides?

Bernard Henrissat: There is a little bit of this. It is estimated that about 10% of the calories that the microbiome is deriving from the food we eat is utilized by the (human) host. It is only 10%. But I believe that of the main "raisons d'être" of the microbiome is simply that we have a lot of carbohydrates, because we eat a diet that is complex and we just derive very little ourselves. When it (food) gets in the distal gut, there is temperature, water, heat and microbes can flourish. So I think that the main "raison d'être" of the microbiota is just for itself more than for the host. There has been a good level/amount of co-evolution (we accept this or this bacteria because they do not harm us), there are some very good things, but the (bacterial) flora is there mostly for itself.

John Gerlt: So they could be producing signaling molecules or other molecules to interact or so that we can interact with them, so as to receive some positive benefits, that is not just a carbon source.

Gebhard Schertler: Can I ask a question here? You said we have realized that many bacteria form films. Are carbohydrates actually a substrate to sustain films of bacteria and make them more specific because we want to retain certain bacteria and not others. Maybe that is a way to look at it?

Markus Aebi: No the carbohydrate is the matrix that creates the biofilms mediated by lectin-carbohydrate interactions, and even carbohydrate-carbohydrate specific interactions.

Discussion among all attendees

Ian Wilson: We now can open this up to the general discussion for any of the speakers. Maybe I will start with asking Chi-Huey (Wong) about the effect of the glycosylation. You talked about stabilization but that was for a particular sequence, where you had to have a phenylalanine upstream. Do you think it is general or for specific cases? And how can we use this information to think of stabilizing glycoproteins in general?

Chi-Huey Wong: I think phenylalanine is of small significance, but in some cases you see glycosylation will affect folding, but not of every protein. Hemagglutinin is one of them. There is no phenylalanine there, but Asn27 is essential for folding. If there is no glycan there, it is not enough. The metal transfer I mentioned, the antibody, none of them has the aromatics but they still affect the folding pathway.

Ian Wilson: But you said for the hemagglutinin that there was a certain residue for which glycosylation is required for folding. We have bacterially expressed hemagglutinin and it seems to fold just fine. So how do you reconcile that?

Chi-Huey Wong: I didn't see that. Did you express the same sequence?

Ian Wilson: Yes, with the bacterially expressed form of the hemagglutinin. We haven't published that yet.

Chi-Huey Wong: We expressed the hemagglutinin in human cells with a mutation at Asn27 but couldn't get it. We use for example PNGase to remove the glycan, we also see the structure fall apart. That one (Asn27) is essential for folding and there is another one, Asn142, that is close to the binding site for the sugar (sialic acid). If you don't have the glycan there, you miss out on the stereospecificity because hemagglutinin will recognize both the α(2-6) and α(2-3) linked sialic acids. These are the two glycosylation sites that we identified. The other four are not really essential for stabilization but I think they are involved in function.

Dario Neri: I have a question for Ian Wilson that may be also relevant for the session tomorrow. When you change the process in a production of an antibody or when you go from a product to a biosimilar, the main changes that you may expect, not the only changes, are at the level of the glycosylation. There are some examples in the pharmaceutical history where changes in production made a big impact in terms of pharmacokinetics and *in vivo* and dose potency, for example the famous XOMA's antibody (note: Raptiva^{TM}) for the treatment of psoriasis that was transferred to Genentech and had different pharmacokinetics and potency. My question is whether there are other examples that we should be aware of where changes in the glycosylation of antibodies result in different pharmacokinetics, extravasation and *in vivo* potency.

Ian Wilson: I am not sure I am the best person to answer that. I will transfer that to Chi-Huey Wong.

Chi-Huey Wong: I think there are limited studies on that, so it would be interesting to see how that would effectively affect the behavior *in vivo*. But as I mentioned, if we take the known antibody, which is usually a mixture, and we analyze all the glycoforms and see how each of them behaves in animals, you will see that the sugars have a dramatic effect on ADCC (antibody-dependent cell-mediated cytotoxicity). If the antibody is against cancer and has also an anti-inflammatory activity, it is a chimera. I think the Fc receptor somehow interacts with the glycan. We need to have a structure of the complex to better understand how the sugars of the antibody play a role in interacting with Fc γ receptor.

Richard Lerner: There is a sort of famous experiment. As you might imagine in the biosimilars arena, there were the "erythropoietin wars". The Irish wanted to go into the business of biosimilars and so they built their production plants. What they found was that the commercial reference erythropoietin produced was more different batch to batch than their own erythropoietin. The good news for the pharma companies was that under the new "biosim" rules, it is good to have a mess because you can't replicate a mess. If you have a homogeneous thing, you can make it and it becomes like a small molecule. But if you put out a product, let's say a peptide product, and there are many components for which you don't quite know what the active principal is, even if they all have the same sequence, you cannot make a "biosim" unless you copy that complete mess. It is a very interesting dynamic because in the small molecules world, one attempts to make the most pure compound you can get your hands on. In the "biosim" world, it would be interesting to hear from the people who worry about fermentation: can you actually by fermentation get the same sort of carbohydrate derivitization twice in row?

Chi-Huey Wong: I mentioned about the N-glycosylation from yeast to human. So the core trisaccharide would be the same from yeast to human. If we take a glycoprotein made from yeast to create the sugar to keep the (trimannose) core and then add the one of interest to the core, we create the human version of the glycoprotein. But it depends on if you want a mixture or a single molecule...

Richard Lerner: But this is because you developed an elegant method! What would you expect is going on with all the other products out there?

Chi-Huey Wong: Maybe Markus (Aebi) will answer that. I think my expectations are in the future. Biosimilars will be one business, but people will try to get a single molecule out of biosimilars as a new drug. That would be a new drug instead of a biosimilar.

Markus Aebi: I think for a single site like a protein this is possible. You can mimic this microheterogeneity, but as soon as you go to multiple sites, this becomes very difficult and the reason for that is that we lacked until now analytical tools to quantify structures per site. I think now we have these tools and this means that we can start addressing the question: what is the function of a specific carbohydrate on a specific site?

Until now we always had handwaving: it has no effect, but we didn't have the tools to generate glycoproteins with a defined structure at the defined sites. Now we can start asking these questions and I think this will be important in the future.

Peter Seeberger: Along those lines, Danishefsky *et al.* have just finished the first total synthesis of chemically made erythropoietin. This is entirely synthetic and basically made like a pharmaceutical drug. It is a single molecule, completely defined. I wouldn't say it is a drug at this moment because it is too expensive to produce, but it gives us a chance to really understand at the atomic level what is the exact composition. Then you can find of a new product with a composition based on that. I think the delineation between biologicals and small molecules is starting to fade.

Richard Lerner: We could have an endless discussion on this. It is true, scientifically speaking, with modern analytical chemistry, you can really characterize your product down to the level of the proton. But it is a practical issue, and scientific issues and FDA practical issues are different. And the FDA says that biosimilars must be the same, and that is just the way it is. When you start to ferment multiple kilogram quantities of a protein like EPO, you cannot have two batches that are the same, it just doesn't work that way. That, in the end, is the real issue. We, as scientists, tend to go for purity, while the regulatories (FDA) go for sameness.

Tom Muir: I have a question for the microbiome folks, maybe Dennis would be the best person to answer that question. I remember my late colleague Ralph Steinman often told me that by many measures the gut is the largest immune organ in the body. It is stuffed full of dendritic cells, T-cells and so forth that keep a certain level of activity all the time, it is not quiescent. There is activity to some level and the idea is that there must be some kind of communication between the microbiome and the immune system to maintain homeostasis. But as far as I know, we don't know what the signaling molecules are. It seems to me that it might be a really interesting class of molecules to try to identify using methods such as the ones you talked about. Can you comment on that?

Dennis Wolan: You are absolutely right. The roles that the levels of the immune system has and the basal levels that are there is brought about by the commensal healthy microbiome. And exactly what kind of metabolites and/or interactions or peptide interactions are being sampled by dendritic cells in the Peyer's patches and things of that nature are still in progress. As these become elucidated, we will definitely be using those as starting point to identify bacterial proteins and/or human expressed proteins that interact with them.

Richard Lerner: Tom, you could put a really fine point under that question. When you have something like a mycoplasma, that is an obligatory parasite, it has to live on the surface of our cells for life and escape the immune system.

A) How does it do it?
B) If we knew how it does it, could we use that strategy to make immunosuppressive drugs?

Because apparently as far as we know certain strains of mycoplasma will live inside you, for life, and totally exposed.

Brian Roth: Related to that question, it looks like this would be a tremendous opportunity for sort of unbiased small molecule based screening; if collections of these can be obtained, they can be screened in a high-throughput fashion against families of proteins. We had a small collaboration with some group at UCSF, which was sending us small collections. But it would be great to get larger collections of thousands (if not potentially millions) of these compounds secreted by bacteria that potentially affect human gut homeostasis and immune cells and so on.

Dennis Wolan: There are absolutely groups that are focused on identifying the metabolites that are produced by both the bacteria as well as the human, and the interactions that result. Michael Fischbach from UCSF for example is trying to determine the natural products that are produced by the bacteria.

Brian Roth: He had synthesized a small number. I can't recall offhand what the data are. This work might still be in progress. But yes, I am aware of that.

Markus Grütter: I have a general question on practicality. How different or similar is it to study the bacteria that exist for example in the gut and what is the diversity and the (number of) different strains. Does one have to study thousands of different strains? Are there similarities? What is the situation there? I would be interested.

Bernard Henrissat: I can begin to answer your question. Everything depends on how we define a species to begin with. There is ample evidence of a large amount of horizontal transfer between bacteria in the dense environment that is the gut. And therefore you can imagine that if you have a bacterium that incorporates a piece of DNA from another, its 16S RNA will not change. But the functional properties of that bacteria will change. So it really depends on how we define species. We try to understand and we would like to answer this question with the current knowledge, which is not enough to derive everything. That is all I can say.

Markus Grütter: But how about the practicality of working with these organisms/microbacteria compared to our laboratory animals?

Adam Godzik: I think with there are a few strains of specific bacteria which emerge like a model system in *in vitro* studies that are being used. For the question related to the gene transfer, we don't know how relevant these strains are to the counterparts in the real system. On one hand metagenomics will study the distribution of strains like we have in a normal gut. And on the other, we have model systems but there is still ongoing debate on how related they are. I know a few cases where for instance some models used did not incorporate interesting genes and therefore for *in vitro* studies turn out to be irrelevant. But we are just developing this library of relevant organisms that could be used for models.

Dennis Wolan: There is the clear example of a situation where you can't just use a single model organism. *B. theta* (*Bacteroides thetaiotaomicron*) for instance, that Ian mentioned, encodes 400 or so glycosidases. And it is not until *B. theta* had been co-inhabited into a mouse with some probiotic (I don't remember the name) that all the glycosidases become expressed. So the interactions between the individual bacteria and how they interact with their host is another layer of complexity.

Christine Orengo: Can I just ask another practical question? I think you alluded to it before about those metadata. If we are trying to understand the functions that these communities have, it is very important to capture the phenotypes that we see in either human health or human diseases or, if you are talking about soil bacteria, then pH and different characteristics. Are there any initiatives to derive ontologies of the type of information that should be collected along with the sequencing data?

<u>Adam Godzik</u>: The answer is yes, there is a consortium for developing standards for metadata from marine organisms. This was most related to microbiomes as I mentioned until this fiasco where all samples were collected without any information about water temperature. It was very difficult later to compare them to anything because the samples arrived "naked". For humans, it still doesn't exist. We have problems analyzing certain data sets because we don't know about the medical history of people or about antibiotic use for instance. I think at this point, the concept is to collect all information that is available, just put it in an unstructured way and worry about it later.

<u>Ian Wilson</u>: I would like to move on and talk about synthesis of glycans. Is this now conquered? Can we get gram quantities? Are there other sequences, in particular of ones that are put on in human cells? Can we get any amount pretty readily?

<u>Peter Seeberger</u>: We can make up to kilogram quantities of certain glycans — of certain mammalian type — but also of bacterial type. We typically on the instrument I showed you make 25 micromoles of material which is about somewhere between 10 and 50 milligrams of what you need. I think this is sufficient for most studies we would like to do. The question that of course comes up: Can you make all structures? The answer to that is clearly no. Another question is how many building blocks do we need to make for example a mammalian glycome? Theoretically, that is in the hundreds of structures but from bioinformatics studies, we found out that we can get away with probably 15 building blocks to make most of the human glycome. When you go to bacterial glycomes, the number is larger. But within certain types of bacteria you can get away again with very fewer glycan building blocks. So I think we are still ways away from making everything, but if you really want to make something, we can do that.

I give the example of a compound called "Prevnar", a 13-valent vaccine against *Streptococcus pneumoniae*. It sells itself for billions dollars worth from Pfizer every year. We have now succeeded in making this entire thing using some synthetic chemistry. Of course the *Streptococcus pneumoniae* has more than 13 serotypes, there are 96 serotypes. We haven't made all the 96 yet! It is a long way away, but if you really want a carbohydrate, you can make it. And I think like in DNA, people couldn't make poly-G for many years. In peptides, you could not, after Merrifield (pioneered solid-phase peptide synthesis) make everything right away. I think there is room for improvement, but again, if you really want something, I really do believe you can make it.

<u>Chi-Huey Wong</u>: I agree. I think if there is a demand the synthesis can be solved. The question is: What to make? The information has to come from biologists.

<u>Kurt Wüthrich</u>: As a follow up to Markus Aebi's talk, are there any systematic studies of polypeptide chains with and without glycosylation? I mean unfolded

polypeptide chains and studies on their folding behavior? It may be difficult to keep some of these polypeptides unfolded in solution, but it is feasible. Has anything of this been done?

Markus Aebi: I think there are folding studies on RNAse A and RNAse B but I am not aware of others, I must say.

Chi-Huey Wong: I think there are many examples (from Patel, from Barbara Imperiali's and many other groups). We had one from about 15 years ago: what we did was make the dodecamer of the RNA polymerase C-terminal repeat that is in random coil. But if you add the mono-sugar, it becomes a type-II β-turn. So the glycosylation is able to induce the conformation from random coil to type-II β-turn. This of course will initiate the folding. But there are many examples using a peptide as a model, among which synthetic peptides with and without sugar to see how the glycosylation affects the folding.

Tom Muir: I think the erythropoietin Peter eluded to from the Danishefsky group, actually did see quite dramatic differences in the folding yield of the protein, as a function of how many sugars were added to it. And it wasn't until they had the full complement on that the folding became reasonably efficient. Without them, it was really a mess due to the folding reaction. That is specific of the *in vitro* folding reaction.

Peter Seeberger: I think the answer is only as good as the models are. Early on those were relatively short peptides with relatively few sugars. And then it is difficult to come to a general conclusion based on that. As we see larger molecules with larger glycans, we will get more information. Of course what this means biologically is less clear. Because some of these glycans are put on during the synthesis of polypeptides in some cases, there may be immediate effects on that. So I think that answering to the question is not that straightforward.

Judith Klinman: I haven't heard anything about lifetime of proteins in the cells as a function of glycosylation. It is something I remember from years ago but I don't know what the current status is on that thinking.

Markus Aebi: So glycoproteins are secreted proteins, so the relevance is either on the plasma membrane or they are secreted in the serum. In the serum, we know very well that the glycan structure determines the lifetime. The asialoglycoprotein receptor that retrieves non-siaylated glycoproteins from the serum and thereby determines the half-life. On the plasma membrane it is an interesting question. How is the quality controlled with respect to the glycocalyx on the cell surface? There are some ideas around of how this is managed and what determines the half-life. But until now we still don't have the tools to analyze the half-life of a glycoprotein *in vivo*. This is not something you just do on a Saturday morning!

Markus Grütter: As follow up on Kurt Wüthrich's question to Markus Aebi: It seems that you have the machinery studied and you know the components that are responsible for the glycosylation on the polypeptide chains. Do you see a possibility to engineer these components sort of semi *in vitro* or directly glycosylate something?

Markus Aebi: So one way is you separate the two things, namely the N-glycosidic linkage, to make the N-glycosidic linkage from the structure of the carbohydrate. This is an engineering way. You can actually separate the two processes. Making the N-glycosidic linkage is at a specific point in the polypeptide chain. For this, you have to rely on enzymes; you cannot do that chemically except with big efforts. So there are *E. coli* systems that can do that, you can then combine it with chemically synthesized carbohydrate and ligate the two things together. That is possible, now whether it is commercially attractive I don't know. I think as soon as the regulations are such that you require really pure components as we discussed before, then it becomes commercially interesting because then it can compete with the production from eukaryotic cell lines.

Chi-Huey Wong: I have a question for Markus. You mentioned that it is maybe the kinetic barrier that gives the mixtures of glycans. You know glycosyl transferases are membrane-bound in the Golgi. I wonder if there is lectin involved to move the newly synthesized glycoproteins around. Can you further elaborate on why we get mixtures?

Markus Aebi: There is a Nobel prize just awarded for this transport phenomenon in the cell on this vesicular transport. There is still some debate as how proteins are transported in the Golgi, whether there is retrograde transport or recycling. I don't want to go into this debate but it is clear that the protein cargo travels through this Golgi and the membrane-bound glycosyl transferases stay within the stacks creating individual gradients and therefore a microenvironment in these micro-reaction vessels, so to say, that then determine the structure of the carbohydrates. That is at least my view. I think it's a given set of glycosyl transferases in the cell that is the hardware. And then you determine the expression levels individually, and these determine in what direction glycosylation goes. And then the speed of transfer that then ends up in specific glycan structures on a protein. That is how I view it.

Ian Wilson: Can I ask you Markus about the site-specific identification? Because that is something I have been extremely interested in. This has really held our work up very badly when trying to think of immunogen design for an HIV vaccine. It is complicated with HIV, due to the 27 glycans per monomer (that is 81 per trimer), so we really would like to know the specific glycans at particularly positions. How can we do that with such a large protein, because with each tryptic peptide you get an average of 4 glycans.

Markus Aebi: We are using analytical tools that were developed at the Academia Sinica to exactly address this question. These are MS-based analysis HCD fragmentation that allows you to identify first glycopeptides and, at the same time from the same fragmentation, use the information to identify the polypeptides. Then we use as a second dimension sequencing of the carbohydrates. This new technology allows us to exactly address this type of question and quantify the corresponding results.

Ian Wilson: You were suggesting in your talk there wasn't perhaps as much variation as one might think at a specific position and that there are actually relatively few isoforms/glycoforms there. Did you ever see mixed high mannose and complex sugars at the same position or is that never found?

Markus Aebi: In this system (a purely experimental system), we are using yeast proteins expressed in insect cells. This is as reductionism as it can be. There we did not see high mannose structures expressed except for those in the ER. But I would expect that as soon as you have very strong interactions between a carbohydrate and the surface of a protein then you get high mannose structures on the final protein. And I would be eager to measure these interactions. We can model it by molecular dynamics, but I think NMR would be an ideal system to see how specific these interactions actually are and whether we are able to predict them for the outcome with pure protein.

Jason Chin: I have a question about glycosylation in the Golgi in two parts. You talked about a series of reactions happening as you move through the Golgi. And then possibly co-translational initial glycosylation, which possibly precedes the folding event. My first question is about the co-translational folding: To what extent is it possible to recapitulate these things *in vitro* and to make detailed biophysical measurements? My second question is about the subsequent series of glycosylation and how we should think about those. Is there any directional flow where you have a series of glycosylations and any error you make in any step is propagated to the next? Or is it more sophisticated than that it that and there might be retrograde transport coupled to deglycosylation and error correction? How do we think about this process and how can the analytical techniques that we have allow us to distinguish between those possibilities?

Markus Aebi: I think first that the term "co-translational" you used is relevant. It is more "before folding" or "after folding". To couple the glycosylation to the translocation event is one way to make sure that glycosylation occurs before folding. But it can also occur much later on in the process, in the folding pathway.

We cannot mimic, but we can do *in vitro* glycosylation and we couple it to the translocation event. These studies have been done extensively, also to monitor other things of the translocation events and glycosylation was used as a marker. We can detail how far the (poly)peptide has to reach into the ER lumen to get glycosylated and we know quite well from *in vitro* studies how this processes work. In terms of retrograde transport, we take it off again. We lack to my knowledge the enzymatic machinery in the lumen of the ER to do that. To deglycosylate proteins, they do have to exit the ER again in retrograde translocation. Then they are degraded in the proteasome. But before the sugar has to be taken off from the polypeptide.

Jason Chin: So this means there is no error correction?

Markus Aebi: No. You just degrade the protein that is not properly folded.

Jason Chin: OK. But you never correct the glycosylation on the protein?

Markus Aebi: As far as I know, no. Because taking of the sugar generates an aspartic acid instead of an asparagine, which wouldn't make much sense.

Ian Wilson: Can I ask some questions about identifying the glycan specificity? The glycan arrays are fantastic inventions of the last few years in order to decide what the specificity of the glycan is. But we have encountered a number of situations where we actually don't find much specificity on the array. The binding is just too weak. So you either need higher density arrays or you need mixed glycans. The particular examples that I showed you are antibodies that actually have 2 or 3 binding sites for a glycan plus interaction with the polypeptide. So the individual interactions are likely to be probably very weak (probably in the millimolar range), but can I ask about the detection and the specificity. How can you actually detect so weakly binding glycans?

Peter Seeberger: I think all the data you get out of your glycan arrays is only as good as your input, as far as content of the array is concerned. I think we are only at the very beginning. The Consortium of Functional Glycomics I think has 600+ compounds. We have now hundreds of compounds on our arrays. But it depends on what you put on there. We cannot believe that with 600 glycans, we really even begin to cover either the human glycome or the microbial glycome. And I think particularly the arrays that are out there right now are weak on the bacterial glycome. There are almost no bacterial structures on there. That is what my department has been addressing these recent years because they feel there is a real need. They are now doing the same thing for plants. So I think the content has to be improved, we need more glycans. It is not going to be pretty but someone has to make those molecules, or isolate them.

The second question is the interactions between proteins and glycans. You are implying most of these are weak interactions. I think this is only partially true. There are some known interactions where the binding is in the low nanomolar level. Of course people have dogmatically written about it that this is not the case, but I think we see very tight binding, particularly in the case of glycosyl-glycans. Heparins and proteins bind exceptionally tightly to each other. There is charge, so these things are possible. I think it depends on what the glycan should do: if the glycan is at the surface and it has to temporarily engage the receptor, it makes sense that we get a weak binding. But then binding is temporarily and it opens up again. If you want a signaling event in the cell, I think you want to have a tight and specific binding event that leads to the transmission of information. And I think that depending on what the function of the glycan is, we can have different types of binding that range from very tight to not tight at all, where multivalency and other things plays a role, but I think people have been quick in making broad ranging statements about the functions of glycans based on one or two examples in some cases. So I do not say we do not need some sort of identity arrays, but I think we need both and we should be aware that in some cases with glycans we see very tight binding.

Ian Wilson: I mean the situation I am talking about is: we actually know roughly what the specificity is, we know the sort of glycans that are present on the array, so that is not the issue. We know that we can get crystal structures. We actually have sometimes crystal structures even with the individual glycans and with the glycosylated protein. We actually see these interactions that we are not detecting on the array, and we know that the glycans are actually on the array.

Peter Seeberger: I think this is a matter of context. Because I think people have focused only on the protein or only on the glycan. It would be really interesting to look at the whole system, because glycosylated proteins have both components: the protein and the glycan. And I think it has been difficult in the systems that have been made by Markus and others, and through synthetic means we are getting towards this end now.

Chi-Huey Wong: I can add some points about this. The antibodies you discovered from HIV patients, some of them, like 2D12, PGT16, PGT9, they don't recognize just one glycan. They recognize two different glycans. How do you pick that up from the array? The current array is made from glass slides activated by N-hydroxysuccinimide ester (NHS). We know it is unstable and very often you see that the active ester is not evenly distributed on the surface. So the array you have got is not very homogeneous. I showed that aluminium oxide-coated glass slides will be very even because the chemistry used for such arrays is different. It is based on the use of a phosphonic acid tail, which will react with aluminium oxide spontaneously. That will create a much better controlled array with a pre-arranged distance between the sugars. With that we are able to pick up the heteroligand binding and multivalent binding with very high dissociation constants there. And I think this could lead to the design of carbohydrate-based vaccines in the future.

The other problem with arrays, as Peter mentioned, is that we don't know how many sugars we have to put on. I think the NIH commission has several laboratories to make 10,000 glycans to be put on the array. But whether that is enough, we still don't know. If we get into the microbiome system, it would be enormous. So

I think it is probably better to make the array for special purposes: we have the array for influenza, for cancer, for gp120 (HIV) (because it has about 85 different glycans that are put on the array), and that is how we use it to understand the specificities of the carbohydrate-binding antibodies.

Aled Edwards: I think this experiment must have been done? How much success has there been in reconstituting in *S. cerevisiae* by knocking them all out one by one, getting a minimal glycosyl transferase-ome and then adding back and looking what effects they have. And are you going to do that with the baculovirus with the CRISPRs — start knocking out one by one so you can have an *in vivo* reconstituted system as it were?

Markus Aebi: I know from one example where they have engineered insect cells (I forgot the cell line) for the specific production of antibodies with fucosylated glycan structures, for which they have knocked out and put in the specific glycosyl transferases. I think it was a Japanese company and they were able to generate a pure glycan structure on the produced antibody. So this is possible.

Aled Edwards: I was more thinking of reducing (I don't know if there are 50 of them) down to five. The cells will be barely alive and then add them back (glycosyl transferases) one by one to see exactly what happens.

Markus Aebi: So when you fiddle around with ER glycosylation machinery, you are dead! When you take the early enzymes in the Golgi, you hardly survive. And for the later ones, the sialyl transferases, there is no serious phenotype in mice.

Chi-Huey Wong: The knock-out experiment was not very successful, particularly in mammalian cells, because they (glycosyl transferases) are all essential. One glycosyl transferase could be responsible for many different glycoproteins. But one successful approach is the Genentech approach that uses a fucosyltransferase knock out to make antibodies without fucose. That has been produced on larger scales for development.

Ian Wilson: May I ask Chi-Huey Wong about the vaccine approach? A lot of the earliest thoughts on trying to get better interactions/responses for neutralizing anabolism, better immune response against viral antigens, was actually to cover up the surface with additional glycans, not to take glycans off. And then you actually only leave particular sites (such as receptor binding sites or fusion machinery) unexposed, but you are covering up everything else including all the hypervariable loops. So you have done completely the opposite, that is you have taken all the sugars off and you found you can get good responses. But assuming that you must get a lot of non-neutralizing response mixed in with neutralizing response, how do you boost the neutralizing response?

Chi-Huey Wong: That is a good question. Where we have evidence so far is that once we remove the outer part of a sugar, you generate more different kinds of antibodies. We used a single B-cell to get about 400 clones and some of them are very interesting. We even obtained antibodies that recognize the sequence covered by the sugars. I think we need to get a complex to better understand that. I think what happens is that (you know antibody-protein interactions are much stronger than antibody-glycan interactions) the antibody pushes the sugar away and thermodynamically you get the antibody-protein interaction there. We need to get that evidence by structure, but numerous studies showed that we get more different kinds of antibodies and they target new epitopes. The antibodies generated are better in terms of neutralization. We have to do more studies to confirm that.

Richard Lerner: I think this is important — more than a gem of an idea — this dialogue between Ian Wilson and Chi-Huey Wong. Because the neutralizing antibody people have pointed to the fusion area, there has been a race between the vaccine makers to sort of make the fusion piece, and jettison everything else. From a synthetic point of view, that turned out to be very difficult to do, because once you take those pieces out, they don't fold right. So the alternative would be to do what Ian Wilson just said: that is leave the thing intact but cover everybody else with sugars. You have the cake and eat it too! You have got the epitope you want in the right conformation more or less, and you have blacked out everything else. That would be a way to focus vaccination. You (to Chi-Huey Wong) could take the same technology you have had, turn it on its head and use it to cover everything.

Chi-Huey Wong: I agree. We have just begun to understand how carbohydrates are taken up by B-cells and by dendritic cells, and how these antigens are processed, how they are cleaved and then presented to T-cells. This is unknown. We begin to understand because several carbohydrates make small changes there. What we know is that in the case of proteins, of course protease is involved in the processing, but for carbohydrates we don't know about it. People think it has to do with nitrogen oxide radicals cleaving glycosidic bonds but there are some other glycosylases involved. We don't quite understand the specificity of that, and that is why often one carbohydrate by chance sometimes creates different kinds of antibodies, which can recognize the internal part of the epitope.

Richard Lerner: But by the end of the day, the immune response is a binding energy competition amongst 10^8 combatants. And anything you can do to lower the binding energy potential of the rest of the molecule in terms of the initial event, never mind the protease later on, will focus the immune response. You can probably focus on everything that you want.

Ian Wilson: There is evidence of what you are suggesting that certainly for these antibodies to the hemagglutinin, that the glycans are pushed out of the way. You can see that the best antibodies are actually navigating through the glycans. If they weren't there, you might get at the surface better, but if you think of trying to get an immunogen that is going to raise the same type of antibodies, you might need to start with something that has less glycans on it, then add something later as a boost with more glycans. You should use this combination approach. There is good evidence in the HIV field, this is also true, that you can start with something which has less glycans on it. There are usually less glycans on the transmitted virus than on the virus that ends up after a few rounds of replication in your body. Many times, the germline antibody doesn't even seem to recognize the viral antigen. And one way that it sometimes can recognize is if you take the glycans off. There is thought now of, again, taking some of the glycans off to get initial priming of the germline response and then add more glycans back to make it more native-like.

Richard Lerner: But nobody would have ever done that, if you had not shown that the loop can penetrate. That is what changes everything because why would you take a sugar off to expose an epitope underneath, unless you knew what you know now. Because the natural system would cover it up again. You have to consider that, in the context of the idea, that there is enough energy in the loop as it were to push the sugars away and get at what is underneath. If you can't do that, there's no sense doing it that way.

Kurt Wüthrich: We all seem to agree that glycosylation is extremely important and that we don't know the basis for its important role. I have a specific example: Peter Seeberger talked about prion proteins and the GPI anchor. There is another story related to the glycosylation of the prion protein in the cellular form. There are two glycosylation sites and based on the determination of the glycosylation levels, it has been determined that some people, mostly in England, were infected with a new variant of Creutzfeldt-Jacob disease. Just simply from the numbers, it seems that this may not have been a correct interpretation of the data. I can hardly imagine that there would have been new variant of Creutzfeldt-Jacob disease in just 250 cases and never again, considering the extent of the BSE crisis in England. I suppose that since you worked on the GPI anchor site of the problem, you must have looked into this as well.

Peter Seeberger: Very simple answer: No, we did not yet. One of the limitations for us actually was to get the protein properly. You pointed out during the break that it is not easy to work with this protein. I think the expertise of working both with these complex proteins that also have to be handled taking into account certain safety regulations in some countries as well plus the required glycosylation expertise for us took a long time. So your plans now are that after we have made the protein plus the GPI anchor to put on glycans as well. There is a group in my department that works only on that problem. They made progress but they haven't yet made the whole thing.

<u>Ian Wilson</u>: Do we have general questions that anybody wants to ask?

<u>Christine Orengo</u>: It is a change in topic completely. Just a brief question to Bernard Henrissat: You talked about your subfamilies you are characterizing in the CAZyme classification and you explicitly said you don't use any sequence information. But I imagine that within some families even from the tree you can see clustering and specificities associated with the different branches. So I imagine there must be some sequence signals. I just wondered if you had a look at that at all and why you choose not to use that sequence information.

<u>Bernard Henrissat</u>: The situation is very difficult. In a family-by-family case, we have figured out that there is no universal threshold to determine these subfamilies. Sometimes in some families of glycosyl transferases, especially those that convert the blood groups A and B, just a couple of residues are enough to change specificity. In other families, we can go as low as 15% identity and still have the same function. So really we need to have expert knowledge on each and every family. When we see subfamilies, we don't know what is the driving force or whether we will discover new functions. They just provide opportunities for discoveries, but we don't have a 100% hit on discovery. Most of the time we discover that what was the signal turned out to be taxonomical drift, that sort of things. This is getting more and more complicated because the trees get to be saturated by the sequence information we have these days. So it is easy to explain on a slide, it is a lot harder to do in real life.

<u>Don Hilvert</u>: Maybe to return to a Kurt's question about chemical synthesis of carbohydrates. What are the remaining chemical challenges to be able to make the entire universe of carbohydrates? Is it the building blocks or the particular linkages that are difficult to accomplish? Are there strategic changes that need to be made in the approach that one takes?

<u>Peter Seeberger</u>: Two things that you mentioned. The first is you need the building blocks. For human systems that is pretty easy, but for some bacterial systems that is difficult because you cannot pull out the monosaccharides. So we have to synthesize the monosaccharides from first principles. It is doable, but in some cases there are as many as 15 chemical steps required to get the building block. That has been worked out by a number of different laboratories. The more complex question is: the chemical glycosylation reaction has been known for 110 years now. And to this day, we know it depends on the electronics, sterics and conformation of both the nucleophile as well as the glycosylating agent. In many cases, we can tune that very nicely, but in some particular cases you get steric mismatches, and you do not get what you want. It has been so far difficult to predict and so a few cognoscenti know how to do this and others don't. We have tried to create an operator-independent database that compares many different pairs of coupling partners and therefore derive some empirical evidence to predict glycosylations.

This is growing now and I think this will make it easier for people to decide what building blocks to use. Based on that, we have come up with what we call a list of approved building blocks. We know they can be made, they are stable and upon activation, and they give you reliably high activity at the site of glycosylation-linkage. Their number is growing and I think this is what people who would like to make a glycan and are not experts need. They can either buy these building blocks and make the glycan themselves or go to a commercial supplier that will make a glycan for them. I strongly believe that the glyco-field needs to integrate more with the people who make proteins because you cannot get a molecule pure, you can't ask the (right) questions. Not everything can be made yet, but it can be made in many areas.

Don Hilvert: On the other hand, the catalysts do exist in Nature. What about the chemo-enzymatic approaches? Is it just really the availability of the enzymes? Or the stability?

Chi-Huey Wong: I think that would be the way to go in the future. Of course it depends on the product. If you want to make glycoprotein or heparin sulfate or more complex structures, I would not just go around with chemical methods. If you want to scale up, I think the enzymatic process could be the choice, because you cannot tolerate any stereoisomer in the product, particularly if you are dealing with a vaccine. So I think both chemical and biological methods have to be considered and of course it depends on what you want to make. The problem with the enzymes is that they are not available, but they are there! If there is a demand, I think they will become available. This is very similar to the story of aspartame, the sweetener. When it came out, it was made by mixed anhydride chemical synthesis. The reason for that is that thermolysin was expensive and now it is as cheap as sodium chloride. When there is a big demand, I don't think enzyme is a problem.

Ian Wilson: A lot of what we're having to do now is trying to think of not doing all glycoproteins but designing and synthesizing some sort of epitope scaffold. So you have maybe one peptide or multiple peptides and you want to put say Man$_5$ in one place and Man$_8$ in another place, and a complex sugar in another place. How can we do that?

Chi-Huey Wong: I think that this can be easily done by chemical means. Peter Seeberger is one who set up a solid-phase synthesis of target molecules. These are small enough to be dealt with chemistry.

Peter Seeberger: I think one has to be pragmatic. As Chi-Huey Wong said before: depending on the issue or what you have to make, you can use chemical means where you can create quickly diversity. If you go to large scales then you might do it enzymatically. But I think it is some of a "chicken egg" problem because materials are not available. People don't even dare to ask about them. The technologies exist but they are still expensive and people don't dare to do it. At some point, you have to break from this vicious cycle.

Craig Townsend: I direct this question to the bio-informaticists. Is there any example of a drug that has its effect against the primary target, but whose beneficial effects are amplified by normalizing in some way the microbiome? I understand it is hard to separate cause and effect here.

Adam Godzik: I am sure there are but I don't know myself.

Dennis Wolan: There is the example that appeared in Science a few years ago: CPT-11 which is a chemotherapeutic that is processed by glucuronidases of the bacteria. So not necessarily what you are saying, but we have to take into consideration the metabolic processes on drugs in our PK-ADME studies for sure.

Ian Wilson: Any other burning question you want to ask to the panel? I am going to ask Markus one question: in HIV, we have this high mannose patch on gp120 and we are not quite sure how it arises but it seems to be conserved. Somehow the virus wants to hold on to this high mannose patch and we would like to understand why? Maybe for interactions with other molecules on the cell surface? How can you control holding onto this high mannose patch with a hypervariable virus?

Markus Aebi: I believe it is relatively easy to maintain a high mannose patch because you simply have to prevent processing by mannosidases that are in the ER and in the Golgi. You have to hide the non-reducing end of the carbohydrate somehow and you can do that by burying into a pocket of a protein. If that hypothesis is correct, we should be able to measure that by these interactions with a protein. I think the only way is to do that by NMR and more people have to work on this. I think it should be possible but Kurt might know more about if we could measure carbohydrate interactions with such a large molecule using NMR.

Kurt Wüthrich: How large?

Markus Aebi: gp120?

Kurt Wüthrich: No problem.

Markus Aebi: So you can model or you can measure. I would say it is the way to go.

Kurt Wüthrich: Maybe this is a very naïve question. Why did we exclusively talk about glycosylated peptides or polypeptides and not at all about carbohydrates today? For example, carbohydrates as substrates for enzymes or the microbiome, especially in the gut?

Bernard Henrissat: I did touch on this. I explained that we are only able to digest three sugars and the rest is entirely done by the microbiome. Lactose, starch and sucrose, that is the basic food and that is what gets absorbed by the small intestine before it gets into the bowel.

Peter Seeberger: I think also it is interesting to see how these biofilms form. What glycans are involved? How do they interact and how do the communities that live together in the different parts of the body, how do they decide who is interacting with whom. I think glycans and carbohydrates are involved in these interactions. The problem, at least from my perspective, is that nobody knows the answer to that and I am not sure we actually have the tools at this moment to resolve this sort of question. But to me that's the real key question: why do certain types of bacteria decide to be in a certain place and deal with each other. And how do they communicate? Are glycans involved in that? Probably yes partially. Maybe someone has the answer here? I don't think anybody has.

Markus Aebi: Kurt, you are absolutely right. You are touching a problem that glycans are not only here to decorate proteins. They are part of the cell wall, of the fungal and plant cells wall. You name it. But this is the general rule that carbohydrates are made to interact with the outside of a cell and so you can touch whatever you like. You will end up with carbohydrates at least in our view here. This is a totally new field. We are the wrong experts.

Chi-Huey Wong: I have a question related to high mannose cluster in human viruses. I don't know why most human viruses have glycoproteins with the high mannose type: hepatitis, HIV, dengi? Does this high mannose glycan escape from the ER and never get into Golgi? Or do they go the Golgi and for some reason are not further glycosylated?

Markus Aebi: I am not an immunologist but in my view, high mannose is "self", a self entity recognized by the immune system. (N. Callewaert, you might correct me)

Nico Callewaert: The answer is multiple fold. HCV for example is packaged from the ER membrane directly. It doesn't do that from the plasma membrane. The HIV does, as far as we know, but there are many viral glycoproteins that are densely packed with one another, so steric occlusion away from the Golgi glycosyl transferases after that packaging has happened, maybe an important factor in that.

Ian Wilson: I think we have come to the end of this session. I thank the speakers and panel members, and all of you for your questions.

Session 4

GPCRs and Transporters: Ligands, Cofactors, Drug Development

Phylogenetic tree of GPCRs and published GPCR structures. See Figure 1 contributed by Raymond Stevens on page 187.

GPCRs AND TRANSPORTERS: LIGANDS, COFACTORS, DRUG DEVELOPMENT

GUNNAR VON HEIJNE

Department of Biochemistry and Biophysics, Stockholm University
106 91 Stockholm, Sweden

My view of the present state of research in the field

We are currently witnessing an almost exponential increase in structural studies of integral membrane proteins. While a low-resolution structure of bacteriorhodopsin, obtained by electron crystallography, was published as early as 1975 [1], the first high-resolution X-ray structure of an integral membrane protein, the photosynthetic reaction center, was published only in 1985 [2]. For many years there was but a trickle of new structures coming in — as late as 2000, there was only about 50 unique structures in the PDB. Today, however, more than 400 unique structures have been published (see http://blanco.biomol.uci.edu/mpstruc/) and new ones appear at an increasing rate.

What lies behind this small explosion of membrane protein structures? Many more crystallographers are active in the field today than 10 years ago, and a number of methodological advances have increased the success rate of structure determination. Perhaps the most important first breakthrough was the adoption of a structural genomics approach: instead of working for decades on a single protein, most researchers today start from a large collection of homologous proteins collected from different organisms (mainly prokaryotic, but now also eukaryotic). The genes are cloned, proteins are expressed in easy-to-use organisms such as *E. coli* or yeast, and rapid screens to single out those proteins that are produced in high amounts and behave well during purification are applied. Finally, the best-behaved candidates are subjected to high-throughput crystallization screens in a wide range of detergents or lipid cubic phases, possibly with the addition of various ligands, lipids, or antibodies to help stabilize the protein. Proteins can also be stabilized by mutagenesis strategies. Finally, crystals are screened at synchrotrons, and phases are solved either by molecular replacement (if a similar structure is already known), by the introduction selenomethionine residues into the protein, or by classical heavy-atom derivatives.

But not all interesting membrane proteins have bacterial homologs. In such cases, the only open avenue is to tackle the difficulties of working with eukaryotic membrane proteins head-on. This realization has given rise to a dramatic improvement in techniques, both in terms of high-throughput approaches and in different ways of stabilizing these often fickle proteins to increase the probability that they will crystallize.

In parallel, new methodology has increased the number of membrane protein structures solved by NMR — a technique that also allows studies of protein dynamics — and there are also new structures being determined by electron crystallography of two-dimensional crystals or by single-particle cryo electron microscopy.

This session is devoted to two particularly important classes of integral membrane proteins: G protein-coupled receptors (GPCRs) and small-molecule transporters. Both classes contain centrally important drug targets, and play critical roles in many signaling and metabolic pathways. Both fields have seen great progress during the past few years, opening up new possibilities for drug screening and design.

My recent research contributions to the field

My own research deals with the question of how integral membrane proteins are targeted to, integrated into, and fold in the membrane during biosynthesis. These processes are common to all membrane proteins (including GPCRs and transporters), and look very much the same in prokaryotic and eukaryotic cells. In short, ribosomes translating a membrane protein are quickly targeted to so-called translocons, protein-conducting channels found in, e.g., the cytoplasmic membrane of bacteria, the endoplasmic reticular membrane in eukaryotic cells, and the inner membrane of mitochondria. The translocon then mediates the orderly membrane integration of hydrophobic transmembrane segments as they emerge from the ribosome.

We have focused primarily on the energetics of the membrane-insertion process as it is played out in live cells. Recent contributions include detailed measurements of how much each of the twenty natural amino acids (and some non-natural amino acids to boot) contributes to the overall free energy of membrane insertion of a transmembrane segment [3–5], as well as measurements of the forces exerted on the nascent polypeptide chain during membrane integration [6, 7].

Outlook to future developments of research in the field

Given the rapid pace of structure determination, representative structures for all major subclasses of GPCRs and small-molecule transporters will no doubt become available within a few years. The next step is to obtain structural information on protein complexes between GPCRs and other proteins; indeed, a structure of a GPCR in complex with its cognate G protein has already been solved [8]. Beyond the static structures, the conformational changes involved in GPCR signaling and small-molecule transport will become increasingly well understood, thanks to structural, biochemical, and biophysical studies. New drug candidates will be developed, based on high-throughput screening approaches and structure-guided design.

Acknowledgments

Grants from the European Research Council, the Swedish Foundation for Strategic Research, the Swedish Research Council, and the Knut and Alice Wallenberg Foundation are gratefully acknowledged.

References

1. R. Henderson, P. N. T. Unwin, *Nature* **257**, 28 (1975).
2. J. Deisenhofer, O. Epp, K. Miki, R. Huber, H. Michel, *Nature* **318**, 618 (1985).
3. T. Hessa, H. Kim, K. Bihlmaier, C. Lundin, J. Boekel *et al.*, *Nature* **433**, 377 (2005).
4. T. Hessa, N. M. Meindl-Beinker, A. Bernsel, H. Kim, Y. Sato *et al.*, *Nature* **450**, 1026 (2007).
5. K. Öjemalm, T. Higuchi, Y. Jiang, Ü. Langel, I. Nilsson *et al.*, *Proc. Natl. Acad. Sci. USA* **108**, E359 (2011).
6. F. Cymer, G. von Heijne, *Proc. Natl. Acad. Sci. USA* **110**, 14640 (2013).
7. N. Ismail, R. Hedman, N. Schiller, G. von Heijne, *Nat. Struct. Mol. Biol.* **19**, 1018 (2012).
8. S. G. Rasmussen, B. T. DeVree, Y. Zou, A. C. Kruse, K. Y. Chung *et al.*, *Nature* **477**, 549 (2011).

STUDIES OF GPCR CONFORMATIONS IN NON-CRYSTALLINE MILIEUS

KURT WÜTHRICH

Cecil H. and Ida M. Green Professor of Structural Biology
The Scripps Research Institute, La Jolla, CA 92037, USA, and
Professor of Biophysics, ETH Zürich, Zürich, Switzerland

My view of the present state of structural biology with GPCRs

The main goal of GPCR structural biology is to obtain insight into the mechanisms by which signals elicited by ligands in the extracellular "orthosteric" binding site are transmitted over a distance of about 35 Å to intracellular partner proteins (Fig. 1). This includes that one tries to establish correlations between the chemical structures of drug molecules and the signaling pathways, as well as to gain insight into the mechanisms of action of allosteric effectors, as indicated in Fig. 1. Present-day GPCR structural biology can work from a highly promising, recently established platform, since crystal structures of more than 20 different human GPCRs are available. These include the first structure of a human GPCR, β_2AR, which was determined in 2007 [1], a high-resolution structure of $A_{2A}R$ [2], a β_2AR–G-protein complex [3], and numerous GPCR complexes with pharmaceutical ligands (surveyed in [4]). The availability of the high-resolution GPCR scaffolds provided by X-ray crystallography enables precise design of complementary experiments for studies in non-crystalline milieus, for example, with NMR spectroscopy or single-molecule spectroscopy.

Our recent contributions to GPCR research

Using NMR spectroscopy in solution we pursued two different lines of research. Firstly, in support of subsequent structural studies, we characterized solutions of detergent-reconstituted integral membrane proteins (IMP). This included the screening of a wide array of newly designed detergents [5], the use of microcoil NMR equipment for measurements of translational diffusion coefficients to characterize the size of the IMP-containing particles [6], and applications of the results from [5] and [6] to optimize IMP solutions for NMR studies and for crystallization trials [7–9]. Secondly, we inserted fluorine-containing labels in strategic locations of GPCRs (Fig. 1). Observation of the NMR signals of these ^{19}F-NMR probes enabled the observation of conformational equilibria between long-lived inactive and activated states of GPCRs, as illustrated in Fig. 2 for β_2AR. This resulted in the

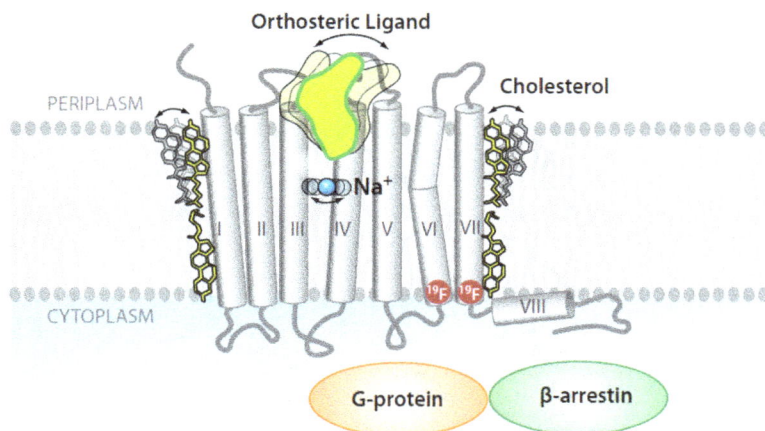

Fig. 1. Survey of GPCR structure and function. GPCR structures are characterized by seven trans-membrane helices. On the periplasmic surface there is an "orthosteric" ligand binding site (the arrow indicates mobility in the ligand binding site). Allosteric binding sites have been located along the periphery of the seven-helix bundle (cholesterol) and in the center of the bundle (Na^+). Binding of drug molecules to the orthosteric site elicits signaling to partner proteins in the cytoplasm, for example, G protein and β-arrestin. The two red circles indicate locations for fluorine-19 labels that have been used for ^{19}F-NMR observation (Fig. 2) of conformational changes associated with trans-membrane signaling between the orthosteric site ligand binding and the cytoplasmic protein surface. (Reproduced from [10]).

demonstration that signal transfer to G proteins and to β-arrestin are transmitted through different signaling pathways [11, 12].

Outlook to future developments of NMR spectroscopy with GPCRs

Exploratory experiments with several different GPCRs showed that the results on conformational transitions between long-lived active and inactive states, as obtained with β_2AR, cannot readily be generalized for other receptors. It will therefore be exciting to pursue comparative studies of different GPCRs with the ^{19}F-NMR probe approach. In addition, we have established an expression system that enables labeling of GPCRs with stable isotopes, so that the ^{19}F-NMR probe experiments can be followed up with high resolution structural information, which may include *de novo* structure determination of new GPCRs. For me it is particularly exciting that the glucagon receptor in its interactions with glucagon is one of our primary targets. Several decades ago, NMR studies of glucagon in aqueous solution [13] and in the water/lipid interface on the surface of detergent micelles [14, 15] yielded interesting observations on the conformations of this polypeptide hormone in different milieus. It will be exciting to compare these data with new results on the behavior of glucagon interacting with its receptor.

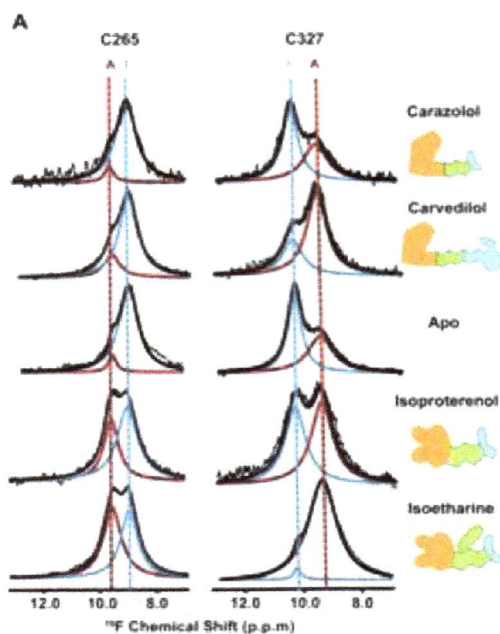

Fig. 2. ^{19}F-NMR observation of conformational equilibria in β_2AR. ^{19}F-NMR signals of ^{TET}C265 and ^{TET}C327 in the apo-form and in four drug complexes of single-residue TET-labeled β_2AR recorded at 280K. The experimental spectra (thin black line showing noise) have been deconvoluted into signals of an activated state (A, red) and an inactive state (I, blue) of β_2AR. The thick black line represents the sum of the signals I and A. The chemical structures of the bound ligands are schematically drawn on the right. (Reproduced from [10]).

Acknowledgments

GPCR research in my laboratory is supported by the NIH Common Fund grant P50 GM073197 and the NIH PSI:Biology grant U54 GM094618. I thank the colleagues and research associates listed in the references 4 to 15 for their courageous and devoted efforts in our membrane protein research program.

References

1. V. Cherezov, D. M. Rosenbaum, M. A. Hanson, S. G. Rasmussen, F. S. Thian *et al.*, *Science* **318**, 1258 (2007).
2. W. Liu, E. Chun, A. A. Thompson, P. Chubukov, F. Xu *et al.*, *Science* **337**, 232 (2012).
3. S. G. Rasmussen, B. T. DeVree, Y. Zou, A. C. Kurse, K. Y. Chung *et al.*, *Nature* **477**, 549 (2011).
4. R. Stevens, V. Cherezov, V. Katritch, R. Abagyan, P. Kuhn *et al.*, *Nat. Rev. Drug Discov.* **12**, 1 (2013).
5. Q. Zhang, R. Horst, M. Geralt, X. Ma, W. Hong *et al.*, *J. Amer. Chem. Soc.*

130, 7357 (2008).

6. P. Stanczak, R. Horst, P. Serrano, K. Wüthrich, *J. Amer. Chem. Soc.* **131**, 18450 (2009).

7. R. Horst, A. L. Horwich, K. Wüthrich, *J. Amer. Chem. Soc.* **20**, 16354 (2011).

8. R. Horst, P. Stanczak, P. Serrano, K. Wüthrich, *J. Phys. Chem. B* **116**, 6775 (2012).

9. R. Horst, P. Stanczak, R. C. Stevens, K. Wüthrich, *Angew. Chem. Int. Ed.* **52**, 331 (2013).

10. T. Didenko, J. J. Liu, R. Horst, R. C. Stevens, K. Wüthrich, *Curr. Opin. Struct. Biol.* **23**, 1 (2013).

11. J. J. Liu, R. Horst, V. Katritch, R. C. Stevens, K. Wüthrich, *Science* **335**, 1106 (2012).

12. R. Horst, J. J. Liu, R. C. Stevens, K. Wüthrich, *Angew. Chem. Int. Ed.* **52**, 331 (2013).

13. C. Bösch, A. Bundi, M. Oppliger, K. Wüthrich, *Eur. J. Biochem.* **91**, 209 (1978).

14. W. Braun, C. Bösch, L. R. Brown, N. Gō, K. Wüthrich, *Biochim. Biophys. Acta* **667**, 377 (1981).

15. W. Braun, G. Wider, K. H. Lee, K. Wüthrich, *J. Mol. Biol.* **169**, 921 (1983).

THE SEVEN TRANSMEMBRANE SUPERFAMILY

RAYMOND C. STEVENS

Department of Integrative Structural and Computational Biology, The Scripps Research Institute, 10550 North Torrey Pines Road, La Jolla, CA 92037, USA and the iHuman Institute, ShanghaiTech University, Shanghai, China

My view of the present state of research on the 7TM superfamily (GPCRs)

The 7-transmembrane (7TM) superfamily is the largest membrane protein family in the human genome, and the target of more than 30% of all therapeutic drugs that treat diseases afflicting mankind. At more than 800 members and responsible for 80% of all human cell signaling, the family includes well-known receptors often referred to as G protein-coupled receptors (GPCRs). In 1976, Henderson and Unwin were the first to observe the 7TM protein architecture with bacteriorhodopsin [1]. Palcewski and co-workers then reported structure-function studies of bovine rhodopsin in 2000 [2]. In 2007, structures of human β_2-adrenergic receptor (β_2AR) were determined by Kobilka, Schertler, and Stevens combining different approaches highlighting the first structures of a human GPCR and the first receptor controlled by diffusible ligands [3–5]. Since then, structures of more than 24 different 7TM receptors have been determined [6], each one impacting a different important area for basic and applied discovery. Furthermore, the structure of a GPCR G protein complex has been determined with an intracellular antibody [7]. At the current pace, it is highly likely that representative structures of all subfamilies will be determined in the coming few years enabling new discoveries in molecular recognition and illuminating the incredible diversity of these receptors. They should also provide deeper insights into structural basis for activation. Collectively, the structures and related follow-up biochemical and biological studies are dramatically changing how we think about the 7TM receptors. For example, with the observations of cholesterol and sodium binding to receptors and complemented by functional studies, we are now starting to view these molecules as allosteric machines controlled by many different molecules in the membrane beside just endogenous or synthetic pharmacological molecules [8].

My recent contributions to research on the 7TM superfamily (GPCRs)

Like many others in the field, my laboratory started pursuing structure-function studies of this superfamily more than 20 years ago, with no success until fairly recently. Combining our novel membrane protein technology developments created specifically for GPCR structure-function studies including miniaturization and automation, alongside Brian Kobilka's extensive knowledge of the protein chemistry of β_2AR, the structure of β_2AR was determined at a resolution that enabled the characterization of correct ligand orientation and many of the receptors novel features. Given our large investments in technology development, we were able to quickly follow-up with the structure of the human A_{2A} adenosine receptor ($A_{2A}AR$) [9], and also establish the human GPCR Network (http://gpcr.scripps.edu) that has successfully breached barriers to routine structural studies of human GPCRs [10]. Efficient determination of GPCR structures requires the development of a robust approach to enable a high likelihood of success and to provide a platform that is amenable to optimization and cost reduction based on accumulated experience. The approach used by the Scripps GPCR Network in the initial successful structure determinations of the β_2AR and the $A_{2A}AR$ led to the development of a pipeline infrastructure with feedback loops that could be used as a template for the determination of other GPCR structures. This pipeline is constantly being optimized using a 'family learning approach', which is similar to the approach used by the Structural Genomics Consortium [10] most notably with protein kinases [11].

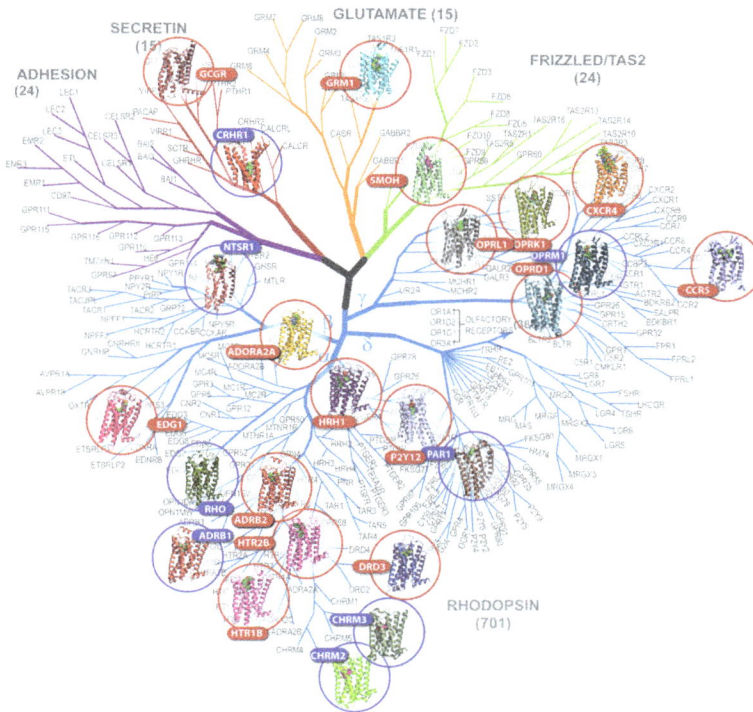

Fig. 1. Phylogenetic tree of GPCRs and published GPCR structures (24), those human GPCR structures published by the Scripps GPCR Network (16) are shown with red circles.

Since 2007, our process has determined the structures of two thirds of the available unique receptor structures (16 of the 25) and provided follow up insight into their structures and functions (Fig. 1). Each of the structures has been studied in collaboration with a scientific community leader who has extensive knowledge of the specific receptor system from a chemical, biological and pharmacological perspective. This highly collaborative approach is beneficial not only to ensure that the science is at the highest quality, and that insightful follow-up experiment and discovery are done, but it also establishes an avenue for extensive dissemination and communication of the data and results beyond the structural biology community. In addition, computational docking challenges with the community, specifically GPCR Dock 2008 [12], 2010 [13], and 2013 are also helping to monitor the field of computational GPCR modeling and docking, to provide us with feedback for prioritizing target selection and thus ensuring continued high impact to the biomedical community.

Diversity and modularity of GPCR structures [6, 14]

Although all GPCRs are characterized by a similar 7TM topology, the five major families of human GPCRs share little sequence identity (SI) and possess different extracellular N-terminal domains (SI < 10% in the 7TM domain). The largest and most diverse *Rhodopsin* family (also called Class A) consists of approximately 700 GPCRs in humans, and its subgroups α–δ have an SI of 25% or more. Each subgroup contains numerous subfamilies, and GPCR subtypes within these subfamilies share a higher SI ($\geq$ 30%) and often a common ligand selectivity. A survey of all solved GPCR structures is improving our understanding (and correcting misunderstandings) of the architecture of these receptors. GPCRs exhibit modularity in the extracellular module, which is responsible for binding diverse ligands and has much

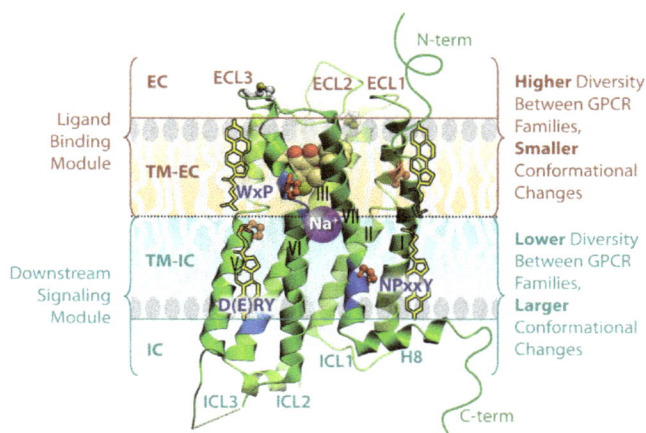

Fig. 2.   General architecture and modularity of GPCRs.

higher structural diversity than the intracellular module, which is involved in binding a limited number of downstream effectors including G proteins, arrestins, and kinases and hence shows higher sequence/structure conservation (Fig. 2).

A highly diverse repertoire of structural features in the ligand-binding pockets of different GPCR subfamilies apparently reflects evolutionary pressure to selectively recognize ligands that vary greatly in shapes, sizes and electrostatic properties. Even within the preserved overall 7TM bundle architecture, there is a remarkable diversity of shapes and features of binding pockets between GPCR subfamilies, manifested in both side-chain diversity and variations in backbone conformations of TM helices and extracellular loops.

As an example, the structures of opioid and serotonin receptors involved in binding of psychedelic drugs have been studied. In collaboration with Bryan Roth at UNC, these structures highlight the diversity in the binding pockets and now provide a molecular basis for answering the question as to why the nociceptin receptor (NOP) does not bind morphinans, unlike the other classical opioid receptors, or how molecules like LSD elicit distinct and biased responses by closely related serotonin receptors. The structures show that replacement of only a few key residues in the ligand binding pocket region results in conformational changes in the structure including shifts in helices V/VI leading to different pharmacological behavior.

Crystal structures of GPCRs also provide a robust three-dimensional structural framework for computational modeling of receptor dynamics and oligomerization state, as well as for ligand docking and virtual ligand screening (VLS). The growing number of structure-based VLS studies demonstrates encouragingly high hit rates (20–70%) in the identification of new ligand chemotypes as lead compounds for $A_{2A}AR$, CXCR4 chemokine, D_3 dopamine, and H_1 histamine receptors, as well as in lead optimization. Following their successful applications to kinases, proteases and other protein target families, structure-based ligand screening technologies are now becoming an important part of the GPCR drug discovery process in pharmaceutical and biotechnology companies. For example, the $S1P_1$ structure [15] is now being used by the biotech startup Receptos in their drug development efforts targeting multiple sclerosis and irritable bowel syndrome with two different Phase II/III clinical trials currently underway.

7TMs as allosteric machines [8]

Pharmacological responses of GPCRs can be fine-tuned by allosteric modulators. Structural studies providing such observations have been limited by the medium resolution of past GPCR structures. Thus, we reengineered the human $A_{2A}AR$ by replacing its third intracellular loop with apocytochrome $b_{562}RIL$ and solved its structure to 1.8 Å resolution. This high-resolution structure allowed us to identify 57 ordered water molecules inside the receptor comprising three major clusters. The central cluster harbors a putative sodium ion bound to the highly conserved aspartate residue $Asp^{2.50}$. Additionally, two cholesterol moieties stabilize the conformation of helix VI. These high-resolution details shed light on the role of water molecules, sodium ions, and lipids/cholesterol in GPCR stabilization and function.

Biased signaling pathways in 7TMs characterized by ^{19}F-NMR [16]

Extracellular ligand binding to GPCRs modulates G protein and β-arrestin signaling by changing the conformational states of the cytoplasmic region of the receptor. In collaboration with the Wüthrich laboratory at TSRI, we placed site-specific ^{19}F-NMR labels in the β_2AR in complexes with various ligands, and observed that the cytoplasmic ends of helices VI and VII adopt two major conformational states. Changes in the NMR signals reveal that agonist binding primarily shifts the equilibrium toward the G protein–specific active state of helix VI. In contrast, β-arrestin–biased ligands predominantly impact the conformational states of helix VII. The selective effects of different ligands on the conformational equilibria involving helices VI and VII provide insights into the long-range structural plasticity of β_2AR in partial and biased agonist signaling. The use of ^{19}F-NMR may thus prove to be a widely applicable tool for structure-activity relationship studies of GPCR ligands.

Outlook for research on 7TM superfamily (GPCRs)

Over the past 40 years, pharmacologists have treated GPCRs as black boxes that recognize and respond to certain ligands, and narrowly defined them as either coupled to G proteins, or more recently coupled to arrestin. However, we now know there are more than 200 other intracellular proteins that interact with the 7TM receptor family and we are only beginning to understand signaling. Furthermore, with only 24 known structures, this represents only a few percent of the total number of receptors, each one characterized revealing great new discoveries no one expected.

Our ultimate goal is to understand how complexes, cells, organs, and species work at a molecular level. To accomplish this endeavor and better model how organisms work, wide ranging types of data need to be integrated together including computational approaches. Human cell signaling with 7TMs at the center, are a model system to investigate, with broad reaching impact on many different fields of basic and applied science to help mankind.

Acknowledgments

This work was funded by the NIH Common Fund for Structural Biology grant P50 GM073197 and NIH PSI:Biology grant U54 GM094618. This work could not have been accomplished without the incredible teamwork effort by Enrique Abola, Vadim Cherezov, Vsevolod Katritch, Peter Kuhn, Angela Walker, Kurt Wüthrich and all of the students and postdocs we have worked with over the past 20 years. I am also very grateful to our collaborators Ruben Abagyan, Chris de Graaf, Tracy Handel, Adriaan IJzerman, So Iwata, Ken Jacobson, Jonathan Javitch, Brian Kobilka, and

Bryan Roth. Lastly, I would like to thank Richard Lerner, Peter Schultz, Ian Wilson, and Jim Wells for being so consistently supportive of high risk/high reward research after more than 20 years of effort.

References

1. R. Henderson, P. N. Unwin, *Nature* **257**, 28 (1975).
2. K. Palczewski, T. Kumasaka, T. Hori, C. A. Behnke, H. Motoshima *et al.*, *Science* **289**, 739 (2000).
3. V. Cherezov, D. M. Rosenbaum, M. A. Hanson, S. G. Rasmussen, F. S. Thian *et al.*, *Science* **318**, 1258 (2007).
4. S. G. Rasmussen, H. J. Choi, D. M. Rosenbaum, T. S. Kobilka, F. S. Thian *et al.*, *Nature* **450**, 383 (2007).
5. D. M. Rosenbaum, V. Cherezov, M. A. Hanson, S. G. Rasmussen, F. S. Thian *et al.*, *Science* **318**, 1266 (2007).
6. V. Katritch, V. Cherezov, R. C. Stevens, *Annu. Rev. Pharm. Toxicol.* **53**, 531 (2013).
7. S. G. Rasmussen, B. T. DeVree, Y. Zou, A. C. Kruse, K. Y. Chung *et al.*, *Nature* **477**, 549 (2011).
8. W. Liu, E. Chun, A. A. Thompson, P. Chubukov, F. Xu *et al.*, *Science* **337**, 232 (2012).
9. V. P. Jaakola, M. T. Griffith, M. A. Hanson, V. Cherezov, E. Y. Chien *et al.*, *Science* **322**, 1211 (2008).
10. R. C. Stevens, V. Cherezov, V. Katritch, R. Abagyan, P. Kuhn *et al.*, *Nat. Rev. Drug Discov.* **12**, 25 (2013).
11. S. Knapp, P. Arruda, J. Blagg, S. Burley, D. H. Drewry *et al.*, *Nat. Chem. Biol.* **9**, 3 (2013).
12. M. Michino, E. Abola, J. S. Dixon, J. Moult, R. C. Stevens *et al.*, *Nat. Rev. Drug Discov.* **8**, 455 (2009).
13. I. Kufareva, M. Rueda, V. Katritch, R. C. Stevens, R. Abagyan, *Structure* **19**, 1108 (2011).
14. V. Katritch, V. Cherezov, R. C. Stevens, *Trends Pharmacol. Sci.* **33**, 17 (2012).
15. M. A. Hanson, C. B. Roth, E. Jo, M. T. Griffith, F. L. Scott *et al.*, *Science* **335**, 851 (2012).
16. J. J. Liu, R. Horst, V. Katritch, R. C. Stevens, K. Wüthrich *et al.*, *Science* **335**, 1106 (2012).

NANOBODIES FOR THE STRUCTURAL AND FUNCTIONAL INVESTIGATION OF GPCR TRANSMEMBRANE SIGNALING

ELS PARDON and JAN STEYAERT

Structural Biology Brussels, Vrije Universiteit Brussel, Pleinlaan 2, 1050, Brussel and Structural Biology Research Centre, VIB, Pleinlaan 2, 1050, Brussel

My view of the present state of research on GPCRs and transporters: ligands, cofactors, drug development

Many membrane proteins including GPCRs and transporters are conformationally complex molecules. These are exiting times because we are starting to collect high resolution structures of the key functional conformers for a number of model systems. Most important, we are also witnessing the development of appropriate biophysical methods to study the (thermo)dynamics of the conformational transitions and the way ligands, cofactors, lipids or other proteins disturb these conformational equilibria. Ultimately, we will have to incorporate the contribution of the electrochemical membrane potentials to fully understand transmembrane transport and signaling. No doubt that the remarkable progress that we are making in these fields will lead to better drugs.

My recent research contributions to GPCRs and transporters: ligands, cofactors, drug development

Polytopic membrane proteins such as GPCRs and transporters are dynamic proteins that exist in an ensemble of functionally distinct conformational states. Crystallogenesis typically traps the most stable low energy states, making it challenging to obtain agonist bound active-state X-ray structures of GPCRs. Stabilization of an active conformation of a GPCR can be achieved in different ways. The most physiologic approach is to use a native signaling partner such as a G protein. An alternative to using a G protein is to identify another binding protein that can stabilize the same conformational state.

Antibodies evolved to bind to a diverse array of protein structures with high affinity and specificity. Last years, we generated Nanobodies (Nbs) that selectively recognize the active states of the $\beta2$ adrenergic receptor ($\beta2AR$) [1] and muscarinic receptor 2 (M2R) [2]. Such Nbs that faithfully mimic the effects of the G proteins Gs or Gi were used to obtain diffraction quality crystals and to solve the very first structures of the active agonist-bound states of $\beta2AR$ or M2R. More interesting, we also identified nanobodies that stabilize the $\beta2AR\cdot Gs$ complex [3]. One of these Nbs was used to obtain the high-resolution crystal structure of this complex, providing the first molecular view of transmembrane signaling by a GPCR.

Our work illustrates the power of the Nanobody platform for GPCR research [4]. Nanobodies are the small (15 kDa) and stable single domain fragments harboring the full antigen-binding capacity of the original heavy chain only antibodies that naturally occur in Camelids. Because of their unique three-dimensional structure, nanobodies have access to cavities or clefts on the surface of proteins. Our Nanobody discovery platform has the competitive advantage over other recombinant crystallization chaperones that the cloned Nanobody library represents the full collection of the naturally circulating humoral antigen-binding repertoire of heavy chain antibodies, contrary to combinatorial libraries of conventional antibody fragments. Because Nbs are encoded by single exons, the full antigen-binding capacity of *in vivo* matured antibodies can be cloned and efficiently screened for high affinity binders, allowing one to fully exploit the humoral response of large mammals against native antigens.

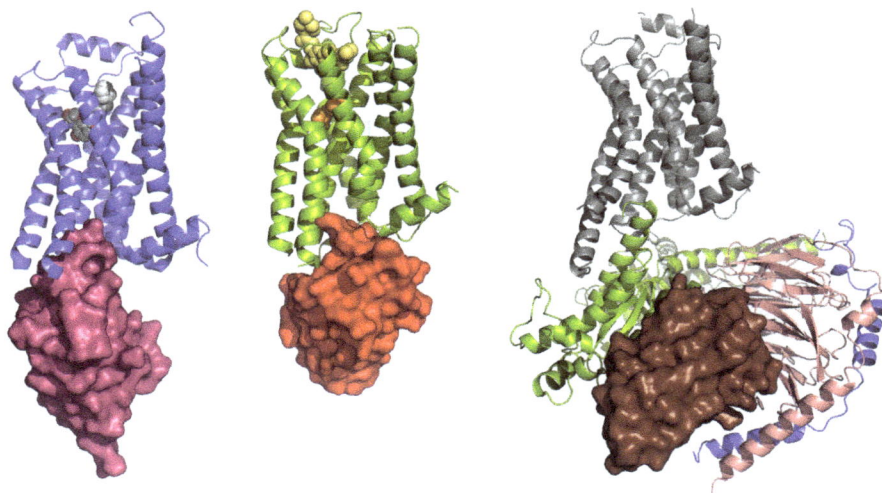

Fig. 1. Nanobody-assisted X-ray crystallography of GPCR active states. (Left) Agonist bound active state of β2AR (ribbon representation) stabilized by Nb80 (surface representation); (Middle) Active state of M2R (ribbon representation) bound to an agonist and a positive allosteric modulator that is stabilized by Nb9-8 (surface representation); (Right) β2AR-Gs transmembrane signaling complex (ribbon representation) stabilized by Nb35 (surface representation).

Outlook to future developments of research on GPCRs and transporters: ligands, cofactors, drug development

Last years, we demonstrated that Nanobodies are exquisite tools for the structural investigation of conformationally complex membrane proteins. Future work will focus on Nanobodies that stabilize different GPCR conformational states including active, inactive and biased conformations of the receptors, aiming at understanding how different ligands elicit other biological responses. Our work on the β2AR-Gs complex indicates that nanobodies can also be used to stabilize protein complexes. In the coming years we will develop methods to select nanobodies that stabilize transient protein-protein interactions and focus on GPCR-G protein (Gs, Gi, Go, ...), GPCR-arrestin and GPCR-kinase complexes and other signaling complexes involving β-arrestin, ERK2 and C-Src amongst others.

Acknowledgments

The Steyaert lab was supported by the Fonds Wetenschappelijk Onderzoek-Vlaanderen through research grants G011110N and G049512N, Innoviris Brussels through Impulse Life Science program BRGEOZ132, the Belgian Federal Science Policy Office through IAP7-40 and by SBO program IWT120026 from the Flemish Agency for Innovation by Science and Technology.

References

1. S. G. Rasmussen, H. J. Choi, J. J. Fung, E. Pardon, P. Casarosa *et al.*, *Nature* **469**, 175 (2011).
2. A. C. Kruse, A. M. Ring, A. Manglik, J. Hu, K. Hu *et al.*, *Nature* **504**, 101 (2013).
3. S. G. Rasmussen, B. T. DeVree, Y. Zou, A. C. Kruse, K. Y. Chung *et al.*, *Nature* **477**, 549 (2011).
4. J. Steyaert, B. K. Kobilka, *Curr. Opin. Struct. Biol.* **21**, 567 (2011).

THE HIDDEN PHARMACOLOGY OF THE HUMAN GPCR-OME

BRYAN L. ROTH

Department of Pharmacology and Division of Chemical Biology and Medicinal Chemistry
University of North Carolina Chapel Hill Medical School, 4072 Genetic Medicine Building
Chapel Hill, NC 27599, USA

My view of the present state of research on GPCR drug discovery

With nearly 900 members, the human GPCR-ome comprises the largest family of druggable targets in the genome although most members have not been chemically interrogated.

G protein-coupled receptors (GPCRs) constitute the single largest class of druggable targets in the human genome [1]. The human genome includes genes for ~ 900 olfactory and non-olfactory GPCRs and careful analyses estimate the number of non-olfactory GPCRs to be between 342 and 356 (http://www.iuphar-db.org/index.jsp) [2]. Of these, approximately 38% are classified as "orphans", *i.e.*, their natural (endogenous) ligands remain unknown (see IUPHAR GPCR Database at: http://www.iuphar-db.org/index.jsp). Additionally a significant proportion of the non-orphan GPCRs remain incompletely or inaccurately characterized with respect to the ligands that modulate their activity.

My recent contributions to the area of GPCR drug discovery

Over the past several years, my colleagues and I have shown that approved medications and their metabolites exert both known and unknown effects via unanticipated interactions with GPCRs [3–5]. In some cases, such as the infamous 'fen/phen' diet drug fiasco, an unanticipated interaction of a drug metabolite (*e.g.*, norfenfluramine) with a GPCR (*e.g.*, the 5-HT$_{2B}$ serotonin receptor) led to tremendous both human suffering and disasterous economic consequences [6]. In recent work done in close collaboration with Ray Stevens' group we were able to provide a molecular explanation for how drugs can interact with these receptors and induce serious side-effects [7].

Most recently my group has been interested in using cheminformatics and other *in silico* approaches to facilitate GPCR drug discovery and genome-wide technologies to chemically interrogate the GPCR-ome and these will be discussed separately.

Identifying ligands for orphan and non-orphan GPCRs

Considerable effort has been devoted to finding both endogenous and synthetic ligands for non-orphan GPCRs. For instance, we provided the initial validation data

[8] and later explored the utility of the **S**imilarity **E**nsemble **A**pproach (SEA; [8]) for identifying GPCRs as unanticipated targets of approved medications and chemical probes [5, 9–11] — all of these studies in collaboration with Brian Shoichet's group at UCSF. We have also used *in silico* approaches for drug design and lead optimization. The most recent example of this was the successful design of multi-target drugs for aminergic GPCRs [12] using Baysean-based approach. In general, we have found these particular approaches to be useful for predicting small molecule interactions with molecular targets for which there is a large amount of annotation as to small molecule interactions. Unfortunately, for much of the GPCR-ome, little chemical matter is annotated (see Fig. 1).

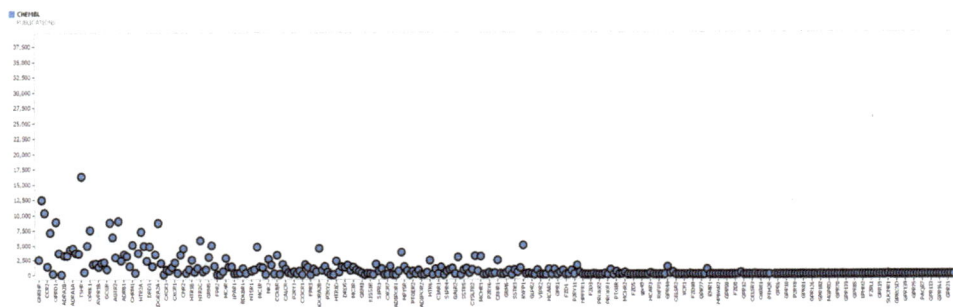

Fig. 1.   Annotated small molecules for the GPCR-ome. Shown is a scatter-gram listing of representative members of the GPCR-ome on the X-axis and # of small molecules annotated from Chembl. As can be seen, most of the GPCR-ome has relatively sparse annotation.

My lab has had some success with identifying ligands for orphan receptors [13–15] and this remains an active area of investigation.

Genome-wide technologies to chemically interrogate the GPCR-ome

Orphan GPCRs represent a particularly challenging target class because conventional GPCR screening approaches require either *a priori* knowledge of the particular G protein(s) to which the orphan GPCR preferentially couples or the availability of positive control compounds (either agonists or antagonists) Various means to circumvent these challenges to provide so-called 'universal' GPCR screening platforms include the use of chimeric (or universal) G proteins, label-free impedance-based approaches, arrestin-based screening platforms, melanosome-based approaches and others (see [16] for review), although these have not been applied on a large scale to identify ligands for orphan GPCRs.

We have been able to recently adapt an arrestin-based platform initially for D2-dopamine receptor screening [17] and later for other receptors including k-opioid [18], serotonergic [7] and now nearly every GPCR in the human genome (see online resource at: http://pdsp.med.unc.edu/PDSP%20Protocols%20II%202013-03-28.pdf; Kroeze *et al.*, *manuscript in preparation*).

Because the bulk of the GPCR-ome is not annotated from a chemistry perspective, developing genome-wide platforms to interrogate the GPCR-ome will greatly accelerate our ability to develop drugs and small molecule probes for the 'hidden GPCR-ome'. Additionally, as new chemical annotation will accelerate *in silico* approaches such information will allow us to reveal the subterranean (or hidden) pharmacology of known drugs and bioactive compounds. Revealing such information will have a tremendous impact on not only drug discovery but also safety pharmacology and, ultimately, the health and well-being of humanity.

Outlook to future developments of research on GPCR drug discovery and chemical biology

Illuminating the pharmacology of the hidden GPCR-ome will transform drug discovery, although this will require novel and highly scalable technologies.

Acknowledgments

The work described here was supported by grants from the NIH (RO1MH61887; DA017204; an NIDA EUREKA Award; U19MH82441 and the NIMH Psychoactive Drug Screening Program)

References

1. A. L. Hopkins, C. R. Groom, *Nat. Rev. Drug Discov.* **1**, 727 (2002).
2. R. Fredriksson, H. B. Schioth, *Mol. Pharmacol.* **67**, 1414 (2005).
3. R. B. Rothman, M. H. Baumann, J. E. Savage, L. Rauser, A. McBride *et al.*, *Circulation* **102**, 2836 (2000).
4. B. L. Roth, D. J. Sheffler, W. K. Kroeze, *Nat. Rev. Drug Discov.* **3**, 353 (2004).
5. M. J. Keiser, V. Setola, J. J. Irwin, C. Laggner, A. I. Abbas *et al.*, *Nature* **462**, 175 (2009).
6. B. L. Roth, *N. Engl. J. Med.* **356**, 6 (2007).
7. D. Wacker, C. Wang, V. Katritch, G. W. Han, X. P. Huang *et al.*, *Science* **340**, 615 (2013).
8. M. J. Keiser, B. L. Roth, B. N. Armbruster, P. Ernsberger, J. J. Irwin *et al.*, *Nat. Biotechnol.* **25**, 197 (2007).
9. E. Gregori-Puigjané, V. Setola, J. Hert, B. A. Crews, J. J. Irwin *et al.*, *Proc. Natl. Acad. Sci. USA* **109**, 11178 (2012).
10. P. N. Yadav, A. I. Abbas, M. S. Farrell, V. Setola, N. Sciaky *et al.*, *Neuropsychopharmacol.* **36**, 638 (2011).
11. C. Laggner, D. Kogel, V. Setola, A. Tolia, H. Lin *et al.*, *Nat. Chem. Biol.* **8**, 144 (2011).
12. J. Besnard, G. F. Ruda, V. Setola, K. Abecassis, R. M. Rodriguiz *et al.*, *Nature* **492**, 215 (2012).

13. B. L. Roth, S. C. Craigo, M. S. Choudhary, A. Uluer, F. J. Monsma Jr. *et al.*, *J. Pharmacol. Exp. Ther.* **268**, 1403 (1994).

14. T. Nguyen, D. A. Shapiro, S. R. George, V. Setola, D. K. Lee *et al.*, *Mol. Pharmacol.* **59**, 427 (2001).

15. F. Oury, G. Sumara, O. Sumara, M. Ferron, H. Chang *et al.*, *Cell* **144**, 796 (2011).

16. J. A. Allen, B. L. Roth, *Annu. Rev. Pharmacol. Toxicol.* **51**, 117 (2011).

17. J. A. Allen, J. M. Yost, V. Setola, X. Chen, M. F. Sassano *et al.*, *Proc. Natl. Acad. Sci. USA* **108**, 18488 (2011).

18. H. Wu, D. Wacker, M. Mileni, V. Katritch, G. W. Han *et al.*, *Nature* **485**, 327 (2012).

STRUCTURES AND REACTION MECHANISMS OF ABC TRANSPORTERS

KASPAR LOCHER

Institute of Molecular Biology and Biophysics, Department of Biology, ETH Zürich;
Schafmattstrasse 20, HPK D17,
8093 Zürich, Switzerland

My view of the present state of research on the structure and mechanism of ATP-binding cassette (ABC) transporters

ABC transporters are primary active transporters that have an important role in the physiology and pathophysiology of many organisms. Fueled by hydrolysis of ATP, these proteins allow nutrients, metabolites, and noxious substances to be moved across lipid bilayers [1]. In bacteria, ABC importers facilitate the uptake of nutrients from the environment via high-affinity uptake pathways, thus enabling pathogenic bacteria to acquire essential (micro-)nutrients even when they are scarce. ABC exporters catalyze the extrusion of components required for the biosynthesis of the bacterial cell envelope and often contribute to drug resistance, which can pose problems in combatting pathogenic bacteria. In eukaryotes, ABC transporters are generally exporters that move diverse substrates across cellular or organelle membranes. Various hereditary human diseases have been associated with the dysfunction of ABC transporters, and their over-expression in tumor cells can cause multi-drug resistance, a serious impediment for the treatment of cancer [2].

A great deal of progress has been made over the past 15 years in the structural and mechanistic investigation of ABC transporters. Specifically, X-ray crystallographic studies have revealed their folds and have visualized distinct conformations. The structural insight was useful for interpreting functional, biophysical, and biochemical studies and proved invaluable in understanding the coupling mechanisms of several model systems. While the first structures described ABC transporters of bacterial or archaeal origin, advances in over-expression and purification of eukaryotic membrane proteins have allowed the mechanisms of ABC transporters related to human diseases to be tackled.

My recent research contributions to the structural and mechanistic investigations of ATP-binding cassette transporters

Research in my group at ETH Zürich has focused on the structure and mechanism of various membrane proteins, with a strong emphasis on ABC transporters. We have spent considerable efforts at establishing and optimizing the over-expression and purification of ABC transporters in bacterial host cells, and more recently also in eukaryotic cells. Using X-ray crystallography, we have determined the first structures of type I and type II ABC importers, as well as that of the first ABC exporter [3–5]. The structures of these bacterial proteins were useful for studying the transport mechanisms not only of bacterial, but also of eukaryotic ABC transporters. Whereas the transmembrane domains of transporters belonging to different subfamilies revealed distinct folds, the "motor domains" (nucleotide-binding domains) share a conserved fold and therefore also some similarity in binding and hydrolyzing ATP. We identified an essential structural motif termed "coupling helix", which is present in the transmembrane domains of all ABC transporters whose structures were solved to date. This short α–helix is nestled into a groove formed at the subdomain boundaries of the nucleotide-binding domains. The coupling helices provide the interface between the motor and the transmembrane domains, which is critical for the transmission of conformational changes and thus for coupling ATP binding and hydrolysis to substrate transport (Fig. 1).

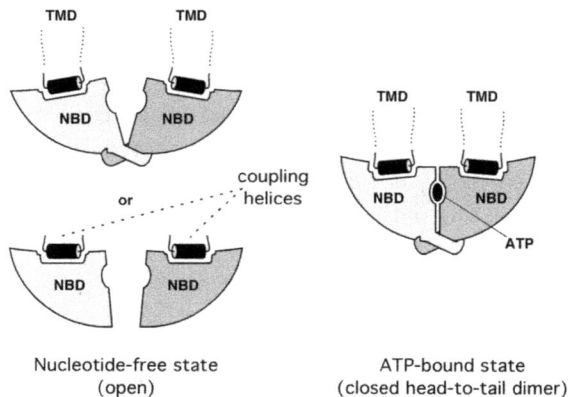

Fig. 1.   Transmission interface in ABC transporters. The coupling helices are a conserved architectural motif of transmembrane domains. They are pushed together during ATP binding and pulled apart upon release of the hydrolysis products. This motion constitutes the power stroke of ABC transporters. The nucleotide binding domains form a closed, head-to-tail sandwich in the ATP-bound state. In the nucleotide-free state, they may either be spatially separated (lower left) or held together by C-terminal motifs or domains (upper left).

In 2007, by comparing distinct structures of transporters trapped in the absence of nucleotide or in the ATP-bound state, we deduced a coupling mechanism that could rationalize substrate uptake by type I ABC importers and extrusion by ABC exporters [6]. The mechanism predicted that in the nucleotide-free state, ABC transporters adopt an inward-facing conformation, in which they release their substrates into the cytoplasm (importers) or recruit them from the cytoplasm or from the inner leaflet of the lipid bilayer (ABC exporters). In the ATP-bound conformation, the transporters adopt an outward-facing conformation, either releasing substrates (exporters) or acquiring them from the cognate binding proteins (importers). Alternating access was thus coupled to an ATP-dependent, motor domain-driven reaction.

We have recently captured the vitamin B12 transporter BtuCD-F from *E. coli* (a type II ABC importer) in an intermediate state with bound AMPPNP, a non-hydrolysable ATP analog. Combined with functional data, this allowed a full transport cycle to be formulated. To our surprise, the structural rearrangements in the transmembrane domains of BtuCD-F are distinct from those of type I ABC importers or ABC exporters, suggesting that while the coupling helices move in a similar way, the conformational rearrangements in the transmembrane domains of ABC transporters differ [7]. In BtuCD-F, two distinct cytoplasmic gates (compared to only one in other ABC systems) operate, allowing large substrates such as cobalamin (1355 Da in mass) to be transported without ion leakage to occur.

Outlook to future developments in the structural

Despite recent progress, the reaction mechanisms of ABC transporter–catalyzed processes are only understood for select substrates, all of which are water-soluble. For hydrophobic substrates, the transport mechanisms are poorly understood. This includes the transporters of toxic compounds, prescription or environmental drugs, or (glyco)lipid flippases. Most of these reactions are highly relevant in pathophysiology. For example, ABC transporter–catalyzed drug extrusion from endothelial cells of the blood-brain barrier protects the brain from toxic insult, but simultaneously impedes the delivery of drugs for neurological diseases. The study of structures and mechanisms of ABC transporters of substrates that have a low hydrophilicity (drugs, lipids) will therefore likely unravel novel transport mechanisms. Such studies should have a potential value for future diagnostic or even therapeutic applications besides providing insight into elusive biological reactions.

Acknowledgments

Research in my laboratory was generously funded by the Swiss National Science Foundation (SNF), NCCR Structural Biology, NCCR TransCure, Swiss Cancer League OncoSuisse, and ETH Zürich.

References

1. I. B. Holland, S. P. C. Cole, K. Kuchler, C. F. Higgins, *ABC Proteins: From Bacteria to Man* (Academic, London, 2003).
2. M. M. Gottesman, T. Fojo, S. E. Bates, *Nat. Rev. Cancer* **2**, 48 (2002).
3. R. J. P. Dawson, K. P. Locher, *Nature* **443**, 180 (2006).
4. K. Hollenstein, D. C. Frei, K. P. Locher, *Nature* **446**, 213 (2007).

5. R. N. Hvorup, B. A. Goetz, M. Niederer, K. Hollenstein, E. Perozo *et al.*, *Science* **317**, 1387 (2007).
6. R. J. P. Dawson, K. Hollenstein, K. P. Locher, *Mol. Microbiol.* **65**, 250 (2007).
7. V. M. Korkhov, S. A. Mireku, K. P. Locher, *Nature* **490**, 367 (2012).

SESSION 4: GPCRs AND TRANSPORTERS: LIGANDS, COFACTORS, DRUG DEVELOPMENT

CHAIR: GUNNAR VON HEIJNE

AUDITORS: C. GOVAERTS[1], M. PARMENTIER[2]

(1) S.F.M.B., Université Libre de Bruxelles, CP206/2, Campus Plaine, Bd du Triomphe, 1050 Brussels, Belgium
(2) WELBIO and IRIBHM, Université Libre de Bruxelles, Campus Erasme, 808 route de Lennik, 1070 Brussels, Belgium

Discussion among panel members

<u>Gunnar von Heijne</u>: All right, thank you. That was very intense. I don't think there's a problem for the people in the panel to now come up with a load of questions to discuss but maybe I can just kick off by going back to one of the questions I put up on my initial slide which is — if we start with the GPCR field — where would you say that, you know, think about the basic mechanism of the GPCRs, what could you say that we kind of know for sure now about the mechanisms and where are we kind of speculating and hypothesizing?

<u>Bryan Roth</u>: We don't know anything (laughs). So, I think we have some snapshots of some of the initial potential complexes but I think, as was mentioned, we really have no idea at the level of detail that we'd like to, about how the arrestin-GPCR complex is organized and how it functions. And then downstream from that there are, as Ray mentioned, potentially hundreds of other interacting proteins, which also are very important for signal transduction. I don't think there are any complexes of those yet that have been reported with the GPCRs, so we'd like to see those from a structural perspective. And then, there's this whole web of interactions inside the cell which is really a seamless web, which changes on a sort of millisecond basis and that is, I would say, completely unknown in terms of what is going on. So I would say, despite magnificent progress in structural biology, we still are just at the very earliest stages of getting any sense of what's really going on. So, I would say the Golden Age is yet to come. But we're getting there.

<u>Gebhard Schertler</u>: So for me it's quite exciting. You know, a few years ago, we were dreaming of seeing a ligand at all in a GPCR, and all we had were models, so now we have really... every day we can see ligands, if we put the money in. So in terms of drug discovery cycle, we are really in a much better situation. If you want to try that, if we look at the understanding of the system, I showed you that we very well understand what the role of some of the conserved residues is, I think we also have seen there is a kind of general activation mechanism but there's also

another very puzzling thing about GPCRs. They are actually very small, right? They're tiny. But they interact with arrestin, G protein, can have different signaling outputs. They will interact with the kinase, responding on the ligand you have and then you will get a different phosphorylation on the receptor. So there's something going on and so far the answer is that the different conformations that are possible, are the answer. But you know, let's look at this a bit more complicated. If it's chewing gum, then nothing is determined anymore, all right? So, how could it be a bit more systematically analyzed? So there are different answers. I showed you very conserved residues, so are there distinct states that we stabilize? Sometimes they have been called microswitches, so I think there are elements which are fixed but because there are several of them they can be combinatorially switched, so I think in this way there's some underlying principle which allows us to give different outputs. But it is not that every ligand induces a completely independent signaling pathway, it uses various sub-elements, and sub hydrogen-bond networks in the GPCR to actually give you this new conformation. And this needs to still be proven, you know, we didn't see so many states in detail that we really can see that, I mean. We have an impressive number of structures and sometimes it's also a little bit hard to correlate the structures with what state it is. So we get the structure, we kind of look afterwards: What state it is? I have shown you a tryptophan on helix five, which is sometimes in and out. That's a little bit an indicator for at which point you're in the activation cycle, or which state you have, but it's actually not very strong and not 100% correlated but it be can used a little bit for that. So very often, after the structural biology, we imply what state it was. It's not — even the complexes — it is not so clear what states exactly are relative to the other measurements. And I think this is a big problem. I think what would be very great is to know much more about the interactions with the kinases, there we have no structures, obviously we have a weaker interaction. Will it be really a full complex or is it a transient complex, can it be stabilized in the future? That would be very interesting. I think we will have very soon arrestin structures and we will see how they trigger actually other elements or the same element in the GPCR, but I think that will be resolved actually very, very soon.

Ray Stevens: I have a question for Brian, and this stems from yesterday's discussions on the enzymes and stability. Gebhard just talked about stability. We don't really characterize, we don't consider stability to be a pharmacological property. But stability, how long a GPCR or a transporter resides in the membrane is incredibly important. We look at PK/PD effects of drugs. Stability obviously is very important for the actual protein targets. So why is receptor stability ignored?

Bryan Roth: I think because GPCRs are generally in terms of their residence in the membrane, and signaling, they're not regulated by, so much by ubiquitination and intracellular proteolysis, that sort of things. It's more a vesicular recycling mode of regulation.

Ray Stevens: Well, that's been our view. It's all about trafficking, it's about insertion, trafficking. But maybe it's also about residence time.

Bryan Roth: Yeah, that's an interesting idea and I don't know what to say about that. I can sort of give you an anecdote from my past. So Anne Lesage is here as an auditor and she was a post-doc with me when I was at Stanford. I had a failed postdoc there and was very lucky to get an academic position, but the reason the postdoc failed was that I thought that GPCRs were regulated transcriptionally and this was sort of early in the era when GPCRs were being cloned. And basically what I found out was that transcriptional regulation is not the way they are regulated and all the regulation apparently is post-translational and as far as is known, mainly, basically turning the reactivity on and off, but not so much stabilizing their activity. Now it's possible that there are things out there we don't know about, related to GPCRs. But, I'll keep an open mind. It's possible that this is an area that has never been interrogated appropriately, but I'm not aware of anything related to that with GPCRs.

Gebhard Schertler: So, another problem we have from the structural biology: we work on very strongly reduced systems. So we have removed N-termini, C-termini, inserted loops, and we don't think that they are biologically not important. Some of them, have clear features from intrinsically disordered structures, like the C-terminus of rhodopsin. And I think in some GPCR situations they will be important to assemble or co-assemble or bringing proximity, other factors, which are important in this signaling network. And we know very little about it, and actually we know also not so well how to study this. All what people have done is: they have found hundreds of things that actually co-precipitate with GPCRs, so there are full lists of them. But there's actually not a knowledge created from that. And this is potentially a great sea of new waves of pharmacological intervention. So because we are not knowing or understanding these signaling pathways and the signaling routine, we really also can't address this so well. Biased-signaling is great but, at the moment, it's a molecular phenomenon. We are not linking it very well to human disease phenomenology, all right? So we don't have fantastic examples where biased-signaling is clearly the cause of a side effect or clearly the cause of a cure, or clearly the cause of a disease. So it is much more complicated. However, finding this effect might explain certain unexplained things so far. We might think in another way about it. But we're not immediately at this stage where we can say: we want to bias something and then we get this medical outcome. So a lot to learn there. And it has to be partially done with patients or animal models and it cannot be done in cells or on the molecular side at all.

Bryan Roth: So there is one example of where biased signaling is unambiguously important. And this relates to the niacin flush. I think I'm probably the only physician here right? But maybe some of you have taken niacin for lowering cholesterol.

It is very effective for lowering cholesterol and fortunately as effective actually as statins. The problem with niacin now is that many people when they take it they get a very severe flush, and it turns out that the flush is actually due to arrestin signaling so Bob Lefkowitz and colleagues published a beautiful paper in the Journal of Clinical Investigation a few years ago where they were able to dissociate the flush side effect of niacin from the lipid lowering effects of niacin, and for actually what was previously an orphan GPCR. And so the idea, sort of going forward, has been to basically create a G protein-biased agonist for this particular GPCR, that clusters all lowering agents which don't have this effect of flushing. So that's one that's key. There are a number of compounds in clinical trials, now from Travina, which are G protein-biased compounds for the angiotensin receptor and we will probably know about them shortly. And then, in full disclosure, my lab has a large collaboration with Pfizer pharmaceuticals to create arrestin-biased compounds for treating schizophrenia and those, at least in animal models are extraordinarily effective, and work only through arrestin. So I think these compounds certainly are percolating up and this is a really active area of drug discovery right now, and it won't be too long I think before we have this compounds. That being said, it's just one aspect of bias. So there all these others signaling pathways beyond arrestin that are not...

Ray Stevens: So that was going to be my question to you, Brian. Everybody keeps talking about arrestin or G protein. Why is the field so obsessed with those two signaling pathways? It's seems that it's actually the name GPCR that has held back the field because we have two hundred other intracellular proteins they interact with.

Brian Roth: Yes, I agree 99.9 percent. The reason basically is for convenience that are, for those of you that have been involved in sort of lead optimization, things at the later stage of drug discovery need sort of very facile assays, and it's simply because the assays are available currently for G protein signaling and arrestin signaling and they are extraordinarily robust and selective and so...

Ray Stevens: Related to Al Edward's comment — if you make a tool compound, all of a sudden everyone will start studying it. The appeal with G protein and arrestin signaling is the simplicity of the assay. It is what is available, and so that's what everybody does.

Brian Roth: Right. Those are the assay, that's what people do. We are trying to expand the universe, but yes there is no good reason other than those are the tools currently available that are in wide distribution.

Gebhard Schertler: I have a question for Kurt. First, I mean the NMR labeling studies are really fantastic because it directly showed that there are different states, which we can observe in solution. What is your view of how NMR could look at the dynamics of GPCRs in more general terms in the next fifteen years?

Kurt Wüthrich: Well, as you know, NMR can do everything you want (laughs), as long as you have a sufficiently big magnet. To be serious, in β_2AR we saw two states of two different sites in the protein, which means that we must have seen at least three, possibly four states, depending on whether the inactive state is identical or different in the two. And if we added more probes, the chances are that we would see additional states. So it would appear that we have a manifold of relatively long-lived states with locally different conformations. I did not mention this before: we have measured the lifetimes of the two states which we called A and I, and found a lower limit of about 1 second. So these are really long-lived states, meaning that the conformational change must involve major rearrangements, very likely including backbone segments. You mentioned before that in order to get diffracting crystals, you had to eliminate loops and truncate chain ends. In most cases, fusion proteins were used. Then you have to add CHS, and it is very critical which kind of detergent you use, and so on and so forth. And I mean, considering the amount of structural information that has been obtained, it is amazing that you are not telling us how these GPCRs work. I'm surprised, you see, since we now have, at least, a hundred structures of 25 human GPCRs with different ligands.

Ray Stevens: I'd actually say we've learned a tremendous amount. You're right. So right now there are more than two dozens different types of GPCRs solved, there are over a hundred co-crystal structures. We have now started to decipher sodium effects and other solvent ion effects. Cholesterol is a big influence. We are beginning to understand molecular recognition. So I disagree with your disappointment in the amount of what we've learned.

Kurt Wüthrich: Oh no no no, it wasn't disappointment, it was surprise (laughs).

Ray Stevens: OK I'm surprised by your surprise (laughs). I think we've learned a tremendous amount. Now, we look to you and NMR, in terms of understanding the dynamics, because we do think that these receptors, the way that they function is not from their static pictures. We think that it's how they move and at what timescales, but I think we've learned a tremendous amount in the last five years.

Kurt Wüthrich: Yes, we heard that one structure has been solved at room temperature.

Ray Stevens: That's correct.

Kurt Wüthrich: You see, in solution NMR experiments we don't have all these stabilizing interactions. We heard that a nanobody has to be added to stabilize states of GPCRs that cannot otherwise be seen. Now, are these artifacts, or are these interesting states? It has also been pointed out that you have to bind the G protein or a nanobody to stabilize the activated form. We don't have any of this, and in the absence of a cytoplasmic partner protein we get up to 90% population of what we consider to be the activated state. So there are clearly data that need to be combined to sort things out.

Jan Steyaert: I think we all agree that GPCRs are highly cooperative things so that means that if you...

Kurt Wüthrich: Highly?

Jan Steyaert: Highly cooperative molecules, so if you want to look at these highly cooperative systems, you cannot do it just with one, with two of the components. You at least need three components. You need a ligand at one end and you need the cooperative partner at the other end. And so far we are always looking at the receptor with one of the partners, so to really understand how these things work we will need at least three molecular complexes: we will need a ligand at the outside and we will need the other partner at the inside. That's the only way to look at cooperative systems. And the GPCR by default is a cooperative system. So as long as you look at it from a two components system, I mean, you are missing, you always miss the downstream partner being it the G protein, being it the arrestin, so to really understand a cooperative system you need at least three components. And so far most of the structural information has been gained on two components systems.

Kaspar Locher: Can I ask, Kurt, how do you measure your NMR spectra? What is the GPCR in? And is that a good model for a lipid membrane that contains up to 20% cholesterol?

Kurt Wüthrich: Well it contains 20% CHS. It's a mixed micelle of 80% detergent and 20% CHS. It's the same mixture that yielded crystals that diffract. Is that the answer to your question?

Kaspar Locher: So now the question is, if you're asking the crystallographers or expressing your surprise that no clear consistent mechanism has come out and implied that NMR would solve that, my concern is that if you conduct these experiments at reasonably high temperatures I assume with 80% detergent, the dynamics could be altered and in ABC transporters a lot of dynamics are altered when you go into detergents so I guess there are question marks there too.

Kurt Wüthrich: Yes, but I mean we don't work at liquid nitrogen temperature. We work at body temperature, not at "high temperature". So which dynamics would be relevant?

Ray Stevens: You know, this has been a debate, and all of us in the room have at least heard of it this debate for forty years: NMR versus crystallography, solid state versus solution state. I heard Kurt talk 30 years ago against the evils of crystallography, and I wish I would have recorded that lecture. I think the reality is, you use multiple techniques to try to understand what's going on. We know that with crystallography, we are going to get a static snapshot, that's the best that we can do. But we get a pretty accurate static snapshot. Then we use NMR and the dynamics, it's the best that we can do. We would like to be able to do, in more sort of a lipid-like environment. We like using the lipid cubic phase because it's a little bit more a lipid-like environment, but it's all about combining the pieces together.

Gunnar von Heijne: Okay, so I think now it's time for us to increase our coffee and cholesterol levels. So we will be back at 4.10 for the open discussion. Thank you.

Discussion among all attendees

Gunnar von Heijne: Right, who wants to start?

Kaspar Locher: I actually didn't get a chance to ask a question at the panel discussion that I have of the GPCR fellows to the right of me. I had and remade this comment about not working with anything other than human, and I'm wondering how different is mouse biology in terms of GPCRs from human biology? Because eventually, whatever drug you're going to produce, and I think that's one of your goals, you're going to produce a molecule that you going to have to put into a mouse, and I'm wondering what's known about this.

Bryan Roth: So I preempt the question from Ray because we actually deal with this on a daily basis. In the first of the two serotonin structure papers, we go in this in great details. So there is this huge mouse/human difference in serotonin receptor pharmacology due to a single amino acid in the binding pocket, which actually killed drug discovery for Pfizer because they were developing compounds against the mouse receptor and it turned out that they didn't interact with the human receptor. And this is sort of ubiquitous in GPCRs that there can be big species-specific differences in pharmacology in the binding pocket, usually due to a single amino acid differences. And so, as a practical matter, what's normally done when you're advancing compounds is to have the cognate receptor for every species that you're looking at, for better or worse. So we have this project ongoing on serotonin receptors, which have the added complication that there are editing isoforms. So we have to have editing isoforms for human, mouse, and rat and every compound that advances far enough has to actually be profiled in those before it goes into the experimental animal. So it turns out for this particular family, if you want to be assured that what you're looking at has a pharmacological relevance to the human, you have to have a human receptor, unfortunately.

Ray Stevens: I would add: each one of these structures is tough to solve, each one takes us anywhere from three to four years to solve. So, if you going to work that hard, you may as well go after the human. That's sort of one. The second is: it was mentioned that we use different stabilization techniques. Chris Tate and Gebhard did beautiful work on stabilizing membrane proteins with mutations. We try to stabilize largely through chemical ligands. And again, most of the ligands that have been developed are actually for human receptors, expressed in human cell lines. Then from screening that way, eventually we get them into mouse in terms of studying some of the animal studies. Earlier in my career, we we're working on two different drug discovery programs, and we made great drugs that worked for mouse enzymes and it turned out they didn't work all that well for human. And I decided, at the very beginning, we were going to focus on human.

Gunnar von Heijne: But Kaspar, maybe we can throw the question back to you. So in the transporter field, it's mostly prokaryotic transporters so far. So how do you reason in this?

Kaspar Locher: So, as I tried to point out, ten or twelve years ago when there were no structures of ABC transporters it made perfect sense to use bacterial proteins as model systems to understand anything at all. Folds, coupling mechanisms and we're finishing some of this work and looking at some of the reactions that are still poorly understood, like glycolipid flipping. Apart from that, I would agree with Ray very much and the last two projects have shown how the human multidrug transporters... Because it's a very similar situation there. In mice, multidrug transport is a little different and you do not want to have the situation where you're looking at pharmacological insight gained from structures only to find that it is different in the humans later on. So I agree very much.

Ray Stevens: So, if I can ask one more question to Kaspar. You said that the transporter field is two decades behind the GPCR field. And I wonder why is that? Because, if I think about structural information, we saw a lot of transporters structures before we started see GPCRs structures. Is there something inherent about the transporters? How are they different? Because everybody thinks their protein is the most unstable, most frustrating, annoying protein on the planet. And so what is it about transporters that makes them so challenging?

Kaspar Locher: The ABC transporters are the most frustrating (laughs). Just kidding. So we can't talk about all transporters in one bag. There are many that are relatively easy to study. They are very diverse but I do think that their importance for drug clearance, drugs moving around in the human body has been underestimated. And this is catching up now, and part of the program that we have in Switzerland, which is called TransCure, is trying to remedy that, and looking at much more details of transporters involved in health processes. So, we're behind in terms of interest and manpower of people who study this.

Richard Lerner: I want to raise an issue that Ray raised but went by very fast. And that is regulation by dwell time. The reason we became interested in that is — you'll see tomorrow just briefly — we make agonist antibodies to receptors. And sometimes, you take a given cell clone, a stem cell let's say, and you add a normal ligand, and it will make a white cell let's say. And then you add the agonist antibody and instead it will make a brain cell, and out of the same cells. And so incidentally using the same downstream signaling pathways. So our model for this is a kinetic model, and that is that a dwell time keeps the signal on for a longer time and the cell can read signal length or signal gradient like happens in development biology or something like that. And the same receptor can be used in the same way depending on the dwell time. And I wonder if, in the case of the GPCRs in particular, what you term bias is a function of dwell time?

Ray Stevens: So if I can make one comment to that Richard — on this whole issue of how long a receptor is sort of around. So I agree, the longer a receptor resides, the longer it can have an effect on different signaling pathways. We need to tease out what's going on. But probably one of the most frustrating aspects at least of the GPCR field is kinetics. You would think with these billion-dollar GPCR drugs available that pharmaceutical companies would have actually studied the kinetics of these different ligands, but if you want to know what the off-rate is of Plavix or some of these other drugs, it's not known. In fact, pharmaceutical companies do not study kinetics of receptor-ligands to the best of my knowledge. And we've begged, and we've asked for just about everything from the different companies. So I think kinetics of binding is poorly understood, receptor residence time is very poorly understood. We understand some aspects of trafficking absolutely. You know there is a really nice seminal work that has been done. C-terminus being a point where the receptor gets pulled out of the membrane. What about GPCR kinetics?

Richard Lerner: But Ray, just to follow up with that, wouldn't you think that as it were dwell time analoging would be an important, you know they're more than happy to put a methyl on it.

Ray Stevens: Well if I look at PK/PD effects of drugs, you would think that the drug companies and anybody that's doing drug discovery, you would think people want to study the on- and off- rates of ligands as well, related to that, but they don't. It's hard. Part of the answer is that it's just really hard to get this information and data and you want to do it on a large library of ligands, it's even tougher. But I think it's an important area of discovery.

Gebhard Schertler: I mean, it is interesting that G protein signaling normally is on the fast side, you know minutes to 20 minutes. And then, biased signaling, if you read it out, for example with kinetics, it does come later. So if you could have something differentiating and not being on long enough you might already get the bias. But it might not explain the whole thing.

Kurt Wüthrich: I'd like to come back to this monobody, two-body, three-body comment of before the break. In the monobody NMR experiments, with apo-β2AR, we have a visible population of the activated state with identical chemical shifts to the states that we see with increased or decreased populations when different ligands are bound. This indicates that the different states that we see are an inherent property of the GPCR and that it is not needed that a partner protein is associated with the GPCR to generate the activated conformation. I think, this is an important starting point for understanding how things happen. You see, you then have the receptor as a monobody, and you can start to add one, or multiple components and see how these components affect the equilibrium between the two conformations that preexist in the GPCR. I just want to emphasize this, when we talk about mono, di- or tri-body systems, that we at least seem to have a starting point with a monobody. Of course, we do not know — now I can expand into the future — what the two states are that we see at each of these probe positions, and we of course have to follow up with whatever method is available. We will try NMR structural work, that means isotope labeling of the GPCRs and defining the structural variations between the two states that we see at each point. You see, with the fluorine labels we have been able to measure the thermodynamics and the kinetics in the system of multiple pairs of two states, but we cannot make statements about the structures behind it. For this, we need to collect additional data. That's the outlook for the future.

Lode Wyns: I have a totally different question, maybe linking this morning session and this afternoon. It's well known there's a number of GPCR-related diseases where you have point mutations. It seems the GPCRs are perfectly happy, but signaling from the endosomes or signaling from the ER to a large extent seems to be OK but they are withheld and they don't get to the surface. What is being done in this field?

Bryan Roth: So, there's a whole area of pharmaco-chaperones, and I think there are actually a couple companies that have this. I think there are compounds that, so just to explain for people, there is this idea that when they're misfolded, they're stuck in the ER and there's a concept called pharmaco-chaperones where typically there are actually drugs that are technically called inverse agonists. So they strongly stabilize the antagonist state of the receptor, and for many receptors, vasopressin receptor for which there are mutations that cause diabetes insipitus, which are due basically to the receptor being trapped in the ER. So vasopressin receptor inverse agonist actually then is able to make the receptor go to the cell surface. So, they're out there. And I think, it's you know, from a big pharma perspective it would probably be a niche field because they're generally rare diseases but for those people that have them they are quite devastating and... So I think it's moving along. I'm not aware of any drugs that are approved but I suspect that sometime in the future they will be out there.

Ray Stevens: So, along those same lines, Brian happens to sit on the advisory board for the center that we have, and one of the things that we proposed last year was we want to start to study these splice variants and disease mutations. The advisory board discouraged us from doing so because they said that the number was too small. Do you remember that?

Bryan Roth: I remember that, yes. I think I said that.

Ray Stevens: For the GPCR family, they must mutate, but I think the bottom line is: it actually is rare. It doesn't happen nearly as much as you would think. Now one of the best examples is the smoothened receptor. So Genentech has this great drug, GDC0449, for basal cell carcinoma. They give it to people, it does a great job taking care of the cancer, but then months later the cancer comes back with a vengeance, massive. And, so what's going on? And Genentech did some beautiful follow-up experiments. My understanding of the sort of results is that we all have these variants in us, it's just that the drug was hitting the wild type sequence, and those other variants were able to then come out and dominate, and so Lilly now has a follow-on drug. So I think there are some receptors, but smoothened again is not your canonical GPCR, it's involved in embryonic development, which is a little bit sort of unusual. So there are some examples but it seems to be fewer than I personally would have thought, I thought it would have been many more.

Marina Rodnina: I have a question about optogenetics. Is it a useful tool for you in fact? And what are the perspectives of using that in pharmacology?

Ray Stevens: One of my graduate students just did his qualifying exam on optogenetics of GPCRs and it was a fantastic, he passed, so I was proud. I think the answer is: absolutely. One of the things we really want to try to understand is more about the sort of signaling, the trafficking, how these receptors are moving, how they are functioning in the cells. I think it's a fantastic area of discovery. And it's just starting to be pursued. I think Dennis was on the committee, any comments?

Dennis Wolan: Yeah, based on his proposal you mean, it was absolutely sound and very great potential to do these discoveries with this optogenetics.

Judith Klinman: I have a very naive question. From a mechanistic point of view, where is the GTP actually hydrolyzed? Is it hydrolyzed? I mean these are G coupled proteins, there is GTP getting hydrolyzed, no?

Ray Stevens: So G proteins are composed of three subunits, $G\alpha$, β and γ, and the enzyme reaction takes place in $G\alpha$. You have the receptor in the membrane, and then you have the $G\alpha\beta\gamma$ complex, that is the structure that Jan showed from Brian Kobilka's lab, that then forms a complex and then you get hydrolysis.

Judith Klinman: OK, so for example Kurt in your experiment, where you're looking at the two conformers, and you're showing an interconversion between them with the agonist or the antagonist binding, if you put in a GTP analog that can't get hydrolyzed, can you start linking these conformational changes to chemical reaction?

Jan Steyaert: I think there's a misunderstanding there. The structure is a nucleotide-free complex, and that's exactly what the GPCR does, it just catalyzes, in enzymatic terms, the release of GDP, and then the complex falls apart. It binds GTP and the complex falls apart, and the hydrolysis happens in the free $G\alpha$ later on, devoid of the G protein.

Judith Klinman: I see, so he can't even do the experiment. Okay, thank you.

Gebhard Schertler: Yes but GTPγS is routinely used as G protein assay, you know.

Gunnar von Heijne: Other questions at this point? Before we leave the GPCRs, so maybe we will come back to them, I'm sure, but, you made a remark that was new to me and that was the suggestion that maybe GPCRs are channels, or transporters, some of them could be. So could you expand a bit on that?

Ray Stevens: When we started to get these high resolution structures, one of the discoveries from 2011, was we saw this sodium binding site, and that was quite novel to us. We really didn't expect it at all. And from that work then, we started to do some follow-up experiments, looking at things that would dock in there, and we discovered amiloride, which is a channel blocker, that can actually bind into that pocket quite nicely. So that started us thinking, and this is a hypothesis right now, that there's enough space where sodium could actually pass through. It's probably not enough space with some receptors to be a channel but perhaps as a transporter, it's possible. So one of the areas that we are now investigating is trying to understand that possibility. We've also started looking at lithium, one of my big scientific curiosity is lithium treatment. How does it work? I asked Brian, he's the sort of expert in this area, and he says "we really don't know". So we started looking at other ions, is there a possibility that lithium is going in this site, controlling the receptors? We find rubidium is incredibly stabilizing and we think that rubidium goes into that pocket. Divalents don't go in. So we are now starting the traditional soluble-protein type of studies now with the GPCRs to understand this pocket and also if there is transporter or channel activity. The last comment on that is: every receptor is different. We've now observed this in the delta opioid receptor and in the adenosine A2A receptor. We think that this pocket is highly conserved in a lot of GPCRs. I think it's now been seen by Gebhard, in the β1 adrenergic receptor. And so we're starting to dig more and more into the exact role, and what role other ions might play with receptors.

Rudolf Glockshuber: I have a question to Kurt and all of you actually. So could your NMR experiments indicate that the drug binds according to a mechanism I would say conformational selection. So, in the free state it exists in two different conformations, and one of them is one that binds the drug. So the question is, are there other examples where one has something like an induced fit mechanism, where the structure of the drug-receptor complex is different from any of the states that are populated in the ligand-free state?

Gebhard Schertler: So the data from Kurt would look like that but there is actually something very curious about the GPCRs. If you put in, one of the very good synthetic agonist, isoprenaline into the receptor, the receptor does not go into an active conformation. So you do the structure of that, and this has been tried by Brian, and also with other agonists, you actually get a very ground-state-like structure back. So the binding site for that is there but that doesn't give you an open G protein binding site. So when you do fluorescence lifetime measurement on a labeled receptor, again an experiment Brian did a very long time ago, and you put isoprenaline in, you can see again a few states, I don't know if this has been repeated with NMR, but what you see is that the active conformation is actually only increased in a tiny little amount. So to actually get activation of G protein and binding of the G protein and an exchange of nucleotide, you do not need all the receptor in an active conformation. And I don't think this is a rare thing. So there I don't know exactly how this fits with induced fit. It might fit still in the ligand binding pocket. But it doesn't give you immediately a complete flip between two conformations. And that is why Brian has now crystallized with the G protein, we have sometimes crystallized with the C-terminus of the G protein. There's only one structure, and that is opsin — and we have also repeated that one — that actually is there with a complete conformational change of helix 6 and that is the opsin structure that was observed at very low pH, in the Berlin group. So there's some complexity also there. There is not a direct link between what you put in the ligand site, and the conformation on the intracellular side of the receptor.

Ray Stevens: Back in 2007 when we published the $\beta 2$ structure, one of our last sentences was: it's going to be a long time before we have an agonist-bound structure, and it's going to take G proteins. Now it turns out, that's incorrect. People are getting agonist-bound structures left and right. And again, I'm not calling it an active state structure, I'm calling it an agonist-bound structure, but when we compared A2A, with the opsin — rhodopsin system, with the structure with G protein with $\beta 2$, what we're finding is lots of similarities. So, you can have an agonist bind, and it preloads that binding site so... I'm just curious as to your thoughts.

Gebhard Schertler: No, I already said, you can bind the agonist but you don't have fixed movements of helix 6 for example, so there is not a direct linking between these two sites. It looks more like a release of the helix.

Kurt Wüthrich: Well, a direct answer to Rudi's question is to say that the "principle of Le Chatelier" is acting here when we talk about binding of a drug molecule. The drug molecule sees both conformations in the apo-protein. It binds to one, and removes it from the equilibrium, and Le Chatelier then pushes the equilibrium around. I think that's the answer to your question as I would see it.

Rudolf Glockshuber: I think that's quite clear. My question was rather are there other examples where one has an induced fit like mechanism, that maybe transmits to the cytoplasmic side.

Gebhard Schertler: So, when we go to larger ligands, there are examples where side chains are rotated. So for example I think in the muscarinic there's one tryptophan that gets put out of the way so that a larger ligand can bind and then actually has a higher affinity. So when we make our ligands in drug discovery bigger, we go out of the orthosteric binding site of the natural ligand. Then some rearrangements are happening, at least of the side chains, and this has been observed in some places and also we have already seen an example in one structure where you had an allosteric ligand bound, and we have kind of seen bivalent ligands which can actually reach both binding sites. So there are interesting detailed things coming about... In these cases, you have clearly also a change of the binding site, at least where you have one real rotamer change that's induced by the ligand.

Tom Muir: So I think, Ray or Gebhard mentioned dimerization of the GPCRs. And now there's a lot of back and forth on this. I'd just want to know what the current thinking was, from all the structures, based on crystal packing, you know, the Kobilka structure, would there be room for two G proteins within a dimer, would they occupy the same space. Where are we with that?

Ray Stevens: Bryan Roth helped organize a meeting at the NIH last summer. So we were just discussing this at the break. What happened in the field was that a lot of people studied GPCR dimerization related to GPCR signaling and they used overexpression systems where they made lots of receptors, and I think the general attitude is that with many of that, much of that data is artifactual. It's overexpression and you force dimerization to occur and you saw what you wanted to see. Now, that being said, I personally believe that there are some cases where there are true dimers. I think the class C receptors, the mGLUrs. I think that data is pretty solid where they are dimers. I think personally something like CXCR4, where we crystallized it in five different forms, different crystal forms, different ligands, different variants, and every time, it was always the same dimer. Now chemokines CXCR4 are chemical gradient systems. Where it's all about concentration of the receptor, and concentration of ligand, and you can have a 1:1, 2:1, 1:2 ratio. I think it's receptor specific. I was just talking to Bryan Roth at the break where, I still like to fantasize that you can have some receptors that heterodimerize, and that make some sense on biology: dopamine with adenosine, but he says that he has data that says: that's an artifact. So it's an open question for the field right now. It's going to be receptor specific. So people have to be very careful. One of the things that we're learning, and I'm glad that we looked at more than one GPCR for the past 5 years, I'm glad that were going to look at all 826, because if you focus on one, and you try then to make correlation to other receptors, you get tripped up. And so I think it's important to study the whole family.

Bryan Roth: So, let me just expand on this. This is something that my lab spent a lot of time unsuccessfully trying to replicate, and I won't state what the papers are but some of the more high visibility heterodimers papers, we were never able to replicate the data. Despite, you know, expending vast resources to do so, multiple knock-out mice. One of these has actually been formally non-replicated by Abbott labs. This was the 5hT2A-mGluR2 that was in Cell and Nature. I mean it could happen but it's not robust enough to be replicated independently by at least one group. So it is, I would say sort of a contentious area, that you could sort of populate this group here with people that believe — I would say believe — in heterodimers strongly, as well as people that don't. So I think it's at that level. It's still, still a bit up in the air, and the thing that I always look for are the independent replications. I think this is the key. One of the nice things about the crystal structures, and I think the opioid receptors were really, really nice examples of this. All four came out basically in the same issue by two independent groups, and they were very similar. And in one case the nociceptin receptor had a different fusion partner, and a completely different part of the receptor, and intact intracellular loops and, but the overall packing, and everything was very, very similar. So, this is the sort of things that you like to see where multiple groups are essentially getting the same data. You don't really see this so much in the dimer field. It's more one-off type things so I'm still waiting for the really strong signal from multiple independent labs replicating something. But I think it's something, certainly something to be aware of. But my personal success is that we can't replicate this stuff.

Bernard Henrissat: How much of the incredibly detailed knowledge that you guys are accumulating can be transmitted to other species? You know, we don't care whether some animals get sick, but just out of curiosity: if you go from mammalian systems to a non-mammalian system, if you go to insects, birds, if you go to plants, are these drugs working the same?

Gebhard Schertler: I think at the end we don't know, there are very famous examples where it doesn't work, like before I would like to mention the $\beta 3$ project, which was done for obesity. It was again done on rodents, and it didn't work in humans, so even small differences can make a difference, but if I look just at the sequence space, and the conserved residues, and the mechanism I tried at least to sketch to you with a movie, clearly all the residues are there, and the principal activation mechanisms are the same, but, you know the ligand sites are very finely tuned, and also the signaling network is very finely tuned, and you very easily get different outputs from that.

<u>Dario Neri</u>: The title of this session covers also drug development. There were some aspects touched, for example by Gebhard where structural knowledge would facilitate for example fragment-based drug discovery. I would be curious if you could expand a little bit on where and how structural knowledge helps drug development and maybe also limitations, in translating structural information into drug development. Maybe one needs to close one cycle of...

<u>Gebhard Schertler</u>: The first thing I want to say is obviously I was one of the people making Heptares, I have shares of it. In addition, I'm on the scientific advisory board; I'm not really allowed to tell you exactly what they do. What they have done is that they have solved many in-house structures. They really have solved first structures with no ligands, then with ligands they made and then also with fragments in about that order. They have now also different classes, and there was a publication of another class of receptors recently. What is very important is they are not using only the information from only one receptor. They have assembled a fantastic modelling team. Every model that they use is derived from more than one structure, so every time we get a new fragment, a new structure, they will integrate that in their modelling effort. It is actually a very strongly model-driven company and they have given quite a lot of talks also about the necessity to either know or model the waters in the binding sites, and they have for example very much this idea that there are two classes of waters, some that are in good energetic positions, and you shouldn't necessarily replace them by the ligand, others which you could well replace. A lot of this is really very high end modelling, you could believe it or not but there have been reasonably successful with this hypothesis. So they have filled the pipeline, and they have a lot of partnership deals, but I think they also have very skilled experts doing this, it's not only the structural information that is pushing this. They have exceptional chemists, they are very fast, you know, always doing a series, and not one compound. So drugs and receptor structure is one driver of drug discovery, it's not a drug discovery pipeline. I really don't believe at all.

<u>Ray Stevens</u>: I will make the same disclosure. About the same time Heptares got started, I started a company called Receptos. Receptos is four years old, and it has a S1P1 molecule in phase III, and did its IPO last May. Structure was not used — the molecules came from Hugh Rosen and Ed Roberts with med-chem design. In this case, structure was helpful in understanding how different molecules bound and what was unique And most importantly in that case, structure was used there were 763 different ligands to look at. How do you prioritize those that go into clinical trials? So structure was used as to try to prioritize things, and use that knowledge combined with med-chem intuition and everything else. So I think, clearly structure is being very helpful. Now, the other point is the use of virtual ligand screening in drug discovery to find new molecules. If I look at kinases and enzymes, we have hit rates of 5%. If you find a hundred different molecules by virtual ligand screening, you're lucky if five of them then get validated. But with GPCRs it's very, very

different. It's a hit rate of 40, 60, 80%. So the question is: why? We've done this with Dopamine D3, we've done it with A2A, Brian Shoichet has done it with some nice systems. I think the reason is, you have these beautiful three-dimensional pockets, unlike enzymes. They are actually more rigid than we thought they were going to be, and because you get these beautiful three-dimensional pockets, to then do your docking, we're seeing great validation. Brian said it earlier, the best is yet to come. You're going to see a gold mine in the structures that are just coming out now. So you'll see a lot of structure-based drugs emerging over the next couple of years.

Bryan Roth: So, let me expand on this. Let me just say I have nothing to disclose, because I don't own a company. But what we've done is mainly published in collaboration with Brian Shoichet, and my own lab. It's something we use every day, basically, and if you have a structure, obviously, it makes things way better. And it greatly accelerates the pace of things. But it's a tool. That being said, the last slide I showed was the discovery of this positive allosteric modulator for this orphan receptor. The thing that may not have been clear from this slide is: that was discovered *in silico*, entirely *in silico*. So this is a receptor for which there is no structure. It was made completely based on homology modelling, and we did some mutagenesis with the very low affinity compound, and then based on docking with the Shoichet group we were able to this extremely potent compound which is actually effective *in vivo*. So, it's possible to go all the way with docking. I think this is probably going to be a rare case, but it does work. So it certainly is a fantastic tool that we have now, that we just didn't have before.

Dennis Wolan: So GPCRs have fantastic ligand efficiency, very well known for that. So what kind of promiscuity is there across these molecules? I mean you have 826 GPCRs, and you have assays for majority of these?

Bryan Roth: Yes.

Dennis Wolan: So, if you have assays for the majority, are they fairly similar assays?

Bryan Roth: So I didn't really get into that cause I didn't want to get into that degree of granularity, but it's basically an arrestin-based assay, which is cross platform. So technically it allows us to screen all the "druggable" GPCRs in the genome in a single 384-well plate. And it had sort of "pennies per well"- it's highly scalable. So what we've done, I didn't show these slides, we looked at the world's most promiscuous compound which is clozapine, which before we had this platform we knew to bind to probably a hundred different molecular targets, and then — I have a slide, so if anybody is interested I can show it to them — then we screened it against the entire GPCR-ome, and what we found was that there were no other targets. Basically it binds to the biogenic amine sort of subfamily of receptors. It hits all of those and then hit some other enzymes but outside of that, interestingly enough, it has no activity. And we screened both highly promiscuous kinase inhibitors and other promiscuous drugs, and you don't sort of see "panGPCR" activity like you would with the kinases, but you do see interesting hot spots. And in many cases these, we can basically deconstruct the information back to a known side effect of the drug or a known therapeutic effect of the drug, that was previously uninterruptable. We were basically able to find a molecular target for that, but it's not, unlike the kinases, or phosphatases or something like that, where we have an inhibitor of one, it hits everything. This does not appear to be the case outside the biogenic amine family.

Ray Stevens: So I showed this phylogenetic tree based on sequence. Brian Shoichet has this really nice chemical version of that map. And if you look at for example muscarinic receptors and chemokines, there's a lot of overlap there. Why that is? Nobody knows. But it's one of the areas that's pretty cool for discovery. We have a table in the lab, of again a lot of ligands do bind multiple receptors, and there are a lot of good examples of it. And so it's another area of investigation.

Kurt Wüthrich: Eventually you want to find correlations between the chemical structure of the ligand, of the drug, and its consequences for what happens on the cytoplasmic end of the molecule. Now, if we had induced fit as Rudolf Glockshuber inquired, then we would have very broad promiscuity. I think that's perfectly clear. If we had induced fit, then the whole system would be highly promiscuous. On the other hand, as far as I can oversee the results of the crystallographic studies, the changes in the global architecture, induced by agonists and antagonists, are very limited. When you go through the series of now 25 GPCRs whose structures are here, the major differences are in the ligand-binding site. And that's what I understand, these are the different sizes of your wine glasses, and the different shapes of your wine glasses in your slides. If we had induced fit, this would all be essentially of limited use.

Gebhard Schertler: Why do you think then, all these residues are conserved if they play no role?

Kurt Wüthrich: I don't understand your comment.

Gebhard Schertler: I showed you, at least three motifs which are dispersed from the binding site towards the intracellular side, and they are conserved in the family A receptors and I also showed you that they build two alternating hydrogen networks, and they are very heavily conserved in the chemokine receptors as in rhodopsin as in amine receptors. And what you're saying is, the binding site, they play no role. I haven't said that, in principle.

Kurt Wüthrich: I think this is a misunderstanding. I understand that there are major variations in the ligand binding site, whereas in the scaffold of the GPCRs there are only rather small changes.

Gebhard Schertler: I agree on that.

Jan Steyaert: Can I comment on this?

Kurt Wüthrich: So this would indicate that specificity is due to the mean structures seen in the crystal structure determinations, whereas an induced fit mechanism would lead to random promiscuity.

Jan Steyaert: Can I comment on that, because if we compare the active structures against the inactive structures, so that meaning, a site occupied by an antagonist or an inverse agonist and compared that to the same receptor loaded with an agonist, you clearly see a compacting, a smaller binding pocket, and you can even observe, in the absence of the G protein... Well, let's turn it around: in the presence of the G protein, the off-rate of ligands really increases dramatically, so the ligand is really binding that small pocket that it's made for, only in the presence of the G protein, and you see factors of 200 even 1000 in terms of slowing down off-rates, and on-rates also slow down, so there's an aspect of induced fit in these binding pockets, upon binding of agonist, natural ligands, and the concomitant binding of the G protein.

Ray Stevens: Some of this is actually semantics because, again, if you look at adenosine A2A, we now have, and I think probably close to twenty co-crystals structures, different ligands, from everything, part- full agonist, partial antagonist and so forth. And it's been remarkable how similar those structures are. One of the things that's been nice is: it's been solved with antibodies, with fusions, with mutations, and the structures are quite similar. The reason I say semantics is those binding sites, within each specific receptor, the differences are not as large as I think a lot of us thought they were going to be. This is all perspective. Myself I was expecting the agonist structure to be very different from the inverse agonist. And it's not nearly as big as I expected. So again; perspective. And if we look at all the different receptors we're seeing a sort of similar trends. The off-rate thing I think is a fascination. I mean, how on earth is it having this effect on off-rates, and changes are pretty small. In the top half we have ligand recognition — very small changes. The bottom half is where you're seeing the bigger conformational changes, and how that's working, and it's quite interesting how it can be few residues making that happen. So I agree with Kurt.

Gebhard Schertler: Can I go in there? I really disagree here, because we have two structures which have bits of G protein bound, we have Brian's, and we have the rhodopsin one, with at least the C-terminal end. And in my lab, we're also doing experiments on the complex. So the first thing you observe, if you make the rhodopsin complex, again with the empty pocket to G protein, that actually the retinal after hydrolysis cannot leave anymore. Okay, so there is a strong change, but that's not on ligand binding, that's when the G protein has bound. And some of the structures you're describing are without the G protein but the agonist bound. They are very similar, and they are all still in a similar conformation as the ground state structure. They have not made that big transition. But when you do that big transition there's no question from the literature that the affinities change dramatically when you add the G protein, and also they change when you add the peptide on it, the C-terminal peptide of the G protein. At that moment, when you really bloom out the cytoplasmic side, you have changes it at the extracellular side, and you're starting to trap the ligand. And if you look carefully at the structures with the antibody and with the G protein complex, it is very clear that the open cleft in the β receptor has closed up, and actually an agonist now cannot come in and replace the ligand anymore. These are absolutely clear measurements, there's no question about this.

Ray Stevens: I think the difference of opinion is: what defines a big conformational change? I'm thinking 8 or 10 Å in the direct binding pocket.

Gebhard Schertler: I think we are talking about different states in the cycle of doing this. You're talking about an early one, which I've called at some point an encounter complex when the agonist is going there, and what we're here talking about is the one that has bound the G protein and at that point we've actually a strong enough conformational change that even the extracellular side is starting to close up. And this you only can see when you have the G protein or a G protein substitute there.

Ray Stevens: Well let's see when we get more structures.

Markus Grütter: I would like to switch to the transporters and ask a question to Kaspar. You have shown in one nice example, the importer, how substrate is transported into the cell. Now there's a whole, large number of ABC transporters that export and that are promiscuous in the way they transport substrates, and we have seen structures of different conformations but not really a convincing site where things bind. How do you see the future with respect of seeing how stuff is transported through this? Apart, I mean, we see the mechanistic movements, but where are the substrates? Do you see a chance to tackle that experimentally and how?

Kaspar Locher: Maybe I should explain first for the non-experts that one of the difficulties of working with ABC exporters is that most of their substrates are so hydrophobic that they compete very strongly, or that detergent competes very strongly for the binding pocket. So whenever we purify them, they are devoid of any substrate, and the substrate don't bind well anymore, and we have to put the transporters back into a lipid bilayers, cholesterol-containing lipid bilayers often, to see the substrate even binding and then usually with lousy activity of tens of μmol. So unlike in the GPCR field, there's a very different mechanism of substrate recognition. And the current ideas are that somewhere on the molecule, and I don't think anybody has a clear idea of where that is, there are multiple residues, could be hydrophobic, aromatic, that contribute to drug binding. How that works and how it's extracting a drug from the favorable interactions that the drug experiences in the lipid bilayer is unclear. But what is clear is that side openings to the lipid bilayer play an important role. How large those are again is a matter of debate. And my personal view is that whenever we take a multidrug or any other ABC exporter out of the membrane, with detergents, it will relax into a non-physiological conformation that's the most stable conformation, that's why we can also crystalize it. And the openings, and the distances of NBDs are as large or larger than those where we can prove inactivation of the transporter. So no clear answers in the near future, there are some systems with good inhibitors and we're currently working with chemists to improve those inhibitors and hopefully get to higher affinities that will allow us to get a structural insight even in detergent solution. Is that the answer to your question?

Markus Grütter: Yes, and how about assays to tackle these questions?

Kaspar Locher: Right, so we approach the idea of assays as follows: until now, assays have mostly looked at the stimulation of ATPase activity. So the idea being, we add a drug into a reconstituted ABC transporter-containing vesicle, it partitions into the lipid bilayer and it binds to a drug binding pocket which is assumed to be a central drug binding pocket thereby speeding up ATPase activity. The problem with this assay is that even DMSO and other solvents will have the same effect. And you can affect ABC transporters even by adding, by changing the lipid composition. What I think we will have to do and what we're working on is to develop very good acceptor proteins that will prevent partitioning of the substrate into the lipid bilayer, and really allow the substrates to be kept outside of the lipid bilayer. After all that's exactly what Nature does. So if you look at the blood-brain barrier, the reason why many of the drugs that we would like to give for brain diseases don't enter the brain is the blood-brain barrier, and there are multiple multidrug transporters expressed there and the reason they work is blood flow and albumin. So the drugs would be transported out of the membrane, they would instantly partition back into the membrane unless there was blood flow and albumin. And similar things we have to do in assays.

Markus Grütter: Thank you.

Gebhard Schertler: Maybe one comment to Kaspar. When we look at rhodopsin in the opsin form, and this is work from Peter Hofmann, he saw really a nice channel for the hydrophobic retinal going in sideways into the G protein-coupled receptor a little bit like you suggest in the transporter. But it's only in one conformation. These openings are really closed in the ground state-like structure.

Kaspar Locher: Right and perhaps I would like to, I didn't get a chance to go to the channel, the potential channel features of a GPCR. I would assume this would be measurable, and I can only assume at the moment that these rates are so slow that they wouldn't be above leak currents of empty membranes. So I doubt very much that GPCRs are transporters of any signal cations.

Ray Stevens: Can we bet on that?

Kaspar Locher: I don't have your purse but I'd be willing to bet (laughs).

Ray Stevens: I'd like to bet beer. But I have sort of insider information so...

Bryan Roth: So, I would not take that bet, Ray.

Ray Stevens: I have a question for Gunnar. So we've been talking about GPCRs and transporters. What about, in terms of their insertion into the membrane. I mean, what is it going to take for us to understand how these receptors, eukaryotic proteins are inserting and folding/refolding.

Gunnar von Heijne: I don't think anything is known about it at this point. Obviously they do get in, that's about as much as we can say I think. They do get in co-translationally in, I think that's clear as well. And they use the SEC translocon but beyond that I don't think we can say anything really. I think we have looked at a few of those just theoretically to look at how hydrophobic these transmembrane segments are. And some are not super hydrophobic. So, there's, you know, although nobody else looked at it, I think the bet is again — if we should start a betting club here — that there's going to be interactions, co-translational interactions between at least some of the helices as they assemble into the membrane. How much of the final fold that's going to be co-translational versus post-translational I don't think anybody can say. But the first elements, I mean we've seen this with another protein, multi-spanning protein, that there is in fact early co-translational interactions between some helices or polar residues on some helices that can help stabilize not so hydrophobic helices across the membrane. But there are very few assays around to look at it, and you know even with purified reconstituted membrane proteins it's very difficult to do unfolding-refolding studies. There are two or three systems where you can do that, basically reversibly and that's of course a far cry from co-translational feeding of transmembrane helices into membranes.

<u>Markus Aebi</u>: Following up on that, nevertheless there's a quality, or let's say a quality control system for proper folding of membrane proteins. And I wonder how this is recognized by this system.

<u>Gunnar von Heijne</u>: That's, as far as I know, that's a post-translational quality control system. So, I mean, clearly many of these proteins don't attain their full three dimensional structure until fairly late. Whether there are chaperones involved, there are a few cases, where it has been shown that there are intramembranous chaperones. The cases I know of, they are very specific, so there is like one chaperone for one protein, but nothing excludes that there could be more general chaperones for instance that we just haven't come across yet. And so, you know, exactly what these states are that are recognized as misfolded we also don't really know. We know that they behave as, in some sense where they can be recognized by the cell as being misfolded, but exactly what they look like is not clear I think.

<u>Kaspar Locher</u>: You have various systems where you can hold the translocation halfway. And in eukaryotes, we heard about the OST complex and it's assumed that there's a direct association with the secretion machinery. Is there an example where a halted peptide that's only semi inserted into the membrane already get glycosylated?

<u>Gunnar von Heijne</u>: Oh yes, I mean you can make truncated ribosome nascent chain complexes where, which have been targeted to the translocon and the nascent peptide is halfway across the membrane and some part is in the lumen, some part is spanning the translocon and some part is stuck in the ribosome. And those can be glycosylated on the luminal side. There is a minimal distance away from the membrane which is basically I think reflecting the distance between the membrane and the active site of the OST in that case. We've been trying to do, to see if we can do the same in bacteria... You know we just don't have a great experiment yet but I think we can see a similar kind of thing, but that's with the substrate, that's not folded, so it's you know quite unfolded in that case.

<u>Dennis Wolan</u>: So, speaking glycosylation, Ray, with your 1.7 Å structure that you have, are you seeing glycosylation, and is there any biological significance?

<u>Ray Stevens</u>: We give these things haircuts, they are glycosylated when they're made, we use PNGase F and we clip off as much as we can, but we see beautiful sugar stems from them. So that's the answer to your question. I want to sort of follow up, with Gunnar, I mean, one of the things we know is: these receptors are glycosylated. You know that if you knock out the glycosylation you don't get expression to the surface. So you know the glycosylation is important. We also know that we can express different parts. I think Brian Kobilka who did the experiment years ago where he co-expressed I think it was the first 4 helices and the last 3 helices, maybe it was 5 + 2 of the different helices. With better knowledge and

questions, we're now repeating the experiment of 4 + 3, where we know, the four helices, the first four helices form this beautiful four-helix bundle. And we're trying to see if we can crystallize just that. It doesn't have a function, but it's a curiosity. And can we start to combine these things together and restore function to try to get this question of how are these things assembling? So is it partial folding?

Gunnar von Heijne: Yes of course those types of experiments have been done with many multi-spanning membrane proteins, starting with bacteriorhodopsin, way, way back. So they can be... Not all combinations work apparently, and I don't think anybody has figured out why certain combinations work and others don't, whether it has to do with insertion or something else but they do seem to work. In many cases they can find each other in the membrane somehow.

Kurt Wüthrich: Well, earlier today, there was the question of the lifetime of membrane proteins in living systems, and this of course is connected with folding and insertion. What is known about the turnover of membrane proteins, of GPCRs, of channels?

Bryan Roth: Yeah, so, I know this. There's a system that we developed, may be or may not be familiar to you, where we use engineered GPCRs to regulate the signaling called DREADDs, and a colleague of mine at Scripps, Mark Mayford did the experiment. So they're expressed using a tetracycline inducible promoter, so he can turn on expression on the brain with TET, and then take it off, and they're gone in two days. So the turnover is actually relatively quick, surprisingly, so this is a basically a muscarinic receptor, sort of prototypical family A GPCR. So at least for that receptor we know quite well. Whether that will be the same for other GPCRs — God only knows — but you know it was actually fairly quick, I wouldn't have suspected that it would be gone so quickly after basically the transcription was shut off. And that's in the absence of ligand. So it's not being activated by any endogenous ligand.

Gunnar von Heijne: So maybe one interesting example of differential stability but within a complex is of course the photosynthetic reaction center, right? You get photo-damage to the central unit, subunit, and that can be picked out of the complex specifically and replaced with a fresh one, without affecting the other subunits. So even within one complex, you know, you can have different stabilities.

Kurt Wüthrich: Would that mean that their individual domains fold independently? And then assemble as a whole domain?

Gunnar von Heijne: Yes, there's some work done on the assembly of these multi-component complexes. They do follow not completely defined but partially defined assembly pathways where, l say, subunits A and B have to assemble before subunit C can come in, but maybe subunit D could come in before and after subunit C so it's not straight linear. But it has certain structure. In the cases that we've looked at — at least these more complex multicomponent membrane proteins systems.

N.B. Some of Professor Stevens' contributions to this section's discussion were added in the proofs.

Session 5
Biologicals and Biosimilars

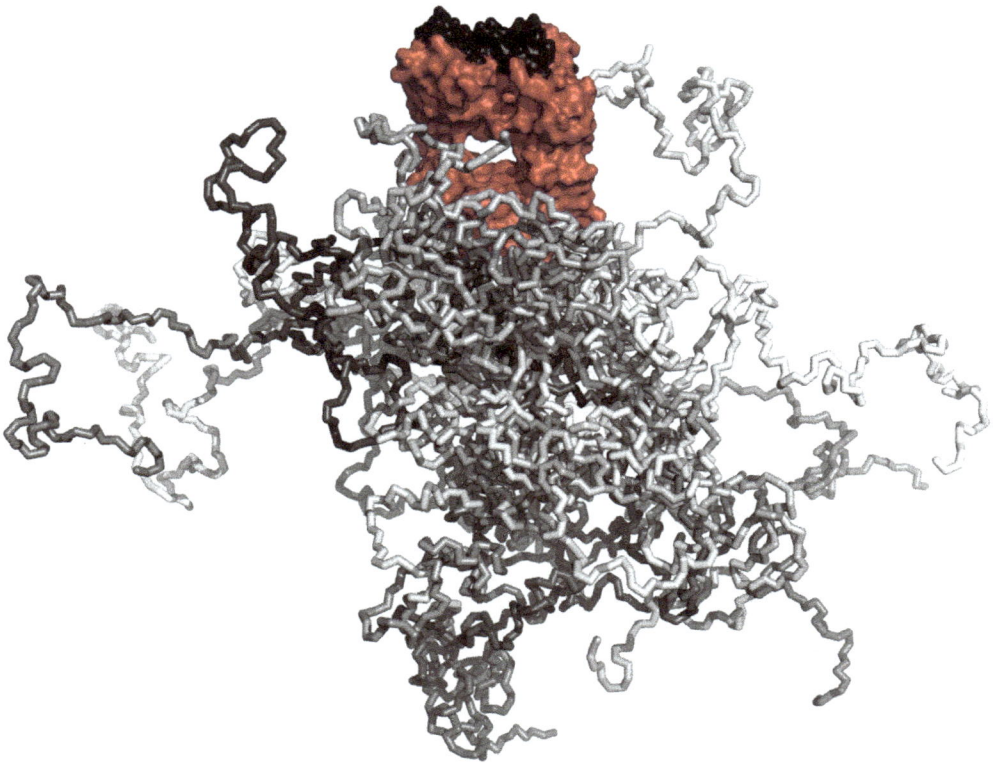

Modeled structure of the PASylated Fab fragment of an antibody. See Figure 2 contributed by Arne Skerra on page 246.

BIOLOGICALS AND BIOSIMILARS

SRABONI GHOSE and MARKUS G. GRÜTTER

Institute of Biochemistry, University of Zürich
Winterthurerstrasse 190, CH-8057 Zürich, Switzerland

Introduction

Biologicals are drugs produced using biotechnology. They may be proteins, including antibodies, or nucleic acids including DNA, RNA or antisense oligonucleotides, that are produced using recombinant technology as opposed to extraction from existing sources *e.g.*, from human blood. Biotechnology has made a wide range of new therapies possible by allowing the large scale production of vaccines, therapeutic protein hormones, cytokines, tissue growth factors, monoclonal antibodies and oligonucleotides for cell or gene therapies. In this discussion we will focus on the categories of biomolecular drugs (biologicals) that are presently managed by the FDA Center for Drugs Evaluation and Research (CDER): monoclonal antibodies, cytokines, tissue growth factors and therapeutic proteins. The first genetically engineered biomolecular drug to receive US approval was human insulin in 1982. Since then, blood clotting factors, growth hormone, erythropoietin, colony stimulating factors, interferon and anticoagulant proteins, all natural proteins or 'first generation' recombinant products, have been brought to the market. Likewise monoclonal antibodies have entered the market. Biologicals today represent a major business segment in the bio/pharmaceutical industry. Based on worldwide sales, seven out of the top 15 pharmaceuticals in 2012 were biologicals [1a]. The therapeutic potential of biologicals is evident in the growing number of product approvals by the FDA and the EMA, 39 in the period 2009–2011 [1b]. Cancer and related conditions remain the main indication. As the first generation biologicals start losing patent protection, and biosimilars enter the market, innovation is seen in the design of next generation drugs as well as in the biomanufacturing process.

First generation biologicals

Two broad categories of protein biologicals are discussed here:

(i) Proteins with an enzymatic or regulatory activity which (a) replace a deficient protein such as insulin, (b) augment an existing pathway such as erythropoietin and (c) modulate an activity such as botulinum toxin

(ii) Proteins with a highly specific targeting activity which (a) directly interfere with an activity such as Trastuzumab (Herceptin) or (b) deliver a therapeutic compound such as a radioligand (^{131}I-tositumomab).

Antibody technology

Besides the ongoing search for high affinity full length therapeutic monoclonal antibodies from libraries, the field has matured through the generation of murine to chimaeric to fully human antibodies. Antibodies in addition are further modified and engineered for higher affinity and stability. Antibody fragments such as Fab, scFv and Fc have improved diffusion profiles due to their smaller molecular weight, thus offering distinct advantages over full length antibodies [2]. Bi-specific T-cell engaging antibodies, Fc domain libraries, dual affinity re-targetting, are all engineered antibody fragment technologies now in clinical trials for different indications. Antibody-drug conjugates (ADC) now offer great potential for development especially since linker engineering has increased stability in the bloodstream and efficient payload release at the target site. As of 2013, 30 ADCs are in clinical trials targeting 24 antigens in oncology indications [3]. Peptibodies are biologically active peptide-Fc fusions that allow for increased affinity as well as efficient manufacturing [4].

Oligonucleotide technology

The first systemically delivered antisense drug mipomersen was approved for the treatment of homozygous familial hypercholesterolemia (hoFH) in 2013. The early promise of oligonucleotide therapeutics was announced with the approval of the first drug in 1992 against genital warts. Drug delivery remains a significant hurdle. RNA therapeutics are particularly suited to restore gene function in conditions where blocking oligonucleotides do not degrade pre-mRNA [5, 6]. Recent encouraging results from clinical trials for Duchenne muscular dystrophy and inherited neurodegenerative diseases are a further example of its potential. Besides RNAi and antisense oligonucleotides, splice-switching and translation-suppressing oligonucleotides as well as siRNA, are being developed for therapeutics.

Biomanufacturing

The manufacturing of biologicals has had to mature as new processes were established on an industrial scale. Developments such as protein engineering, synthetic gene design, recombinant expression technology and large scale fermentation and extraction have all contributed to the scaling-up process. The criteria for biomanufacturing are very different from small molecule manufacturing and new regulatory guidelines had to be defined to guarantee quality standards. The objectives include limiting the immunogenicity of the biological and improving its shelf life.

Research tools have been developed to characterize the biologicals with regard to their aggregation status, three-dimensional structure and post-translational modifications [7]. The majority of biologicals are glycoproteins, the effect of the glycan composition and its engineering has implications for product quality [8]. It has been seen that not only does the quality of the product to be guaranteed but also the quality of the process, thus Quality by Design is now also applied to biomanufacturing.

Biosafety and pharmacovigilance

The safety profile of biologicals is different from small molecule drugs due to their inherent characteristics conferred by the complex composition. These large, sometimes heterogeneous mixtures of isoforms have a greater potential for immunogenicity. These drugs are often used to treat chronic conditions and thus the risk factor is increased. The spontaneous reporting of adverse drug reactions is vital to evaluating the safety of these drugs in the clinic. An example of dramatic and unpredictable consequences is the cytokine storm that occurred with the anti CD-28 monoclonal antibody (TeGenero) study, which was not observed in preclinical studies. The clinical trial design of biologicals has been reviewed after this unfortunate episode.

Biosimilars

Biosimilars are defined as biotherapeutic products which are similar in terms of quality, safety and efficacy to already licensed reference biotherapeutic products. Thus, the biosimilar has the same pharmacological effect without the same chemical identity as the patented drug. It has been argued that biosimilars of monoclonal antibodies will be difficult to generate since each one is unique and that the different process will result in a different product. However, biosimilars have been approved several times both by the FDA and the EMA. The first monoclonal antibody biosimilar was approved in the EU in 2013. The use of biosimilars is set to dramatically increase as the first generation biologicals lose their exclusivity [9]. The clinical studies carried out so far have been positive. New paradigms for testing are required for the comparability exercise of complex drugs such as Avastin as it loses exclusivity in 2017 in the USA. About 20 different biosimilars of Herceptin are now in development.

Second generation biologicals — novel protein scaffolds

Although monoclonal antibodies are very effective therapeutic agents, certain inherent characteristics limit their broader use in the clinic. The large size (150–160 kDa), the poor tissue penetration in case of solid tumours and the structure which does not easily allow binding to enzyme pockets are undesirable features [10]. The next generation biologicals are selected high-affinity binding proteins from rationally designed scaffold protein libraries which offer distinct advantages over a full length antibody. These binders have already been very useful tools in research, in the crystallization of membrane proteins for example and as antibodies. Display technologies have greatly contributed to the generation of these designed libraries. The first scaffold protein biological to be approved was ecallantide, a designed Kunitz domain inhibitor of human plasma kallikrein.

Nanobodies

Nanobodies are single domain antibodies with a molecular weight of 12–15 kDa [11]. Derived from the variable VHH domain of the llama, they have to be humanized to reduce the immunogenicity. Several nanobodies are in clinical trials for different indications such as rheumatoid arthritis and osteoporosis.

Scaffold proteins

DARPins

DARPins are designed ankyrin repeat proteins with a molecular weight of 14 or 18 kDa depending on whether 4 or 5 repeats are present [12a, 12b]. Well expressed in E. coli, DARPins are heat stable and may be highly concentrated. DARPins have been successfully used to target tumours and in diagnostic applications. Several DARPins are in clinical trials in ocular indications.

Anticalins

Anticalins are a class of proteins derived from a human lipocalin scaffold with a molecular weight of about 20 kDa [13]. Lipocalins are abundant in plasma and tissue fluids and well suited for clinical applications due to their low immunogenicity. One anticalin is in an early stage clinical trial in an oncology indication.

Affibodies

Affibodies are high affinity antibody mimetics derived from the IgG binding domain of S. aureus Protein A with a molecular weight of only 6 kDa. The scaffold is stable at extreme pH and high temperature [14]. Radiolabelled affibody-albumin conjugates have been used for diagnostic molecular imaging in oncology. Affibodies are also in clinical trials in an oncology indication.

Monobodies/adnectin

Adnectins are extracellular type III domain of human fibronectin derived scaffold proteins with a molecular weight of about 10 kDa [15]. The domain has a structure similar to antibody variable domains. An adnectin is in a phase II clinical trial as an anti-angiogenic inhibitor in an oncology indication.

Domain antibodies

A domain antibody (dAb) is either the variable domain of an antibody heavy chain (VH domain) or the variable domain of an antibody light chain (VL domain) [16]. Dual targeting human domain antibodies are now in a Phase II clinical trial in a neurology indication.

Fynomers, Nanofitins, Knottins, Avimers, Affilins

Fynomers, derived from the novel, fully human protein scaffold Fyn SH3 domain, have a molecular weight of under 10 kDa [17]. Nanofitins are structurally derived from the DNA binding protein Sac7d, found in *Sulfolobus acidocaldarius*, an archaebacterium [18]. They are unusually heat resistant scaffolds with a molecular weight of about 7 kDa. Knottins are engineered cysteine knot peptides for high affinity molecular recognition against tumor-associated receptors [19]. Avimers are evolved from human LDL receptor domains by *in vitro* exon shuffling and phage display, generating multidomain proteins with binding and inhibitory properties [20]. Affilins are structurally derived from human γ-B crystallin (molecular weight about 20 kDa) or ubiquitin (molecular weight 10 kDa) [21]. All these scaffold proteins are in the preclinical phase of drug discovery.

Outlook

This brief overview shows the extent to which biologicals have developed in the last 3 decades. Antibody engineering has given rise to a new set of binders with innovative characteristics such as bi-functionality and conjugation to chemo- or radiotherapic loads. The future looks bright as the second generation pipeline has an impressive collection of drugs designed to overcome the hurdles of delivery, stability and access to the antigen. Biosimilars have entered the market as the first generation biologicals lose their exclusivity and, having won approval, will be a big player also driving innovation in biomanufacturing. The long term effects of these drugs have to be closely monitored, requiring clear regulatory and pharmacovigilance guidelines.

Acknowledgments

Funding from the Swiss National Science Foundation, NCCR Structural Biology and the University of Zürich is gratefully acknowledged.

References

1. (a) http://www.fiercepharma.com/special-reports/15-best-selling-drugs-2012
   (b) http://www.biopharminternational.com/biopharm/article/articleDetail.jsp?id=775167
2. S. J. Kim, Y. Park, H. J. Hong, *Mol. Cells* **20**, 17 (2005).
3. A. Mullard, *Nat. Rev. Drug Discov.* **12**, 329 (2013).

4. G. Shimamoto, C. Gegg, T. Boone, C. Quéva, *MAbs.* **4**, 586 (2012).

5. R. Kole, A. R. Krainer, S. Altman, *Nat. Rev. Drug Discov.* **11**, 125 (2012).

6. A. L. Southwell, N. H. Skotte, C. F. Bennett, M. R. Hayden, *Trends Mol. Med.* **18**, 634 (2012).

7. S. A. Berkowitz, J. R. Engen, J. R. Mazzeo, G. B. Jones, *Nat. Rev. Drug Discov.* **11**, 527 (2012).

8. N. Lingg, P. Zhang, Z. Song, M. Bardor, *Biotechnol. J.* **7**, 1462 (2012).

9. M. Rinaudo-Gaujous, S. Paul, E. D. Tedesco, C. Genin, X. Roblin *et al.*, *Aliment Pharmacol. Th.* **38**, 914 (2013).

10. T. Wurch, A. Pierré, S. Depil, *Trends Biotechnol.* **30**, 575 (2012).

11. F. Van Bockstaele, J. B. Holz, H. Revets, *Curr. Opin. Investig. Drugs* **10**, 1212 (2009).

12. (a) R. Tamaskovic, M. Simon, N. Stefan, M. Schwill, A. Plückthun, *Meth. Enzymol.* **503**, 101 (2012).
    (b) H. K. Binz, P. Amstutz, A. Kohl, M. T. Stumpp, C. Briand *et al.*, *Nat. Biotechnol.* **22**, 575 (2004).

13. M. Gebauer, A. Skerra, *Meth. Enzymol.* **503**, 157 (2012).

14. S. Hoppmann, Z. Miao, S. Liu, H. Liu, G. Ren *et al.*, *Bioconjug. Chem.* **22**, 413 (2011).

15. M. Ackermann, B. A. Morse, V. Delventhal, I. M. Carvajal, M. A. Konerding, *Angiogenesis* **15**, 685 (2012).

16. L. J. Holt, C. Herring, L. S. Jespers, B. P. Woolven, I. M. Tomlinson, *Trends Biotechnol.* **21**, 484 (2003).

17. D. Grabulovski, M. Kaspar, D. Neri, *J. Biol. Chem.* **282**, 3196 (2007).

18. B. Mouratou, G. Béhar, L. Paillard-Laurance, S. Colinet, F. Pecorari, *Meth. Mol. Biol.* **805**, 315 (2012).

19. S. J. Moore, J. R. Cochran, *Meth. Enzymol.* **503**, 223 (2012).

20. J. Silverman, Q. Liu, A. Bakker, W. To, A. Duguay *et al.*, *Nat. Biotechnol.* **23**, 1556 (2005).

21. H. Ebersbach, E. Fiedler, T. Scheuermann, M. Fiedler, M. T. Stubbs *et al.*, *J. Mol. Biol.* **372**, 172 (2007).

PLATFORM TECHNOLOGIES FOR THE ARTIFICIAL PSEUDO-NATURAL PRODUCT DISCOVERY

HIROAKI SUGA

Department of Chemistry, Graduate School of Science, The University of Tokyo
7-3-1 Hongo, Bunkyo-ku, Tokyo 113-0033, Japan

My view of the present state of research on biologicals and biosimilars

Peptide-based molecules are appealing scaffolds for the development of novel drugs. Their high structural diversity, often combined with a large surface area for interaction with the target, allows for very high affinity to the target with exceptional specificity, as evidenced by the success of peptide therapeutics derived from naturally occurring peptide hormones [1–3]. Moreover, even relatively small peptides are able to bind to target sites that are not the active sites of enzymes and to modulate protein-protein interactions, characteristics which are not generally exhibited by traditional small molecule drugs. However, despite these characteristics, peptides exhibit some features that are undesirable in potential therapeutics. Most notably, their susceptibility to proteases makes them unstable *in vivo*, they are not generally orally available, and they are unable to cross cell membranes making them unsuited for use against intracellular targets.

Interestingly, a number of relatively small peptides (generally of the order of 15 residues or less) derived from natural sources circumvent these apparent limitations (Fig. 1). Perhaps the best example of this is the immunosuppressant cyclosporine, a highly N-methylated, macrocyclic, 11 residue peptide which is orally available, exhibits acceptable pharmacokinetics and exerts its therapeutic effect by binding to intracellular target proteins, cyclophilins [4]. Other clinically useful peptide-based drugs derived from natural sources include the echinocandins (anti-fungal), actinomycin D (antibiotic, but used clinically as a chemotherapeutic), and the antibiotics daptomycin and vancomycin [5]. Taken together, these examples demonstrate that these relatively small naturally occurring peptides can exhibit "drug-like" properties, combining the pharmacokinetic qualities of small molecules with the target binding features of larger peptides.

Notably, almost all naturally occurring peptides with potent bioactivities and favorable pharmacokinetic properties possess non-canonical structural features. The most frequently observed of these is some form of macrocyclic motif, which confers structural rigidity (and thus increased target binding affinity) and resistance to proteases, and also plays an important role in biological membrane permeability [6].

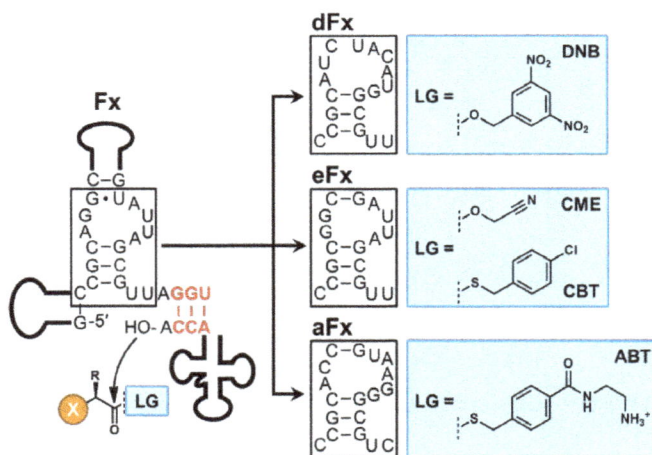

Fig. 1.   Flexizymes. Depending on the substrate kinds, three flexizymes are available.

N-alkylation (commonly *N*-methylation) of the peptide backbone is also commonly observed, and has also been shown to be important for membrane permeability (presumably by reducing the hydrophilicity of the amide backbone) and protease resistance. Additionally, D-stereochemistry, non-canonical side chain structures, backbone heterocycles, and other non-standard moieties are frequently observed in cyclic peptides derived from biological sources, and appear, like macrocyclization and *N*-alkylation, to imbue peptides with improved pharmacokinetic profiles relative to canonical peptides.

The appeal of macrocyclic peptides as drug candidates has lead to significant research efforts to develop screening methods capable of isolating peptides with high target binding affinities from complex libraries. Whilst combinatorial chemistry approaches have shown some promise, several more recently developed techniques have employed biosynthesis of macrocyclic peptides from libraries of combinatorial genetic templates. Such strategies allow for the synthesis of very diverse libraries and the use of iterative selection/amplification-based screening techniques, so that libraries of macrocyclic peptides greater than 10^{13} species can be rapidly screened for binding to a target of interest. Furthermore, the application of genetic code manipulation techniques to these biosynthetic systems and/or post-translational chemical reactions have greatly expanded the range of chemical moieties that can be synthesized, allowing for the screening of libraries of artificial macrocycles that mimic the structural features of naturally occurring bioactive peptides, *i.e.*, artificial pseudo-natural products.

My research themes focus on the development of novel platform technologies that enable us to generate genetically encoded artificial libraries of macrocycles and pseudo-natural products, and the use of selection methods to isolate such molecules with not only high target affinities but also drug-like properties.

My recent research contributions to biologicals and biosimilars

Flexizymes and flexible in-vitro translation (FIT) system: Tools for the genetic code reprogramming

Flexizymes are artificial ribozymes with highly promiscuous aminoacyl-transferase activity. The discovery of these ribozymes through *in vitro* selection experiments, and details of their properties and use has been recently reviewed elsewhere [7]. In brief, flexizymes are small (45 or 46 nucleotides depending on the kinds, Fig. 1) artificial RNAs that catalyze the aminoacylation of essentially any tRNA, using an amino acid substrate that has been chemically activated by a leaving group attached through an ester linkage [8]. Available three flexizymes are capable of charging virtually any amino acid (or any other small molecule with a carboxyl moiety that can be activated) onto diverse tRNAs bearing any anticodon desired.

Importantly, the codon reassignments with non-canonical amino acids using flexizymes are executed by pre-charged non-canonical aa-tRNAs. Therefore, the vacant codons for reassignment can be created by omitting not only the relevant amino acids but also their cognate ARS. Since trace amounts of canonical amino acids are often carried into the reconstituted translation system with the purified ribosomes, translation factors and/or tRNAs, the double exclusion of amino acids and cognate ARSs ensures complete knockdown of any potential competition for elongation. Moreover, aa-tRNA synthesis using flexizymes is far more facile than the chemo-enzymatic method, and flexizymes will accept a much wider range of amino acid substrates than ARSs including *N*-alkyl-amino acids [9, 10], D-amino acids [11], *N*-acylated amino acids [12], and even exotic peptides containing β- and γ-amino acids [13]. Introduction of the resulting non-canonical aa-tRNAs into the custom-made translation system in which specific codons have been vacated thus enables the incorporation of any translatable amino acid(s) into a polypeptide at any desired position(s). Such a translation system integrated with flexizyme chemistry is referred to as a flexible in vitro translation (FIT) reaction. Consequently, it has been possible to generate extremely diverse chemical structures based loosely on a peptide backbone. These have included peptoids (*N*-alkylated polyglycines) [9], (poly)esters [14] and peptides with diverse patterns of *N*-methylation and non-canonical side chains [10].

Furthermore, the diverse chemistries accessible through the FIT system permit the synthesis of peptides containing reactive moieties that can mediate intramolecular cyclization. This allows for the synthesis of macrocyclic peptides without the need for intermolecular chemical cross-linking. As a specific example, the FIT system can be used to synthesize a chloroacetylated N-terminus, which will spontaneously react with a downstream cysteine residue to form a macrocycle through a non-reducible thioether bond [15]. Other examples of macrocyclization through introduction of reactive moieties by FIT synthesis have included oxidative coupling, Huisgen 1,3-dipolar cycloaddition, Michael addition, and native chemical ligation strategies, and a number of these have been combined to produce bicyclic peptides with defined structures [15].

RaPID system: A tool for the discovery of non-canonical macrocyclic peptides

The combination of the FIT system with an mRNA display-based screening method is known as <u>ra</u>ndom non-standard <u>p</u>eptides <u>i</u>ntegrated <u>d</u>iscovery (RaPID, Fig. 2), and has been used to identify a number of bioactive macrocyclic peptides with high binding affinities to therapeutic targets of interest. Like the ARS-mediated genetically reprogrammed mRNA display selections described above, RaPID screening involves translation of a semi-randomized puromycin-linked mRNA library in a genetically-reprogrammed, fully-reconstituted FIT system, such that the resulting non-standard peptides are covalently linked to the mRNA that encodes them. Iterative rounds of selection/amplification are then performed against the desired target immobilized on magnetic beads, with the enriched mRNA libraries translated in the same FIT system at each step.

In their simplest incarnation, such selections can be performed using a FIT system that primarily incorporates canonical amino acids, initiated by a non-canonical D- or L-amino acid bearing a chloroacetyl moiety at the *N*-terminus and

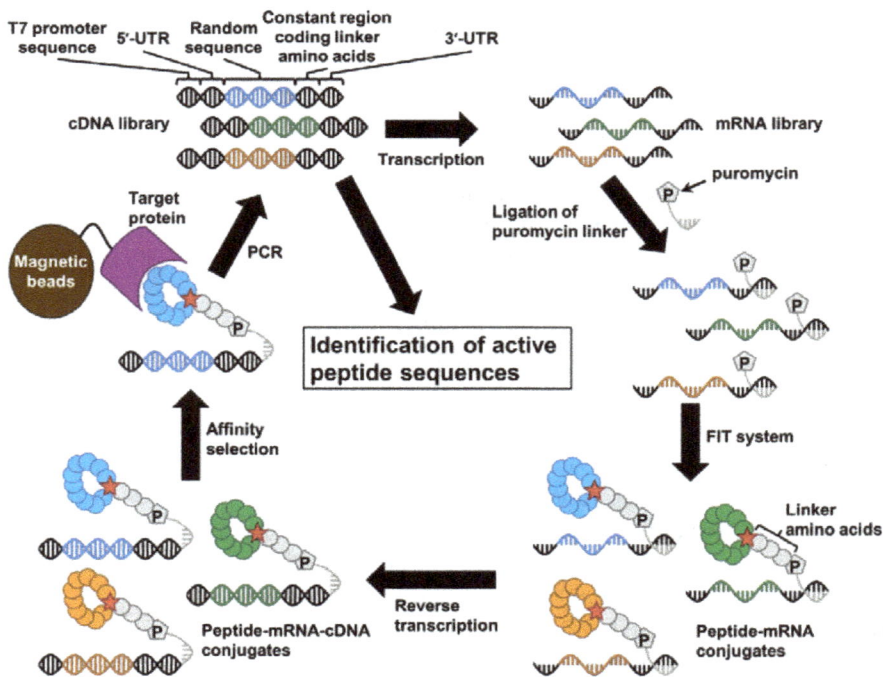

Fig. 2.   RaPID system (mRNA display coupled with FIT system).

a downstream cysteine in order to form a macrocycle through a thioether linkage (as described above). Using such an approach, potent (IC$_{50}$ ~100 nM) and isoform-selective cyclic peptide inhibitors of the human AKT2 serine/threonine kinase have been reported [16]. A similar approach was used to identify macrocyclic peptide inhibitors (again, active at nM concentrations) of the bacterial drug-transporter pfMATE, a finding of particular significance since proteins of this family are responsible for drug resistance in both bacterial pathogens and malignant cells [17]. It is worth noting that this particular work involving 3-dimensional X-ray structures of the macrocyclic peptide-pfMATE complexes provided the first example of how the series of macrocyclic peptides that are obtained by RaPID selection interact with a target protein.

Selections using libraries including more diverse non-canonical amino acid residues have also been reported. For example, cyclic peptide inhibitors of the human Sirtuin2 lysine deacetylase were identified from a library of macrocyclic peptides engineered to include a trifluoroacetyl-lysine "warhead" that mimics the acetyl-lysine substrate of the enzyme [18]. The derived Sirtuin2 inhibitors exhibited high potency (K_D as well as IC$_{50}$ in the single digit nM range) and a high degree of isoform selectivity.

RaPID selection using still more complex non-canonical libraries is also possible. Taking advantage of the capacity of the FIT system to translate N-methylated amino acids, a library engineered to express N-methylated Phe, Ser, Gly and Ala, as well as D-Trp (and macrocyclized through an intramolecular reaction between an N-terminal chloroacetyl group and a downstream cysteine as described above) was screened for inhibitors of the human oncoprotein E6AP [19]. The macrocyclic E6AP inhibitors obtained each included multiple N-methylated residues, and were arguably more like naturally occurring non-standard macrocyclic peptides than any other non-natural macrocyclic peptide inhibitor discovered to date. Moreover, they exhibited extremely high affinity for E6AP, with K_Ds in the sub-nM to single digit nM range, and appeared able to disrupt the protein-protein interaction between E6AP and the P53 tumor suppressor protein.

Outlook to future developments of research on biologicals and biosimilars

The RaPID system is quite robust for selecting macrocyclic peptides against various therapeutic targets. However two major issues remain to be solved. First, despite the fact that selected macrocyclic peptides are generally highly active and selective against the target proteins, they are not necessarily membrane permeable and oral available even though N-methylation of backbone is implemented into the selection scheme. It seems critical to improve the RaPID methodology that involves membrane permeability selection step. Moreover, peptidic natural products contain more diverse kinds of non-peptidic structures, such as azolines and azoles, modified by post-translational modifying enzymes. These non-peptidic modifications into the peptide backbone might significantly contribute to the drug-like properties. Therefore, it is also important for the future methodology development to include such post-translational modification using appropriate enzymes.

Acknowledgments

This work was supported by a JSPS Grants-in-Aid for the Specially Promoted Research (21000005), JST CREST, and MEXT Platform for Drug Discovery, Informatics, and Structural Life Science to H.S.

References

1. A. H. Barnett, D. R. Owens, *Lancet* **349**, 47 (1997).
2. R. L. Koretz, M. Pleguezuelo, V. Arvaniti, P. Barrera Baena, R. Ciria *et al.*, *Cochrane Datab. Syst. Rev.* **1**, CD003617 (2013).
3. A. Takeda, K. Cooper, A. Bird, L. Baxter, G. K. Frampton *et al.*, *Health Technol. Assess.* **14**, 1 (2010).
4. R. E. Handschumacher, M. W. Harding, J. Rice, R. J. Drugge, D. W. Speicher, *Science* **226**, 544 (1984).
5. D. S. Perlin, *Future Microbiol.* **6**, 441 (2011).
6. E. M. Driggers, S. P. Hale, J. Lee, N. K. Terrett, *Nat. Rev. Drug Discov.* **7**, 608 (2008).
7. Y. Goto, T. Katoh, H. Suga, *Nat. Protoc.* **6**, 779 (2011).
8. H. Murakami, A. Ohta, H. Ashigai, H. Suga, *Nat. Meth.* **3**, 357 (2006).
9. T. Kawakami, H. Murakami, H. Suga, *J. Amer. Chem. Soc.* **130**, 16861 (2008).
10. T. Kawakami, H. Murakami, H. Suga, *Chem. Biol.* **15**, 32 (2008).
11. T. Fujino, Y. Goto, H. Suga, H. Murakami, *J. Amer. Chem. Soc.* **135**, 1830 (2013).
12. Y. Goto, A. Ohta, Y. Sako, Y. Yamagishi, H. Murakami, H. Suga, *ACS Chem. Biol.* **3**, 120 (2008).
13. Y. Goto, H. Suga, *J. Amer. Chem. Soc.* **131**, 5040 (2009).
14. A. Ohta, H. Murakami, E. Higashimura, H. Suga, *Chem. Biol.* **14**, 1315 (2007).
15. T. Passioura, H. Suga, *Chemistry* **19**, 6530 (2013).
16. Y. Hayashi, J. Morimoto, H. Suga, *ACS Chem. Biol.* **7**, 607 (2012).
17. Y. Tanaka, C. J. Hipolito, A. D. Maturana, K. Ito, K. Kuroda *et al.*, *Nature* **496**, 247 (2013).
18. J. Morimoto, Y. Hayashi, H. Suga, *Angew. Chem. Int. Ed. Engl.* **51**, 3423 (2012).
19. Y. Yamagishi, I. Shoji, S. Miyagawa, T. Kawakami, T. Katoh *et al.*, *Chem. Biol.* **18**, 1562 (2011).

ANTICALINS® & PASYLATION®:
NEW CONCEPTS FOR BIOPHARMACEUTICAL DRUG DEVELOPMENT FROM PROTEIN DESIGN

ARNE SKERRA

Lehrstuhl für Biologische Chemie, Technische Universität München
Emil-Erlenmeyer-Forum 5, 85350 Freising-Weihenstephan, Germany

My view of the present state of research on biologicals (and biosimilars)

During the past century, since their discovery by Emil von Behring and Paul Ehrlich [1], antibodies (immunoglobulins) have been considered as ideal tools for molecular recognition in the life sciences and biomedical research. The invention of methods for the production of their functional fragments in *E. coli* [2, 3] has boosted the generation of recombinant antibodies with modified amino acid sequences and altered properties.

Since then, human (or humanized) monoclonal antibodies have become a particularly successful class of biologics during the past decade [4], complementing conventional biopharmaceuticals that, altogether, currently represent approximately one quarter of newly approved drugs [5]. High target specificity, therapeutic efficacy, safety and, consequently, low failure rate during preclinical and clinical development contribute to the medical and commercial success of antibodies. Thus, immunotherapy has become an established strategy especially for the treatment of cancer. However, with regard to novel applications in medicine and biotechnology it appears that immunoglobulins are not always optimally suited, in particular due to their composition of two different polypeptide chains and to their very large molecular size.

Members of the lipocalin protein family [6] have shown promise as an alternative scaffold for the generation of ligand-binding proteins via combinatorial protein design [7, 8]. Natural lipocalins are compact proteins which serve for the transport or storage of biomolecular compounds such as vitamins, hormones and secondary metabolites in many organisms, including humans. Their tertiary structure comprises a circularly closed eight-stranded anti-parallel β-sheet. This so-called β-barrel supports four structurally hypervariable loops at its open end, which form the entrance to the ligand pocket. In comparison with recombinant antibodies, lipocalins provide several advantages as they are composed of a single polypeptide chain, have a much smaller size, and their set of four loops can be easily manipulated at the genetic level, allowing the design of alternative binding proteins that have been dubbed 'Anticalins' [7].

Apart from full size immunoglobulins, most biologics that are successfully used in human therapy, for example interferon, erythropoietin, hormones, growth factors, and recombinant antibody fragments, are relatively small proteins. Their molecular sizes are much below the threshold for kidney filtration (about 70 kDa), which leads to rapid elimination from the bloodstream. Accordingly, these therapeutic proteins often have an extremely short half-life in human plasma of just a few hours [9]. This impairs their therapeutic benefit considerably, and in most cases repeated injections at short intervals and high dosing are required in order to achieve a pharmaceutical effect.

A currently accepted strategy to prolong the plasma half-life of biologics is their chemical conjugation with the water-soluble and inert polyethylene glycol (PEG), thereby increasing the hydrodynamic volume of the therapeutic protein or peptide beyond the pore size of renal glomeruli [10]. In recent years several PEGylated biologics have found entry into the clinic, for example Pegasys® or PegIntron®, both PEGylated versions of interferon α, or Pegfilgrastim®, a PEGylated G-CSF.

However, PEGylation of therapeutic proteins is accompanied by additional effort and high cost: suitable reactive PEG derivatives with high purity are expensive, coupling with the recombinant protein often has poor yield and reduces its activity. Also, additional downstream purification steps are needed, wherein product heterogeneity causes further problems. Finally, PEG is not biodegradable and can accumulate in the kidney or other organs upon continued use. Consequently, there is urgent need for other technologies to enable the design of highly active and safe biologics with prolonged action. 'PASylation', which makes use of a conformationally disordered and genetically encodable polymer composed of natural amino acids, provides a biological alternative [11].

My recent research contributions to biologicals (and biosimilars)

Anticalins are a novel class of small and robust engineered protein reagents developed in our laboratory [7, 8], which are usually derived via combinatorial protein design from human plasma lipocalins and offer potential as an alternative to antibodies [12, 13]. The lipocalin protein fold is dominated by a central β-barrel of eight anti-parallel strands, which is open to the solvent at one end where a set of four structurally hypervariable loops forms the entrance to the ligand pocket, similarly to the six complementarity-determining regions (CDRs) of an antibody. While human(ized) immunoglobulins constitute an established class of biotherapeutics today, they show several disadvantages, especially in areas where the full-size molecular format with its immunological effector functions is less relevant. Anticalins have a much smaller size (160–180 residues), providing better tissue penetration, and they comprise just a single polypeptide chain, thus offering the efficient preparation of fusion proteins and/or conjugation with payloads or enzymes, even by employing inexpensive microbial expression systems.

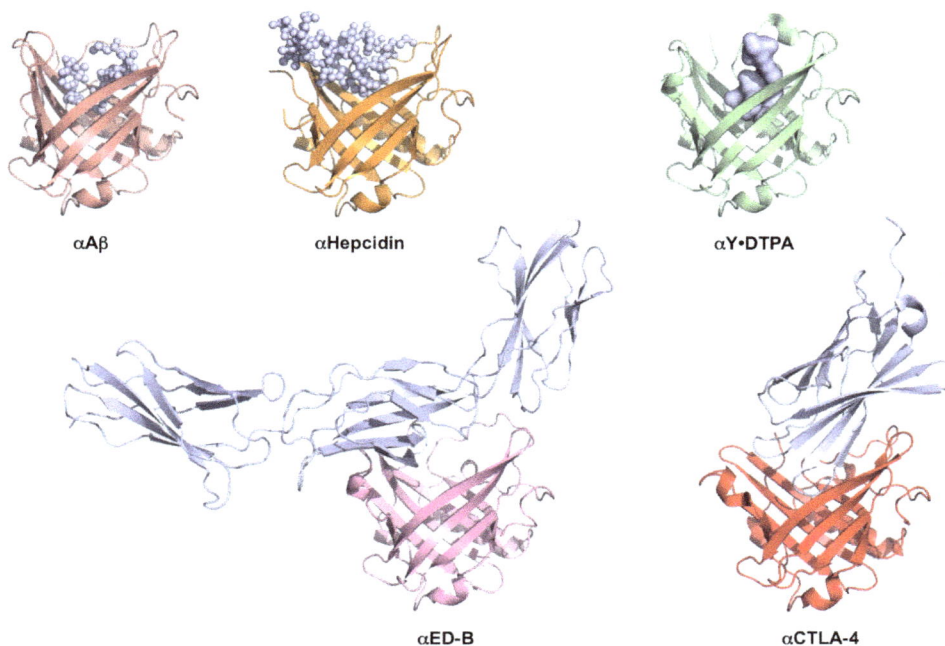

Fig. 1. Panel of the X-ray structures of five Anticalins based on the same human lipocalin scaffold that are directed against diverse medically relevant molecular targets: Aβ, Alzheimer amyloid β peptide; CTLA-4, cytotoxic T lymphocyte-associated antigen 4 (CD152); ED-B, extra-domain B of the extracellular matrix protein fibronectin; Hepcidin, a peptide regulator of iron homeostasis; Y-DTPA, yttrium (III) in complex with the chelator [(R)-2-amino-3-(4-aminophenyl)propyl]-trans-(S,S)-cyclohexane-1,2-diamine-pentaacetic acid.

Anticalins with specificities and tight functional blocking activity toward a series of disease-relevant biomolecules have been conveniently generated by targeted random mutagenesis of natural lipocalins in combination with phage display selection [14–16]: *e.g.*, the pro-angiogenic factor VEGF-A, the tumor-associated extra-domain B of fibronectin and the deactivating T-cell co-receptor CTLA-4. A beneficial feature of Anticalins is their deep binding pocket, which — similarly to the functions of natural lipocalins — also enables the high affinity complexation of small molecule ligands of various kinds, including toxic substances such as digitalis [17] or radiometal-chelate complexes [15] and even the Alzheimer amyloid β peptide (Fig. 1), thus offering interesting applications in biomedical research, diagnostics and therapy as well as in nuclear medicine or as antidotes. The VEGF-specific Anticalin 'Angiocal' has completed Phase I trials for the treatment of solid tumors while the clinical study of another Anticalin, directed against the peptide Hepcidin, is currently in preparation [13].

Except for full size antibodies, which undergo endosomal recycling, rapid clearance from blood circulation by renal filtration is a typical drawback of therapeutic proteins and peptides, affecting most of all conventional biologics such as hormones, growth factors, cytokines, enzymes and almost all clinically relevant antibody fragments (scFv, Fab or single domain/nanobody) or alternative protein scaffolds. At present, chemical conjugation with PEG, which expands the effective molecular size beyond the threshold of kidney filtration, is an established strategy to extend the short plasma half-life of biologicals to a clinically useful range. However, coupling with the chemical polymer PEG requires expensive material and downstream procedures and, in addition, PEG can accumulate in organs due to its lack of biodegradability.

As an alternative, we have developed conformationally disordered polypeptide chains with expanded hydrodynamic volume (Fig. 2) comprising the small amino acid residues Pro, Ala, and/or Ser (PAS). PAS sequences are hydrophilic, uncharged biological polymers with biophysical properties surprisingly similar to PEG [11]. In contrast, PAS polypeptides offer simple fusion to a biologic on the genetic level, thus permitting facile production in *E. coli* (or other expression systems) of fully active proteins in high yields [18] and obviating *in vitro* coupling or modification steps. Also, PAS sequences are biodegradable, thus avoiding tissue accumulation, while showing high stability in blood plasma and lacking toxicity or immunogenicity

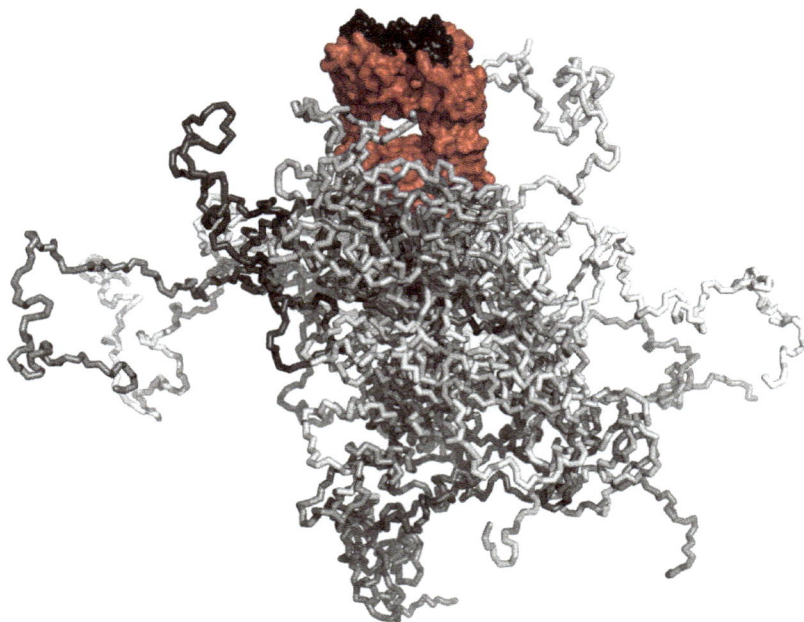

Fig. 2.   Modeled structure of the PASylated Fab fragment of an antibody. The two polypeptide chains of the Fab fragment are colored red while the antigen-binding site is shown in black. 24 arbitrarily selected random conformations of the PAS polymer at the C-terminus of the light chain are presented as snapshots and superimposed in different shades of grey to illustrate its fluctuating random coil-like, space-filling behavior.

in animals. Furthermore, by adjusting the length of the PAS tag, typically between 200 and 1000 residues, the effect on plasma half-life can be easily tuned, allowing extension by one to two orders of magnitude.

Outlook to future developments of research on biologicals (and biosimilars)

The growing field of modern pharmaceuticals is dominated by chemical (small molecule) drugs on the one hand and biologics on the other, which currently contribute approximately one quarter of newly approved substances, with rising overall proportion [5]. Notably, the vast majority of biologics constitutes therapeutic proteins while other biomacromolecules, like carbohydrates or nucleic acids, are confined to niche applications. Peptides may be assigned to both classes, depending on whether they are manufactured by chemical synthesis — which is more costly but facilitates introduction of covalent modifications — or by way of biotechnology. An interesting recent option is the incorporation of non-natural amino acids into biosynthetic proteins via expansion of the genetic code [19]. This opens the route to the site-specific introduction of novel side chain functionalities that can either be useful *per se*, as in the case of spectroscopic probes [20], or may enable the efficient and directed conjugation of a protein with any other chemically functionalized compound, for example radionuclide chelators or small molecule toxins. So far, drug development has focused on the specific recognition and/or tight blocking of disease-relevant cellular/biomolecular targets, which explains the success of antibody technology up to now. Meanwhile, it has become apparent that the careful design of additional pharmaceutical properties, in particular biodistribution, circulation half-life and charging with payloads, will be crucial to provide the next wave of biologics — possibly of a hybrid nature — that offer superior therapeutic effects. Eventually, this development could form a fruitful link between the disciplines of chemistry and biology, which have largely separated since the early days of Ehrlich and von Behring, to yield novel biomolecular drugs for the benefit of human health.

Acknowledgments

The author wishes to thank the Deutsche Forschungsgemeinschaft, the Bundesministerium für Bildung und Forschung, Germany, the European Commission as well as Pieris AG and XL-protein GmbH, both in Germany, for financial support.

References

1. F. Bosch, L. Rosich, *Pharmacology* **82**, 171 (2008).
2. A. Plückthun, A. Skerra, *Meth. Enzymol.* **178**, 497 (1989).
3. A. Skerra, A. Plückthun, *Science* **240**, 1038 (1988).
4. J. M. Reichert, *MAbs* **3**, 76 (2011).
5. G. Walsh, *Nat. Biotechnol.* **28**, 917 (2010).

6. B. Åkerström, N. Borregaard, D. A. Flower, J.-S. Salier, *Lipocalins.* Landes Bioscience, Georgetown, Texas (2006).

7. A. Skerra, *Biochim. Biophys. Acta* **1482**, 337 (2000).

8. G. Beste, F. S. Schmidt, T. Stibora, A. Skerra, *Proc. Natl. Acad. Sci. USA* **96**, 1898 (1999).

9. J. H. Lin, *Curr. Drug Metab.* **10**, 661 (2009).

10. G. Pasut, F. M. Veronese, *J. Control. Release* **161**, 461 (2012).

11. M. Schlapschy, U. Binder, C. Börger, I. Theobald, K. Wachinger *et al.*, *Protein Eng. Des. Sel.* **26**, 489 (2013).

12. A. Skerra, *Curr. Opin. Chem. Biol.* **7**, 683 (2003).

13. C. T. Mendler, A. Skerra, *Drug Future* **38**, 169 (2013).

14. M. Gebauer, A. Schiefner, G. Matschiner, A. Skerra, *J. Mol. Biol.* **425**, 780 (2013).

15. H. J. Kim, A. Eichinger, A. Skerra, *J. Am. Chem. Soc.* **131**, 3565 (2009).

16. D. Schönfeld, G. Matschiner, L. Chatwell, S. Trentmann, H. Gille *et al.*, *Proc. Natl. Acad. Sci. USA* **106**, 8198 (2009).

17. F. Eyer, W. Steimer, T. Nitzsche, N. Jung, H. Neuberger, C. Müller, M. Schlapschy, T. Zilker, A. Skerra, *Toxicol. Appl. Pharmacol.* **263**, 352 (2012).

18. S. Di Cesare, U. Binder, T. Maier, A. Skerra, *BioProcess. Int.* **11**, 30 (2013).

19. T. S. Young, P. G. Schultz, *J. Biol. Chem.* **285**, 11039 (2010).

20. S. M. Kuhn, M. Rubini, M. A. Müller, A. Skerra, *J. Amer. Chem. Soc.* **133**, 3708 (2011).

FROM NATURAL ANTIBODIES TO SYNTHETIC PROTEINS

SACHDEV S. SIDHU

Banting and Best Department of Medical Research & Department of Molecular Genetics
University of Toronto, 160 College Street, Toronto, Ontario, M5S 3E1, Canada

My view of the present state of research on biologicals and biosimilars

The human genome project has been a success in terms of providing basic knowledge of cellular function. Advanced DNA sequencing technologies and technologies for the systematic analysis of genomes, transcriptomes and proteomes have provided unprecedented insights into the inner workings of both healthy and diseased cells. Consequently, the research community is poised to produce tremendous advances in therapies for numerous human diseases, if only this basic knowledge can be translated to the practical level of drug development.

However, it remains unclear how the vast data acquired from genomics and systems biology can be applied to the practical business of drug development. In my opinion, a major cause for this disconnect is the lack of advanced tools for the manipulation of proteins with anything close to the speed and scope that is possible for the manipulation of nucleic acids. Proteins are central to essentially all cellular processes, most diseases are caused by protein malfunction, and most drugs act by modulating the activities of proteins. Thus, it seems clear that basic knowledge from genomics research can only be translated into meaningful therapies if we can develop comprehensive tools for modulating proteins with the same speed and precision that is commonplace for nucleic acids. Such a toolkit would exponentially expand our knowledge of normal and diseased cell function, and most importantly, the tools could be converted into therapeutics that prevent or reverse disease by correcting aberrant protein function.

Unfortunately, despite their undeniable importance, current methods for the generation of protein modulators are not scalable to the genome level. Small molecules remain invaluable as both research tools and drugs, but are limited to a small niche of druggable targets and require extensive effort and expense to produce. Antibodies offer an attractive alternative but they are restricted to extracellular targets and traditional hybridoma technologies that rely on animal immunization are not scalable in a facile manner. In my opinion, there is an urgent need for advanced "synthetic" protein technologies that take advantage of *in vitro* selection technologies and man-made protein libraries. With such designed systems, it would be possible to efficiently scale up affinity reagent production to a genome-wide scale and the binding proteins could be engineered with tailored characteristics that would make them ideally suited for biotechnology applications.

My recent research contributions to biologicals and biosimilars

My group has focused on combining knowledge of protein structure and function with combinatorial phage display technology. Our goal has been to not only develop combinatorial protein libraries capable of generating affinity reagents against diverse antigens, but also, to do so in a defined and controlled manner so that we can gain further insights into library design and develop further improved libraries. The ultimate aim has been to produce synthetic proteins that are simple and highly stable, but are comparable or better than natural antibodies in terms of affinities and specificities. Over the course of the past decade, we have progressed from synthetic antibody libraries, to libraries of synthetic proteins built on alternative scaffolds, to small proteins that can be produced by chemical synthesis.

Our phage-displayed libraries of synthetic antibodies use entirely man-made antigen-binding sites and thus circumvent the need for natural immune repertoires. We have taken a particularly simple approach by using a single, optimized human framework and diversifying only a limited subset of positions within the antigen-binding site that are most important for antigen recognition. Moreover, we have also limited chemical diversity to those amino acids that are most suitable for mediating molecular recognition, including large aromatic residues to provide contacts and small residues to provide conformational flexibility [1, 2]. The resulting libraries have proven to be extremely functional and have generated thousands of high affinity antibodies against hundreds of antigens [3, 4] (Fig. 1(a)). Importantly, the *in vitro* selection technology has enabled controlled selections to tackle molecular recognition tasks that would be extremely difficult if not impossible with animal immunization, including development of antibodies against protein conformational states [5], membrane proteins [6], RNA [7, 8] and post-translational modifications [9]. Also, the *in vitro* nature of the technology has made it possible to scale up the selection process to enable antibody generation against hundreds of antigens in parallel. Thus, synthetic antibody technology has now surpassed both the throughput and quality of hybridoma technology and has enabled numerous applications beyond the scope natural antibodies.

As a further extension of *in vitro* library technology, we have gone beyond antibodies to develop alternative scaffolds adapted to tasks for which antibodies are not suitable. In particular, antibodies are evolved for extracellular function and generally do not fold in cells, thus limiting their use for applications that require protein function in an intracellular environment. To address this issue, we have developed phage display libraries of small proteins that are adapted for intracellular function and have engineered these for enhanced affinity against targets that the natural proteins bind to only weakly. In one example, we have engineered superbinder SH2 domains that bind to phospho-tyrosine peptides with affinities many orders of

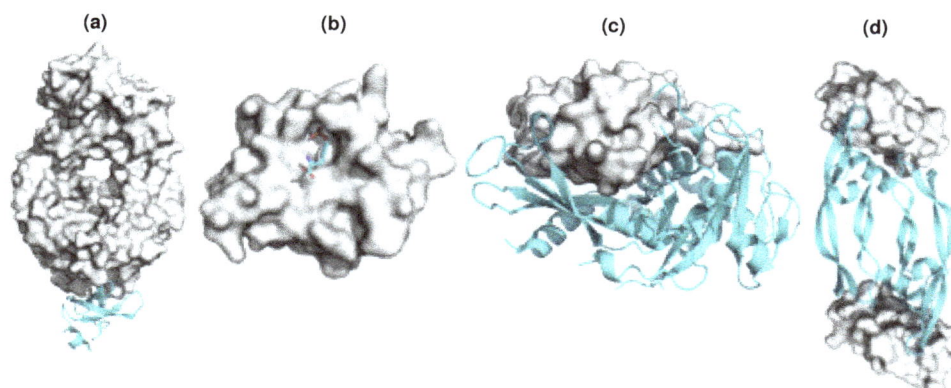

Fig. 1. Synthetic proteins in complex with natural ligands. The structures of synthetic proteins and their natural ligands are shown as *grey* surfaces or *cyan* ribbons, respectively. (a) A synthetic antigen-binding fragment bound to a tudor domain (PDB entry 3PNW). (b) A superbinder SH2 domain bound to a phospho-tyrosine peptide (PDB entry 4F5B). (c) An ubiquitin variant bound to an ubiquitin specific protease (PDB entry 3MTN). (d) A D-amino acid protein bound to vascular endothelial growth factor (PDB entry 4GLS).

magnitude higher than those of natural SH2 domains [10]. This was achieved by optimizing the SH2 domain residues that contact the phospho-tyrosine side chain, and thus, these domains exhibit enhanced affinity without alterations in specificity (Fig. 1(b)). We have used superbinder SH2 domains as inhibitors of natural SH2 domain interactions and have shown that these inhibitors can halt cancer cell growth [10], or alternatively, we have replaced natural SH2 domains with superbinders in the context of whole proteins to alter signaling pathways [11]. In another example, we have used ubiquitin as a scaffold to generate variants that bind tightly and specifically to ubiquitin modifying enzymes [12] (Fig. 1(c)). Using selective ubiquitin variants, we have been able to develop inhibitors of several deubiquitinases and have shown that these inhibitors are functional in live cells. We extended the strategy to target ubiquitin conjugating enzymes and ubiquitin ligase enzymes and, in some cases, showed that ubiquitin variants can also act as enzyme activators. Finally, we showed that the approach can be used to derive binders to non-catalytic ubiquitin-binding domains that function as mediators of protein-protein interactions. Thus, it has become clear that ubiquitin, SH2 domains and other simple intracellular proteins can be engineered as selective inhibitors to modulate protein function inside cells. These advances will open up many avenues for basic research and could also lead to the development of next generation therapeutics.

Taking the concept of synthetic proteins a step further, we have developed small, optimized scaffolds that function like antibodies but are amenable to full chemical synthesis, thus enabling the incorporation of non-natural amino acids. The power of the technology has been demonstrated by the development of D-amino acid proteins that act as inhibitors of natural vascular endothelial growth factor [13] (Fig. 1(d)).

Thus, we showed that D-proteins can recognize natural L-proteins with high affinity and specificity. However, unlike L-proteins, D-proteins are resistant to proteolysis by natural proteases and should be much more stable in serum and cellular environments. Moreover, D-peptides are not recognized by the major histocompatibility complex, and thus, they should exhibit lower immunogenicity. Finally, the use of chemical synthesis also facilitates the incorporation of many other non-natural amino acids with special functions and the incorporation of specialized chemical moieties to enable advanced features such as imaging and drug delivery.

In sum, these advances in the design of synthetic binding proteins extend the applications for affinity reagents well beyond the range of natural antibodies and this should have a transformative effect on many areas of biological research.

Outlook to future developments of research on biologicals and biosimilars

In the post-genomics era, it has become clear that protein science must evolve if we hope to translate genomics information to the protein level. To meet these demands, protein science must adapt new, rapid methods that rely on precisely engineered synthetic proteins, rather than old, slow methods based on animal immunization and natural antibodies. I believe that continued refinements in synthetic protein library designs will continue to accelerate the pace of affinity reagent generation, and developing binding tools against whole proteomes will soon become a reality. Synthetic approaches and *in vitro* technologies are also ideally suited to mesh with emerging technologies in nano-scale devices, microfluidics and arrays. Thus, I predict that not only will we be able to generate synthetic proteins on a proteome scale, but we will also be able to integrate these proteins into next generation systems for cell-based assays and detection. These advances will fuel breakthroughs in basic research and will enable the development of new therapies for unmet medical needs.

Acknowledgments

I thank the Canadian Institutes of Health Research, the Ontario Institute for Cancer Research, the National Institutes of Health and Reflexion Pharmaceuticals for financial support.

References

1. F. A. Fellouse, B. Li, D. M. Compaan, A. A. Peden, S. G. Hymowitz *et al.*, *J. Mol. Biol.* **348**, 1153 (2005).
2. F. A. Fellouse, C. Wiesmann, S. S. Sidhu, *Proc. Natl. Acad. Sci. USA* **101**, 12467 (2004).

3. F. A. Fellouse, K. Esaki, S. Birtalan, D. Raptis, V. J. Cancasci *et al.*, *J. Mol. Biol.* **373**, 924 (2007).
4. H. Persson, W. Ye, A. Wernimont, J. Adams, R. Lam *et al.*, *J. Mol. Biol.* **425**, 803 (2013).
5. J. Gao, S. S. Sidhu, J. A. Wells, *Proc. Natl. Acad. Sci. USA* **106**, 3071 (2009).
6. S. Uysal, V. Vasquez, V. Tereshko, K. Esaki, F. A. Fellouse *et al.*, *Proc. Natl. Acad. Sci. USA* **106**, 6644 (2009).
7. Y. Koldobskaya, E. M. Duguid, D. M. Schechner, N. B. Suslov, J. Ye *et al.*, *Nat. Struct. Mol. Biol.* **18**, 100 (2011).
8. J.-D. Ye, V. Tereshko, J. K. Frederiksen, A. Koide, F. A. Fellouse *et al.*, *Proc. Natl. Acad. Sci. USA* **105**, 82 (2008).
9. K. Newton, M. L. Matsumoto, I. E. Wertz, D. S. Kirkpatrick, J. R. Lill *et al.*, *Cell* **134**, 668 (2008).
10. T. Kaneko, H. Huang, X. Cao, X. Li, C. Li *et al.*, *Sci. Sign.* **5**, ra68 (2012).
11. G. M. Findlay, M. J. Smith, F. Lanner, M. S. Hsiung, G. D. Gish *et al.*, *Cell* **152**, 1008 (2013).
12. A. Ernst, G. Avvakumov, J. Tong, Y. Fan, Y. Zhao *et al.*, *Science* **339**, 590 (2013).
13. K. Mandal, M. Uppalapati, D. Ault-Riche, J. Kenney, J. Lowitz *et al.*, *Proc. Natl. Acad. Sci. USA* **109**, 14779 (2012).

FROM INTACT ANTIBODIES TO ARMED ANTIBODIES

DARIO NERI

Department of Chemistry and Applied Biosciences, Swiss Federal Institute of Technology (ETH Zürich), Wolfgang-Pauli-Strasse 10, CH-8093 Zürich, Switzerland

My view of the present state of research on biologics and biosimilars

Monoclonal antibodies represent the largest and fastest growing class of pharmaceutical biotechnology products [1, 2]. The main application areas of therapeutic antibodies include cancer, autoimmunity and chronic inflammatory conditions. The clinical use of a number of "intact" antibodies (*i.e.*, immunoglobulins in IgG format) is well established and these products are now facing competition by biosimilar products. However, new opportunities are offered by the development of "armed" antibodies, in which the immunoglobulin moiety mainly serves for the selective *in vivo* pharmacodelivery of bioactive payloads (such as cytotoxic drugs, bispecific antibodies, radionuclides or cytokines) to sites of disease, helping increase activity and spare normal tissues. In this article, I present a personal perspective on the opportunities and challenges associated with the use of armed antibody products for the treatment of cancer and inflammation.

Conventional IgG-based therapeutics may act by a variety of mechanisms, which include the blockade of a biological mediator of disease (*e.g.*, TNF in inflammation or VEGF in angiogenesis-related conditions), by the induction of signaling events, by the activation of complement or by the recruitment of suitable leukocytes (*e.g.*, NK cells) at the site of disease (a process also called Antibody-Dependent Cell Cytotoxicity, or ADCC).

In Oncology, monoclonal antibody products are not terribly efficacious when used as single agents for the treatment of bulky solid tumors. This limitation may be related to the rather inefficient antibody uptake into the neoplastic mass, which is a consequence of limited extravasation rate and of blockade by perivascular tumor cells (the so-called "antigen barrier").

In mice, the best tumor-targeting antibodies may exhibit a tumor uptake as large as 50–100% injected dose per gram (*i.e.*, 5–10% of the injected dose localize to a 100 mg tumor mass, after the initial distribution phase). The corresponding %ID/g value on solid tumors in cancer patients may be in the order of 0.1–1%, mainly as a consequence of the different body weight (20–25 g against 50–100 kg) of the two organisms. Thus, only a relatively small portion of the antibody molecule reaches the target following intravenous administration, while virtually all antibody product, at some stage, will pass through an excretory organ (the hepatobiliary system for intact antibodies, the renal system for small antibody fragments). Yet, at suitable time points (*e.g.*, after 24 hours), the best antibody products may exhibit tumor:blood and tumor:organ ratios greater than 10:1, thus being suitable for the preferential pharmacodelivery of payloads at the site of disease [3, 4]. For these reasons, antibodies are increasingly being considered as "vehicles" for the targeting of payloads not only to cancer lesions, but also to other conditions (*e.g.*, arthritis, psoriasis, multiple sclerosis, inflammatory bowel disorders).

My recent research contributions to biologics and biosimilars

Antibodies can be used to deliver bioactive molecules (drugs, cytokines, photosensitizers, radionuclides, etc.) to the tumor environment, thus sparing normal tissues. The targeting of modified sub-endothelial extracellular matrix components using armed antibodies is particularly attractive, because of: (i) the abundance and stability of some of these antigens; (ii) the fact that some of these antigens are strongly expressed in various cancer types and in angiogenesis-dependent diseases, while being virtually undetectable in most normal adult tissues; (iii) the accessibility of these structures from the blood-stream. The use of armed antibodies for the targeting of newly-formed blood vessels is a therapeutic strategy, often referred to as "vascular targeting" [4].

Over the last 20 years, we have demonstrated the feasibility of antibody-based vascular targeting strategies in several mouse models of cancer [4], arthritis [5], endometriosis [6], angiogenesis-related ocular conditions [7], atherosclerosis [8], vasculopathy [9], inflammatory bowel diseases and psoriasis [unpublished]. Five armed antibody products developed in my lab have been moved to Phase I and Phase II clinical trials, in collaboration with industrial partners, for the treatment of cancer and of rheumatoid arthritis.

We have explored the role of cytotoxic agents [10, 11], radionuclides [12] and cytokines [13] as possible payloads for the development of armed antibody products. In particular, antibody-cytokine fusion proteins (immunocytokines) exhibit a considerable therapeutic potential and a number of favorable properties: (i) these products can be very active in patients at low (milligram) doses; (ii) cytokines have often already been studied in the clinic, thus facilitating the development of the corresponding immunocytokine products; (iii) immunocytokines can be used to potentiate or attenuate the action of the immune system at the site of disease and are easy to use in combination with other drugs; (iv) unlike antibody-drug conjugates, these products do not exhibit a direct damage of clearance-associated organs.

Outlook to future developments of research on biologics and biosimilars

Even though biosimilars are not identical to the original product (they may differ in terms of glycosylation patterns and, potentially, also in terms of microheterogeneities and different formulation), the recent approval of biosimilar antibody products indicates that this class of medicinal agents is already a pharmaceutical reality, which may receive a broad indication label after the execution of a limited amount of clinical studies. Pharmaceutical companies may prefer to focus on the development of innovative antibody products, since break-through technologies and novel therapeutic concepts now allow the generation of products of unprecedented activity and selectivity.

Acknowledgments

The research of the group Neri is supported by the ETH Zürich, the Swiss National Science Foundation, Oncosuisse, the European Union (FP Program PRIAT and IMI Project QUIC-CONCEPT), the Swiss Federal Commission for Technology and Innovation (KTI MedTech Award) and by research contracts with the Philogen group.

References

1. G. Walsh, *Nat. Biotechnol.* **28**, 917 (2010).
2. P. J. Carter, *Nat. Rev. Immunol.* **6**, 343 (2006).
3. G. L. Poli, C. Bianchi, G. Virotta, A. Bettini, R. Moretti *et al.*, *Cancer Immunol. Res.* **1**, 134 (2013).
4. D. Neri, R. Bicknell, *Nat. Rev. Cancer* **5**, 436 (2005).
5. K. Schwager, M. Kaspar, F. Bootz, R. Marcolongo, E. Paresce *et al.*, *Arthritis Res. Ther.* **11**, R142 (2009).
6. K. Schwager, F. Bootz, P. Imesch, M. Kaspar, E. Trachsel *et al.*, *Hum. Reprod.* **26**, 2344 (2011).
7. M. Birchler, F. Viti, L. Zardi, B. Spiess, D. Neri, *Nat. Biotechnol.* **17**, 984 (1999).
8. M. Fiechter, K. Frey, T. Fugmann, P. A. Kaufmann, D. Neri, *Atherosclerosis* **214**, 325 (2011).
9. M. Franz, I. Hilger, K. Grün, S. Kossatz, P. Richter *et al.*, *J. Heart Lung Transpl.* **32**, 641 (2013).
10. G. J. Bernardes, G. Casi, S. Trüssel, I. Hartmann, K. Schwager *et al.*, *Angew. Chemie Int. Ed. Engl.* **51**, 941 (2012).
11. N. Krall, J. Scheuermann, D. Neri, *Angew. Chemie Int. Ed. Engl.* **52**, 1384 (2013).
12. M. Steiner, D. Neri, *Clin. Cancer Res.* **17**, 6406 (2011).
13. N. Pasche, D. Neri, *Drug Discov. Today* **17**, 583 (2012).

REGULATING CELLULAR LIFE DEATH AND DEVELOPMENT USING INTRACELLULAR COMBINATORIAL ANTIBODY LIBRARIES

RICHARD LERNER, JIA XIE, HONGKAI ZHANG and KYUNGMOO YEA

Department of Molecular Biology, The Scripps Research Institute

JOEL BLANCHARD and KRISTIN BALDWIN

Department of Neuroscience, The Scripps Research Institute

Recently we developed a method where large antibody libraries are rendered infectious for eukaryotic cells, thereby directly linking the antibody genotype to the phenotype of individual cells. The power of the method derives from the fact that the potential agonists are selected from a large naïve or preselected diversity system that contains about 10^8–10^{11} unique members. In the best case when autocrine screening systems are employed, one can analyze about two million phenotypic events per hour. Using these methods, a variety of agonist antibodies have been discovered that regulate cell fates, including those that are phenocopies of important cytokines as well as those that regulate the lineage specification and transdifferentiation of stem cells. The method only seems to be limited by the availability of selection systems.

One of the most important cellular fates yet to be explored is that of survival from death which is, arguably, the most selectable of all possible phenotypes. We began our studies on the selection of antibodies that inhibit cell death by a study of cell survival after virus infection (Fig. 1).

One can imagine three selection formats for such studies. In the first, one uses prior knowledge to hypothesize which proteins if perturbed would inhibit the pathogenic effects of viral infection. In this case an intracellular library that was enriched by phage panning or binding to the target protein would be used. In this format one simply asks whether the selection of the target was correct and if antibodies expressed inside cells can disable the target. In the second scenario, one does not presume any prior knowledge of the target. Unbiased libraries are employed and the selection becomes a discovery tool for finding molecules that are critical to pathogenic pathways. Finally, when the unbiased format is used, one can follow the discovery rounds by subsequent selection rounds that enrich the libraries against the newly identified protein to validate the discovery and find more antibodies that interact with the target. Here, we will illustrate these principles by showing how unbiased combinatorial antibody libraries expressed inside cells can be used to identify molecules that, when inactivated, prevent virus induced cell death.

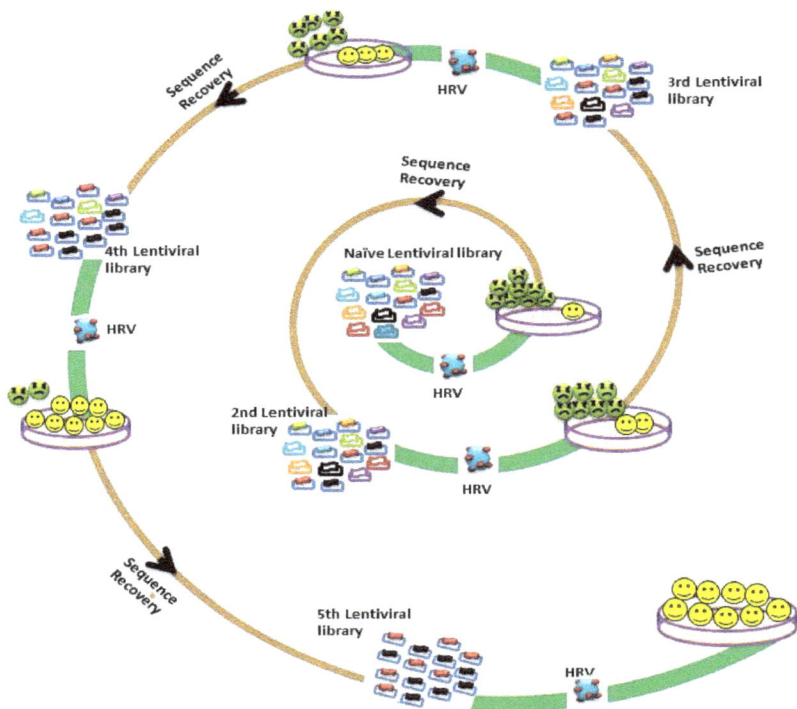

Fig. 1.   Scheme of the phenotype selection based on protection from cell death. The selection starts with a naïve intrabody Lentiviral library that when co-infected with human Rhinovirus in HeLa cell cultures allows selection of antibodies that protect a few cells from death. The integrated antibody fragments (the information) from the cells that survived were recovered and converted to a secondary Lentiviral library which is the starting point for the second round of selection. The selection was carried out for 5 rounds. The ratio of surviving cells (smiley faces) over killed cells (green faces) increases in each round along with functional antibody sequences (red and black hearted particles).

Fig. 2.   Antibodies were used to replace Sox2 and cMyc. These antibodies allowed generation of pluripotent cells and production of live chimeric mice.

While we focused on virus induced cell death as an exemplar, one should be able to extend this method to other systems where there is a selection for survival such as occurs when tumors evade killing by immune cells or small molecules.

In another iteration of the method, we have used intracellular and membrane libraries to generate antibodies that induce differentiation of CD34+ bone marrow cells into red cells, platelets, granulocytes, dendritic cells, brown fat cells, and neural cells from a population of. In the perhaps most dramatic example, antibodies were used to replace transcription factors for the generation of pluripotent cells that were capable of generating live mice (Fig. 2).

NANOBODIES: A UNIVERSE OF VARIABLE DOMAINS AND A TOOLBOX FOR MANY TRADES

LODE WYNS

Structural Biology Brussels, Vrije Universiteit Brussel, Pleinlaan 2, 1050 Brussel, Belgium

My view of the present state of research on nanobodies as biologicals and research tools

The vertebrate immune system can recognize and respond to an enormous number of foreign substances. In doing so and in line with the theme of this conference it expands a universe of proteins, in this case a universe of antibodies. In essentially no time it creates a huge repertoire by V,D,J segment recombination and somatic hypermutation. Over a few days neo-Darwinian real-time evolution by an unprecedented process, harnessed in lymphoid organs creates a vast repertoire [1].

This natural iterative process of mutation/selection has inspired powerful biotechnological approaches: antibody libraries are presented in optimized formats on a range of display systems.

A format presenting unique features is that of camelid VHHs or Nanobodies (Nbs): the variable domain of heavy chain-only antibodies. These antibodies, devoid of light chains were discovered quite serendipitously in the frame of studies on Trypanosomiasis (sleeping sickness) in cattle and camelids. Despite the absence of light chains, Nbs allow the high affinity binding of antigens providing equivalent-sized binding surfaces as bona fide VH-VL constructs do. They allow binding to cryptic epitopes and cavities, present new "non-canonical" CDRs, enlarged CDR1 and CDR3 loops and their germline coding segments present several additional hypermutational hotspots. It is obvious that over the last decade these unique features have found their translation in a wide range of studies, both fundamental and applied [2, 3].

My recent research contributions to nanobodies as biologicals and research tools

Nanobodies as crystallization chaperones

The nanobody format has been successfully exploited to help and crystallize recalcitrant proteins of very different nature.

– In an earliest study [4] a specific VHH allowed structure determination of the intrinsically flexible antidote MazE; a short-lived protein antagonizing a stable toxin mediating programmed cell-death in bacteria. All lattice interactions in the crystal exclusively involve Nb-Nb contacts and only 45% of the polypeptide is ordered.

- Having many independent crystal forms has many benefits for structure-based inhibitor design. Recently the structure of BACE2, an aspartic protease and drug target for the expansion of β-cell mass in diabetes was solved in a series of conformations [5]. A tool chest of BACE2 crystals was created with Fab's, VHHs, and Fynomers. As they bound to many diverse epitopes and give a range of different crystal forms VHHs appear as the best compromise providing bulk to form crystal contacts and sufficient stability.

- S-layers, regular 2D-protein layers constitute major cell wall components in bacteria. They have sparked interest in their use as patterning and display scaffolds in nano-biotech. Structural data however are quite limited to truncated and assembly-negative proteins. The Nbs however, used as chaperones, break the self-polymerizing propensity into 2D lattices. This has allowed the structure determination of full length SbsB (920 residues) and combined with electron density projections from cryo-EM provided the creation of high-resolution maps of the lattices [6].

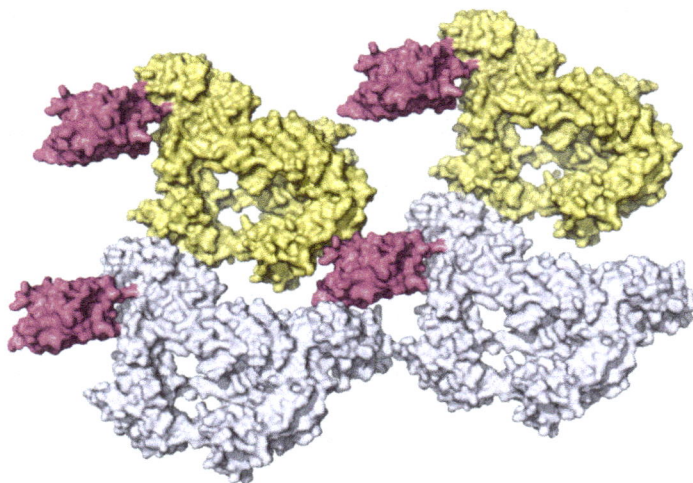

Fig. 1.   Surface representation of the 3D-crystal packing of SbsB S-layer/Nb complex. The inclusion of the Nb (magenta) causes mismatch between crystal and S-layer packing.

Nanobodies in studies of amyloidosis

Nanobodies can protect against the formation of pathogenic aggregates at different stages in the structural transition of a protein in the soluble native state into amyloid fibrils. Studies on β2microglobulin, α-synuclein and amyloidogenic human lysozyme mutants illustrate their value as structural probes to study these processes.

- Nbs against $\beta 2m$ were used to trap and characterize intermediates in the process leading to formation of $\beta 2m$ fibrils, the cause of dialysis related amyloidosis (DRA) [7]. Nbs were selected that block fibrillogenesis of the crucial $\Delta N6\beta 2m$ protein ($\beta 2m$ deleted of its 6 N-terminal residues). The crystal complex with Nbs identifies domain swapping as a plausible mechanism of self-association. In the swapped dimer two extended hinge loops — corresponding to the NHVTLSQ heptapeptide — are unmasked and fold into a new minimal two stranded β sheet. The strands of this sheet are prone to associate and stack perpendicularly to the strands direction to build large intermolecular sheets providing an elongation mechanism by self-templated growth.
- Human lysozyme (HuL) variants (I56T, D67H) are the cause of familial non-neuropathic systemic amyloidosis. Different Nbs mainly directed against either the HuL α or β domain block fibril formation at different stages along the pathway [8, 9].
- Structural studies reveal the epitope for nanobody cAbHuL6 includes neither the site of mutation nor most of the residues in the region destabilized by the mutation. The Nb achieves its effect by binding and fully restoring the structural cooperativity of the wild type protein. It suppresses a local cooperative transition in which the β domain and the C-helix unfold.
- Unlike earlier Nbs, cAbHuL5 binding the α domain, does not restore the global cooperativity and transient unfolding of the mutant proteins. Instead it inhibits a subsequent step in the fibril assembly, involving unfolding and reorganization of the α domain.

Nanobodies in trypanosomiasis

The discovery of heavy chains-only antibodies occurred in the frame of work on trypanosomiasis. In the difficult combat against sleeping sickness they provide new angles of approach.

Trypanosomes evade the immune response by continuously changing the VSG's (Variant Surface Glycoproteins) that cover their entire membrane, which leaves little prospect for a conventional vaccine. Moreover the bloodstream forms are able to turn-over the total surface-exposed VSG pool within 12 minutes through exocytosis in the flagellar pocket. As such this VSG recycling allows antibody shedding and protection.

Nbs raised against VSG's from different *Trypanosome* species were shown to exhibit peculiar and useful activities toward the parasites. Unlike other antibody formats, VHH's penetrate the dense VSG coat and are able to target conserved cryptic epitopes [10]. Normal human serum contains apolipoprotein L-I (apoL-I) which lyses African trypanosomes, except resistant forms. As such *T.b.rhodesiense* expresses SRA protein endowing this parasite to infect and cause disease in humans. A truncated form Tr-apoL-I was engineered, deleted of its SRA-binding domain. A construct conjugating Tr-apoL-I with a VHH efficiently targets conserved cryptic

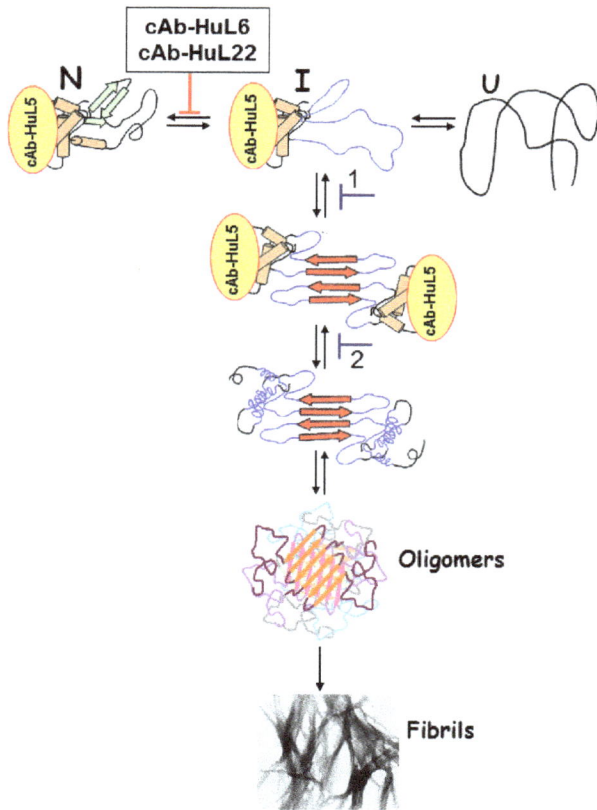

Fig. 2. Possible mechanism for lysozyme fibril formation and its inhibition by nanobodies cAb-HuL6, cAbHuL22, cAbHuL5.

epitopes, common to all trypanosome classes, to generate an immunotoxin with potential for trypanosomiasis therapy [11]. In contrast to larger antibody formats a range of VHH's with nanomolar affinities, both against conserved and class-specific VSG's, show direct trypanotoxic activities and lyse efficiently parasites independent of complement action. By interfering with the early onset endocytosis they circumvent the classic antibody shedding process. Similar morphological features of the process have been reported to occur upon RNAi silencing of genes involved in clathrin-mediated endocytosis. The exact molecular mode-of-action remains to be elucidated [12].

Outlook to future developments of nanobodies as biologicals and research tools

The antibody-based and in particular Nb-based libraries, by excellence products of Nature's use of the "rational design by random design" principle, can for a long time remain formidable tools. Development of Nbs as research tools, Nbs for medical imaging, as diagnostics and therapeutics figures on many research agendas. Innovating, ground-breaking applications are developing at high speed.

- Von Zastrow's team put them at work as "conformational biosensors that reveal GPCR signaling from endosomes". Active-state specific Nbs pinpoint where and when GPCR's signal in cells; video's provide evidence for different G protein and β-arrestin mediated waves [13]. The potential of Nbs to act as conformational biosensors and to modulate protein-protein interactions opens a world of applications.
- In the trypanosome field delivery systems for functional Nbs to the tsetsefly, based on its bacterial endosymbiont *Sodalis glossinidius* have been developed, offering perspectives for interfering with the parasite's life-cycle in the insect [14].
- A system for prophylaxis and therapy against rotavirus disease using transgenic rice expressing rotavirus-specific VHH's was developed recently [15].

Acknowledgments

This work, for a long period, has been supported by grants from VUB, FWO, IWT, VIB and EU.

References

1. S. D. Wagner, M. S. Neuberger, *Annu. Rev. Biochem.* **76**, 1 (2007).
2. S. Muyldermans, C. Cambillau, L. Wyns, *Trends Biochem. Sci.* **26**, 230 (2001).
3. E. De Genst, K. Silence, K. Decanniere, R. Loris, J. Kinne *et al.*, *Proc. Natl. Acad. Sci.* **103**, 4586 (2006).
4. R. Loris, I. Marianovsky, J. Lah, T. Laeremans, G. Glaser *et al.*, *J. Biol. Chem.* **278**, 28252 (2003).
5. D. W. Banner, B. Gsell, J. Benz, J. Bertschinger, D. Burger *et al.*, *Acta Cryst. D* **69**, 1124 (2013).
6. E. Baranova, R. Franzes, A. Garcia Pino, N. Van Gerven, D. Papastolou *et al.*, *Nature* **487**, 119 (2012).
7. K. Domanska, S. Vanderhaegen, V. Srinivasan, E. Pardon, S. Giorgatti *et al.*, *Proc. Natl. Acad. Sci.* **108**, 1314 (2012).
8. M. Dumoulin, A. Last, A. Desmyter, D. Canet, D. Archer *et al.*, *Nature* **424**, 783 (2003).
9. E. De Genst, P. H. Chan, E. Pardon, S. T. Hsu, J. R. Kumita *et al.*, *J. Phys. Chem. B* **117**, 13245 (2013).
10. B. Stijlemans, K. Conrath, V. Cortez-Retamozo, H. Van Xong, L. Wyns *et al.*, *J. Biol. Chem.* **279**, 1256 (2004).
11. T. Baral, S. Magez, B. Stijlemans, K. Conrath, B. Van Hollebeke *et al.*, *Nat. Med.* **12**, 581 (2006).

12. G. Caljon, B. Stijlemans, D. Saerens, S. Muyldermans, S. Magez *et al.*, *PLOS Negl. Trop. Dis.* **6**, e1902 (2012).
13. R. Irannejad, J. C. Tomshine, J. R. Tomshine, M. Chevalier, J. P. Mahoney *et al.*, *Nature* **495**, 534 (2013).
14. L. De Vooght, G. Caljon, B. Stijlemans, P. De Baetselier, M. Coosemans *et al.*, *Microb. Cell Fact.* **11**, 23 (2012).
15. D. Tokuhara, B. Álvarez, M. Mejima, T. Hiroiwa, Y. Takahashi *et al.*, *J. Clin. Invest.* **123**, 3829 (2013).

SESSION 5: BIOLOGICALS AND BIOSIMILARS

CHAIR: MARKUS G. GRÜTTER

AUDITORS: A. LESAGE[1], S. MUYLDERMANS[2]

(1) GrayMatters Consulting, Nachtegalendreef 27, 2980 Halle-Zoersel, Belgium
(2) Structural Biology Research Center, VIB, and Research Group of Cellular and Molecular Immunology, Vrije Universiteit Brussel, Pleinlaan 2, 1050 Brussels, Belgium

Discussion among panel members

<u>Markus Grütter</u>: May I start with a question to Richard Lerner: I was astonished by hearing, although not being in this area, that you can take a particular cell and reprogram it with a single antibody into a fat cell or differentiate it into any other cell. I always thought this involved a whole machinery of reprogramming. How does this go with just one molecule in the cell?

<u>Richard Lerner</u>: Actually in the normal situation, it doesn't require a whole ensemble. Because if you look at the normal hormones like erythropoietin or GCSF or whatever, a single molecule can take a stem cell and cause it to run a particular route. But what is surprising is, which you say is true, a single antibody can cause a stem cell to run a route it is not supposed to run. And I think that has to do with what we were discussing either yesterday or the day before that these receptors can be used in more than one way, because of the dwell time. An antibody is not what they are used to seeing.

<u>Sachdev Sidhu</u>: I continue on a mechanistic level. Do you see most of these dramatic effects come from agonists? Maybe we have been too obsessed with developing antagonists with antibodies.

<u>Richard Lerner</u>: All of the 20 (I did not show them all) are agonists. However there is a paper which is shortly out by McCafferty in Cambridge, and he actually makes an antagonist.

<u>Sachdev Sidhu</u>: That is interesting. Because in therapeutic development we are trying to basically throw away most agonists antibodies and focus on the inhibitors I think.

<u>Richard Lerner</u>: If you had not thrown away all your agonists, Genentech would have been successful!

Sachdev Sidhu: They are turning over as we speak.

Sachdev Sidhu: I have a question for Dr. Suga. Very cool approach. Have you applied it at all to larger proteins, I guess it could be suitable for that?

Hiroaki Suga: Yes, my lab has probably done 20 targets with maybe 5 or 6 of them being membrane proteins. The company that I have actually has done more than 30 different targets. So you can basically target anything. And for membrane proteins, we often have to screen against which one has a biological activity. Enzymes in general are very easy to get inhibitors for very quickly.

Richard Lerner: Professor Wyns, can I ask a question? One of the interesting things I think is, that in infectious diseases, like *Trypanosomes*, the evolution of the surface of the parasite has partly considered "evasion" of the immune system. It is used to evading an ordinary antibody but you showed that with this unusual antibody construct it cannot evade it. But in an organism like the llama, where you got these heavy chain antibodies (nanobodies), its own viruses, its own parasites, would have to have learned to evade that highly specialized system, right? It does not have to evade a human antibody, it has to evade a llama nanobody.

So my guess is if you isolate the viruses or parasites from that organism, they would be able to evade the heavy chain only antibodies. Because they would have learned to evolve its parasites to escape the immune system, just like influenza virus has learned how to do it in man. There must be all kind of infectious agents for that beast.

Lode Wyns: I just skipped one of the slides showing that there are opportunities of multiple types of attack, with the antibodies, as I said: antibodies that sneak in through the VSGs (Variant Surface Glycoproteins) of the *Trypanosome*. You can create antibodies of a whole range now that can be taken up...

Richard Lerner: But that is not the issue. The issue is that you evolve a parasite in part to evade the immune system. The way parasites will evade that scaffold will be different than the way you evade an ordinary antibody. You could test that, if you wanted to.

Lode Wyns: You have always this race between what you do and a what a parasite finds as a solution but...

Richard Lerner: But the parasite is up against the immune system in the organism where it lives. It has got to specialize to evade that particular kind of antibody.

Lode Wyns: But, there is the other question around I think: why did the camels evolve these VHH? And I don't know this.

Richard Lerner: It is obviously a function of the hump!

Markus Grütter: I have a question to Dario Neri. You showed nicely that you load antibodies but also single chains Fv's, shortened pieces with all kinds of interesting payloads and then get it to the target and cure diseases. What is the lifetime of these smaller pieces, these single chain Fv's, in the body?

Dario Neri: We typically work with either bivalent or homotrivalent derivatives, and they all clear very rapidly. Like the half-life in the mouse is about 20-30 minutes, in human it may be 1-2 hours. This is what you want if you have a payload like a pro-inflammatory cytokine that should not stay in blood for a long period of time. So in that particular situation, you want a long residence time at the site of disease — ideally weeks — and you want a rapid clearance from the blood. I think the approach has a potential for certain indications but a more general problem is that antibodies and proteins in general are very slow at getting out of blood vessels. That is even though the pictures look nice at late time points and you have a selectivity, it is still a minute portion of what you inject that reaches the target. This is really a problem because most of the diseases happen outside of blood vessels and you have to get there. So I think part of the fascination is trying to get smaller molecules like Hiroaki Suga showed or by using DNA-encoded chemical libraries. In our experience, when molecules get smaller than 2000 Da, extravasation can be very rapid, it can happen in a matter of seconds, then all of a sudden we are switching the targeting game from hours and days back to minutes. Of course, this will not work for all payloads but it may work for example for cytotoxics with suitable cleavage sites.

Richard Lerner: Can I raise a scale problem because this comes up every time? This could be quite unpopular. Most of the people who worry about making an antibody small have never seen a blood vessel. I don't think there is any capillary that is so small that a full antibody cannot get through it. It is sort of like saying "I am going to build a bridge 60 feet high so a 7 foot mass can get through it". If you look at the scale of the vessel tube relative to the molecular dimensions of the antibody molecule, the ratio is very large. So I would like to challenge the concept of "Small is beautiful" in the antibody world.

Dario Neri: My experience and the experience of many groups is that antibodies eventually get to the perivascular targets, but it is a very slow process. You can measure that very precisely by quantitative biodistribution studies. Unfortunately, actually most companies do not do enough imaging for my taste but quantitative biodistribution studies show very clearly that if you wait 2-3-4 days you get a target to non-target ratio of 10 to 1 (20 to 1), but the process is very slow. So the question is really, does it matter? It depends on the payload. If you have an intact antibody and your antibody is doing something, that is not a problem. It will swim happily in the blood until it gets to the target and does its job. But for certain applications which are very popular today, like antibody-drug conjugates, it may matter whether the bomb is swimming in our body for 2-3-4 days without doing its job. So I think that tailoring of pharmacokinetic properties, and Arne Skerra also showed it nicely, really depends on the therapeutic strategy. Probably different strategies will have different applications.

Richard Lerner: I agree with that 100%. But that is not a direct function of the size of the molecule. Sure if you cut up an antibody into small pieces, it behaves very well. But the idea is that "small" will swim better through the vessels!

Dario Neri: I swear it is true. And you can do a very simple experiment by doing intravital fluorescence microscopy: You inject an antibody labeled with fluorescein and you will see the "swimming" into the vascular structures for hours. You can inject fluorescein as you would do for example for fluorescein angiography. And you will see the molecule going out of the blood vessels in a matter of seconds. I swear it is true.

Arne Skerra: The point is not so much the penetration into small capillaries but more the penetration of the interstitial space, the narrow liquid zones between densely packed cells. Besides, there are some other barriers that are clearly size-dependent with regard to penetration, for example the lung barrier. There are developments going on to administer biologics by inhalation and there it is absolutely clear from published data that the smaller a protein gets the better it crosses the alveolar barrier into the blood. Probably something like that is also true for the blood-brain barrier (BBB). Mark Dennis from Genentech has published that antibodies penetrate the BBB to an extent of may be 1 to 3%, but this fraction is much larger for smaller molecules. One example is leptin which can go through the BBB, and we also have indication that anticalins can cross the BBB, at least as far as we can judge from the therapeutic effects of our anti-amyloid β-anticalins in mice.

Discussion among all attendees

Markus Grütter: We have time for general discussion. I now invite the panel and the audience to start with questions.

Arne Skerra: I have two technical questions to the amazing talk by Dr. Lerner. One is: Do these antibodies act as intrabodies or as secreted antibodies? And the second question is: What may be the influence of lentiviral integration to the genome?

Richard Lerner: Thanks for asking. I prepared this question (showing slides). The answer is that for various phenotypes you have different expressions. You have some that are intrabodies, some that are targeted to the ER or to the plasma membrane. You actually have a library with a diversity of 10^8 integrated into the plasma membrane. (A. Skerra: How exactly is this going?) I will show you how when it comes up.

In terms of the important question about the lentivirus, we are used to a control, using a fluorescent protein-labeled lentivirus (either red or green), and we have never seen a phenotype, ever, that does not mean we never will, but we have never seen a phenotype solely based on lentiviral integration. And every experiment has a parallel lentiviral control.

This is the membrane-bound approach. We call it "nearest neighbor", and what you see (on the slide) is the G-CSF receptor and the antibody (Ab) integrated in the plasma membrane. We call it "nearest neighbor" because it is a first approximation the antibody can only activate what it touches (showing points on the slide). If you look at the fluorescence of the receptor and the antibody, they wind up in the same place. Basically you get an antigen-antibody union in the plasma membrane. It is a powerful way to go because the effect of molarity is very high.

Don Hilvert: Could you clarify what the form of the Ab is? Is it the Fv, the Fab, the entire IgG?

Richard Lerner: It is a scFv dimer. And if I could add: sometimes that doesn't matter. One thing you have to realize is that when you do a lentiviral infection with an MOI (multiplicity of infection) of 3, very often you get more than one lentiviral integration into the genome. So since the Fc is degenerate, you get bispecific antibodies whether you want them or not. And it turns out that sometimes it doesn't matter and sometimes it is required. For example — this one is published — in the case of the EPO agonist antibodies, only the bispecifics work because you need, shown by Lodisch and others, a symmetry to activate that particular signal transduction domain. And you can't get the asymmetry with a symmetrical bispecific. So the bispecific thing is sort of interesting.

Kurt Wüthrich: I would like to lead the discussion towards analytical problems with defining biologicals and defining biosimilars relative to biologicals. One of the participants said yesterday that when looking at post-translational modifications, biologicals are a mess. And biosimilars should be a similar mess to be accepted. We may have analytical tools to show that we have a mess but I think we do not have analytical tools to prove that one mess is similar to another mess. Are there any comments on this point?

Arne Skerra: I am not so sure about that because, for example, the company Roche has launched an entire mass spectrometry department in order to essentially fully characterize the different chemical modifications and glycoforms that occur in monoclonal antibodies. Of course, these are complex molecules so you can only characterize the mixtures and their composition but this to a very detailed extent.

Personally I am convinced that, in the future, if you go to simpler formats, as I explained in my talk, then also the analytics will become simpler because in the end, if we avoid glycosylation or other post-translational modifications, if we also avoid the polydispersity of PEG, then we have mass spectra, for example, that only give one peak and the analytics becomes very simple.

Then the only problem that remains is that there are host-cell related impurities. This is a different topic, but I recently had a discussion with somebody from the FDA about it and I foresee a situation in maybe 5 or 10 years where the current requirements for biologicals characterization may become more similar to the requirements for chemical drugs characterization. Because fundamentally there is not much of a difference with regards to analytical characterization.

Kurt Wüthrich: One should first find out if the mess is useful, or whether a clean, uniformly modified product could do the job.

Richard Lerner: You and I already had many discussions about this. This is mostly a political problem, not a scientific problem. Because modern analytical chemistry, as you know better than anyone, has a resolution that is almost at the level of the proton. But there are several competing forces. You have small biotechs saying "Our product is a mess and if you don't copy the mess forget it". You have Big Pharma saying the opposite. And then you have got the strong political pressure to make generics because the cost of these biologicals is so high. The fact on the matter is that if you want to make up perhaps a fantasy, a scary scenario, about why you must have a mess, you can easily make up that scenario. And there are whole departments in charge with making up a reason why if you clean up the mess, it is dangerous. Because it is "big money".

Arne Skerra: I have a slightly different view. I think there is good reason that in most cases the mess is actually to be avoided because there is one single species which shows the desired activity. For many antibodies, this is for example the unfucosylated species, which may not be the majority in the older product, but nowadays technologies are evolving to make this more pure, even if there will never be 100% purity. I have actually a different perception also from the regulatory authorities. It is not so much the biotechs that are in favor of the product mixtures, it is the Big Pharma because it makes it then more difficult for newcomers in the field to reproduce the spectrum of a composition. In the end, it is only a consideration of the risk for the patient, so the batch-to-batch reproducibility that has to be secured by applying proper analytical means and by ensuring that the composition is always the same within a certain range. But if we come up with designed biological drugs that are single species drugs, then this problem all of a sudden will disappear.

Richard Lerner: You are talking about going forward. But what I am talking about is if you want to make a new "HUMIRA" (Human Monoclonal Antibody in

Rheumatoid Arthritis), what are the criteria that the FDA is going to accept. You are absolutely right in going forward. The scientific community should figure out a way to make the most homogeneous product possible. That is not the situation we are in right now. We have multi-billion dollars antibodies that people want to copy and they will not produce this homogeneous species.

Dario Neri: If I may expand on this topic, the situation at present became very debated because biosimilars were approved for Remicade. Remicade is an antibody (anti-TNF-α) that sells several billions US dollars every year and it is approved for rheumatoid arthritis, psoriatic arthritis, ankylosing spondylitis, psoriasis, IBD (inflammatory bowel disease). Now the situation is the following: each approval for each indication requires the dedicated clinical trials and proof of efficacy. What is not commonly known is that not all TNF blockers, like Remicade, work in all indications. They typically all work in rheumatoid arthritis and they were first approved in arthritis, but not all TNF blockers work for example in IBD. Enbrel (Etanercept) does not, to name one. With the current regulations, the biosimilar is tested in only one indication, typically the easiest, and not at a dose at which the original product was developed, typically a higher dose. But then the product gets automatically approval for all indications, including the difficult ones, for example IBD. So I think this is what will be learned in the near future: either the process is too relaxed and maybe the product does not deserve the approval or maybe the small differences in glycosylation do not have such a big impact.

Sachdev Sidhu: I think it is important to make the distinction, as I think Dr. Lerner said, "going forward" versus the past. I'd say going forward, I think there is a strong evidence that making the best homogeneous product is best because you are going to have better efficacy and hopefully fewer side effects etc. I think a lot of this does affect that to some degree. Looking back though, I think there are two types of biologics, there is the natural EPO, growth hormone, interferons, and there the mess is natural, you try to replicate a natural protein. I still would argue the middle ground, which are existing antibodies, are still human made, they are not entities, and I think there is substantial room there for the "bio better approach" — it is kind of a cheesy phrase! If you can find, an example of Retuxan, the subpopulation that is really effective, you then have a chance of going in, as HUMIRA proved versus Enbrel, then subtle advances can really be picked up and I think there is still a lot of play in the existing antibodies, which are historical entities, they are not EPO, they are not evolved proteins. If you can find the subpopulation that is really effective and show that that it is better even a reduction in cost of goods or some side effects, I think that people would switch to the better medicine. I think even the middle part of the historical stuff is still addressable with more homogeneity, as long as it is a better drug, as a result of that homogeneity.

Lode Wyns: Dr. Lerner was making the problem of the big pharma companies but I really think, through things that I have read in the literature now, that companies are playing the double game. On one hand, they may try to block newcomers but at the same time they themselves want to go for biosimilars. They have a problem there as well. Second I think, if you go through the literature, it is always a mess that is becoming worse, where the mess with new antibodies and new techniques we have, might become a lesser mess. But I have the impression it all concentrates on the immunogenicity of carbohydrates. We should have a lot of people dealing with this immunogenicity of carbohydrates, that is the core of the matter.

Markus Grütter: I would like to switch the topic from this industry-driven discussion to science applications. You have seen all these scaffolds, these new tools that involve a lot of technology libraries with large numbers of molecules from which specific binders to targets can be selected. You have seen technologies that are very useful *e.g.*, in the membrane protein structure biology field or for target validation. My question would be: What is the philosophy to make these technologies available to the community? From C-H Wong, we have heard that one can approach him and cooperate at a certain cost. What is the general view on that?

Brian Roth: I can address that but first I want to get back to the topic you were trying to steer us away from. Just a comment about the point you made, which may or may not be true, that a biological which is highly purified will have the same efficacy as one which has multiple components. Of course, we know that if it is a protein, then fragments may have subtle or profound differences at known as well as at unknown targets, which then could affect the ultimate spectrum of efficacy in a clinical population. That is an experiment or hypothesis which has yet to be fully tested. And certainly with antibodies, that is obviously true, because depending on their conformation, they can have profoundly different effects on the ultimate target. But certainly you will have something, which is more highly characterized, but it may actually have a different spectrum of efficacy. I wouldn't be so strong about that particular point.

In terms of applications, the thing that occurred to me — this sort of makes this answer longer — one of the hats that I wore for a number of years as an adviser for the molecular libraries screening center program (*mli.**nih**.gov/*) which is a huge NIH initiative to discover new small molecules probes, which I think was largely unsuccessful for many reasons. But one of the things they were targeting was protein-protein interactions. These are not so easily targeted by ordinary small molecules but it occurs to me that a lot of the technologies that were discussed are like perfect for that, sort of one of the final frontiers. So I said this as a comment but my question was: some of you have shown a remarkable selectivity within a target class, for example caspases, I guess with the darpins. But have you ever looked outside this target class? Sort of a broader view of the proteome to see what other targets they might be interacting with through chance interactions?

Richard Lerner: That is the whole reason for using unbiased libraries. By definition, if you have a target in mind, the library can't be unbiased because you are panning against that target. The most powerful phenotypic screens do not go into the game with any preconceived notion. It is just: show me the phenotype and I will get the target afterwards. That is how you discover new things. So if that is what you are alluding to, you are correct. Once you with certain exceptions like neurons, but in general, if you go after a known target, you get a known antibody.

Markus Grütter: We don't run genome-wide test of binding but what we do, for example in membrane proteins, is taking a set/a number of different proteins as a control experiment, but these are limited and not done at the level that you were asking. But the interface is such that it is as big and selective as antibodies. Initially, when this system was installed, these questions were addressed on a principle basis. How selective is this on a panel of many different things? You can do selection on entire cells and get specificity to one particular target, so you can fulfill that requirement in principle. It has not been done in the *in vitro* case that I presented but, in general, this test can be done. Maybe somebody else can comment on that also?

Dario Neri: Maybe complementing what you just said. There are experiments published by the group of Hans Lehrach in Berlin and Dolores Cahill in Dublin. They prepared protein arrays with something like 4,000 proteins and then addressed the question: If you throw an antibody in a dot blot type of assay, do you find other specificities? Occasionally you do. So you may find another spot or two, but it is not a general rule. Then you may follow it up and see if the affinity is the same. Typically, these reagents tend to be very specific and you can prove it on an array in the level of thousands of proteins.

Brian Roth: The question didn't relate so much on antibodies because I get the specificity there but more with these synthetic protein reagents that we are talking about. How deeply those have been interrogated? The other point is that I am a big believer in phenotypic screens. I didn't get into this but we actually do a lot of those and in terms of discovering new pathways; that is really the only way to go.

Richard Lerner: If we wanted to be helpful, I'll come back to the biosimilars for a minute. We could have that now, because I think we should discuss science. Given the way the FDA works, this kind of body (the Solvay conference participants) could develop the Solvay criteria of similarity. And if it was endorsed by a certain body or collection or whatever, this is what the FDA would work with. What would you accept? Because nothing is perfect, not even analytical works on small molecules, as we all know. You would accept let's say, an acrylamide gel to start with, then ORD, then D-H exchange, then some sort of phenotypic 2D-NMR, and then maybe even a crystal structure, and of course the mass spectrometry. If you had the Solvay criteria, the 10 Solvay criteria that you have to match in order to be a biosimilar and, if the FDA would accept that, the field could go forward. Because given the fact that these are going to be multibillion dollars entities if they are accepted, the people who are trying to put them on the marketplace, would spring for the cash, to build strong analytical capability to match the Solvay criteria. But somebody has got to sit down and learn to really understand analytical chemistry that applies to protein and say "alright" according to the scientific principles of today, these ten (or 7.5) things are acceptable as a biosimilar. Nobody, no agency, no group has ever sat down and done it.

Hiroaki Suga: I want to respond to your question for the "outside of the target". Because I don't have X-ray structure I didn't show today, but we published an AKT2 inhibitor, and we are now writing a paper on the cMET binders. We collaborated with a MS specialist, his name is T. Natsume. We attached a flag tag and pull it down. He mapped out the whole entire thing. He really found only the target to which it was targeted. He was very surprised — this could be better than an antibody. I think it is the peptide we selected, generally very specific to the target we isolated.

To the question from the chair about collaborations and how to make available the technologies: I don't collaborate with industry people at all, because I have a biotech company that will collaborate with the industry. But I am very open to collaborate with any of the academic people and try to work on some new target together, or some difficult target that people have failed to identify an inhibitor or whatever.

Brian Roth: Let me respond to that. I don't want to sound like a broken record here, but to just put this idea in your mind. I believe what you say if you do "pull down", you only see the target. But I would point out that if the target is a GPCR for instance, you are probably not going to see it (by MS). These targets are present at trace amounts, they frequently do not fly in a mass spectrometer, you are not going to see them, and those can be the ones that really can cause you problems in a big way downstream, I guess it is in human populations. There are other ways to profile these compounds, you might think about this.

Markus Grütter: Just to follow up on the question of selectivity. You said that you believe that the antibodies have the selectivity and these scaffolds don't. There is, from a protein chemistry-based view of these molecules, no reason why these scaffolds shouldn't acquire the same type of selectivity, in principle. Whether all the experiments are always done is another question, but there is no principal barrier for these scaffolds to achieve similar selectivity as antibodies.

Sachdev Sidhu: I would agree with that. I love antibodies as much as anybody else. Scaffolds as long as they are designed correctly, it is biophysics, they can work. The problem is that if they are not designed obviously there: they become hydrophobic and they can't!

But it is not kind of the point, if you really have to adopt the technology to do what you want to do, then it becomes a fantastic technology. We did that with ubiquitin, these things won't work for everything but they work very well for that subset. And if you think about it, in biophysical terms and then design the scaffold, there are plenty of examples where you get the same affinity and specificity, once the design takes the biology and the biophysics into account.

Kaspar Locher: I have a question for Dr. Suga about scaffolds. He made a statement and showed it on his slides that most of the scaffolds and their compounds are membrane permeable. (H. Suga: No!) So the questions is actually: would that be something that is good or bad? Is that tunable or not?

Hiroaki Suga: It is tunable but it requires quite a bit of effort. For instance, what I showed you today was a N-methylated backbone that improved membrane permeability. We had to scan them. We scanned every single residue of the N-methyl group to identify which one retained activity and then we combine them together. So it requires a bit of effort, like very simple medicinal chemistry. If you have the X-ray structure, you can actually down size it a little bit, which helps a lot too. But whatever you selected, it is not necessarily membrane permeable.

Jason Chin: I have a question for Sachdev Sidhu about the antibiome project. I am curious — you are not going to have infinitely specific reagents. So my question is: at what point are you happy that these are good enough? It is great what you are doing in characterizing them. And actually I think having resources where antibodies are properly characterized is fantastic. But how do you decide that there is a hundred-fold specificity?

Sachdev Sidhu: This is very simple. You have to start with a user or application in mind. It is the case with any antibody. Nobody designs it to a specific level. You design it to a level where it becomes useful. What we mean by that is; there is no point in launching an antibiome without a very strong network of users. I don't really care about how tight or specific it is, as long as the cell biologist who wants to use it says "it is working". The same is true for Herceptin once it became usable, it is ok.

If you go back, it might not be 100% specific. You let the application decide that. We try now to convince the NIH and others to engage in large-scale projects to produce 10,000 antibodies. You set up networks for you know somebody is going to test this immediately and you stop when it is good enough because you want to go on the next one. But otherwise, you get with this incredible mess where you may make it too good, that is a waste of money, and if you don't make it good enough, nobody picks it up. So it is defined strictly by what needs to be done with it. A 100 nanomolar binder can be great if a cell biologist needs to do an IP (immunoprecipitation) and then he can put it in the micromolar. And that is fine if what cell biologists say is: "It's working!" and then you move on to the next one. That has to be user-driven.

Don Hilvert: I wanted to follow on Brian Roth comment about the synthetic systems that Hiroaki Suga and Dario Neri presented. What is the optimal set-up building blocks? Do you restrict yourselves to a very small subset of chemical space or is everything open? Because if you have a 10 amino acid peptide, and you use 20 building blocks, you have a number of possible combinations that exceeds what you can typically explore experimentally.

Dario Neri: I guess you are referring to DNA-encoding chemical libraries and, even though the concept has been around for many years, there is still work in progress and there is room for optimization. In my experience, also looking at what other groups do and activity in our lab, the chemistry that is used nowadays becomes problematic when you go beyond 2-3 sets of building blocks. This coincides also with the size: getting bigger than 600 Da, which for medicinal chemistry applications is not good. I would claim, but it is a personal opinion, that the technology at this moment is particularly useful to make good quality libraries based on two sets of building blocks with about 1 million members per library. That is where the technology works very robustly.

As you know I organize every two years a congress on this particular topic and I have seen the field evolving from companies like GSK constructing libraries with billions of molecules with 4 sets of building blocks of very poor quality — going down to libraries of 2-3 sets of building blocks resulting in 1 to 10 million compounds but with better quality and performance. I think a lot of development is still possible. Trying to make these molecules more compact, and smarter in a way.

Hiroaki Suga: In my case, you can put anything, any building block, you can change the side chain of amino acids and stuff like that. But I am not so interested in it and I think it is not really important. I think the diversity of the side chains of amino acids is good enough already. There is more to be gained by changing the "backbone". We are making more efforts with N-methylated backbones or — which I haven't talked about today — we could have oxazole or thiazole kinds of backbones where you are missing the hydrogen donor from the peptide backbone. And that is more important to the development of drug-like molecule in the future.

Chris Walsh: I would like to go back to the interface of small and large molecules with this question. In some sense it is applied rather than basic science, but there is of course tremendous interest in antibody-drug conjugates (ADC) at least in the cancer area. One of the two approved drugs, the one from Seattle Genetics has a peptide as the toxin. I know Dr. Suga wants to get away from antibodies, but at some level one might say: appending some of your optimized peptide scaffolds as antibody-drug conjugates might be interesting. I would like the panel's view on this: Should we really be excited about ADCs for cancer? Otherwise are they just a fad because very little has been said about them here. I understand one side brought them up. We are not about applications but at the interface of small molecules and biologicals, I would be interested. We heard seven presentations, we didn't hear anything about ADCs. Yet in a different context, that is all the big companies would be talking about. Whether it is Ambrex technology or something else. Should we not think that it is an important scientific advance? Is it the future of cancer antibody therapy?

Richard Lerner: I think you are absolutely right. I have been around long enough to watch the evolution of this. Immunotoxins were the original armed antibody right? People fooled around with it. And the reason is that antibodies in general are not very toxic. They have an activator domain and so forth. If you look at most of the multibillion dollar antibodies that are out there, there is a question about these. I believe that the issue you raised is very important for the future. However, if you look at Seattle Genetics, the total chemical armamentarium of the synthetic world has not until now been focused, and there is tremendous room for making very intelligent linkers, tags, and so forth... I think we are going to see some heavy-duty synthetic guys, getting into this business. More payloads, more linkers, smart linkers and so forth... Seattle Genetics is very successful, and rightly so, but for a while they were the only game in town.

Chris Walsh: There are three pieces to antibody-dependent conjugates: one is the antibody. Should we also be thinking about nanobodies or peptide fragments? The second is the linker technology - for better or worse (the third piece is the drug). I started Immunogen in 1981 but this has been a very long time coming. Also the small molecules had to be sufficiently potent. They had to be picomolar. They are very few, but I would propose Dr. Suga's compounds might get to that point. I don't hear anybody talking about using optimized peptides. Both of the ADC that has been approved target microtubules in sort of an old fashion approach.

Richard Lerner: So the first armed antibody was actually taken off the market because it was too hot. So you have to tune the affinity constant of the antibody, with the hotness so to speak of the organic compound. Because if, as is the case of this compound, a single molecule kills the cell, you can't use it.

Dario Neri: I think the discussion is an interesting one. There are I think opportunities for improvement. For example, the first antibody that was on the market in the US (not in Europe and was later taken off the market) was very dirty, like 40% of the product was not even conjugated to the payload. There are some payloads, which are very interesting because they show the direction for the future. Because whatever you inject, as an ADC, at some stage, it is going to get through a clearance organ, typically the liver. So, 100% of what you inject at some stage targets the liver. Maybe it also targets the tumor to a minute extent, but largely the liver. I think we will see developments in self-detoxifying drugs. If you think about maytansine structures, there are esters, epoxides and you can think of course of molecules, which get cleaved and detoxified as they get out of the body. So I think it is an area where modularity will tell us if we are good enough to get better targeting agents, better linkers and I would claim self-detoxifying payloads.

Arne Skerra: I foresee a bright future for drug conjugates once we overcome the dogma of antibodies to be the preferred carriers. And there are two very simple reasons. One is the liver uptake of antibodies, which actually killed Mylotarg (from Pfizer), the product we were talking about. And the other reason is poor tissue penetration and extravasation (we have also talked about in the session). So if we go to either stable small antibody fragments or alternative scaffolds, then we will be able to create a totally new generation of much improved drug conjugates.

Richard Lerner: What is the basis of that statement?

Arne Skerra: I gave two arguments. One is for example liver toxicity, which is the limiting effect (at least it was for Mylotarg) for many preclinical models, as I heard from the biopharmaceutical industry. And we all know there are receptors in the liver, like the asialoglycoprotein receptors, which take up the antibodies after a certain time of circulation. This intoxicates the liver. Once we can avoid that, we can improve the ratio of tumor toxicity to organ toxicity.

Richard Lerner: So if you engineer out the scFv binding piece of an antibody?

Arne Skerra: We are in the process of doing that: we have a collaboration already and it is quite promising.

Sachdev Sidhu: In my opinion it is kind of a middle ground where clearly antibodies have to be the bench mark for ADCs and then you have to look at them and decide whether a particular alternative gives you a significant advantage in one of the three key areas, I would say: clearance, penetration but, as important, internalization and release. And that is a science problem. You have to ask yourself: Can I get something which is designed to either internalize or release better?

But you have to set it up against the benchmark, which is the antibody, and rationally approach that. I think there is a large opportunity there if you can mechanistically understand how to get something in and then the molecule out, but you have to do it in relation to the antibody. I don't think you can do it without a bench marking process.

Chris Walsh: Prof. Suga, any comment?

Hiroaki Suga: Surprisingly, the pharmaceutical companies did come back to think about those. Once they see our peptides... For instance, they can bind extremely well to the target protein but do not inhibit interactions for example with growth factors. And these are really good targets to conjugate with a drug and bring it to the desired target. I think they become very much interested in peptides.

Taking advantage of the small microcyclic peptides; they are very easy to get conjugated, and the purification is also easy, so that may be the bible for that kind of concept, *i.e.*, peptide-conjugates can be potent enough to target the desired cancer.

Arne Skerra: I have a technical question to Dr. Suga. I was amazed of how efficiently you can incorporate methylated amino acids — or even D-amino acids — into your peptide. At which point does the mechanism of the ribosome limit your peptide chemistry?

Hiroaki Suga: We have studied that a lot, we have many papers (like 5) on that. The D-amino acids can be incorporated in the initiation extremely efficiently, no matter what the side chains are. It is during elongation that we get problems. Single amino acids or D-amino acids can be inserted at certain side chains, but not many of them. But the major problem is actually consecutive incorporation, which is completely inhibited. That is why we usually have only a single D-amino acid or having it at a specific position where it is ok with respect to the elongation position. On the other hand, some N-methylated amino acid, like charged N-methylated amino acid (due to charge effects), are very poorly incorporated at the elongation stage. We have all knowledge of what you can do best and what you cannot do well, and we designed a genetic code in that way.

Gebhard Schertler: A question for the future. We can differentiate cells and do a lot with selection. And selection has make such progress. Are we fast enough to take cells from a patient and deliver an agent that is specific for that patient? That is a first question. My second question is: in cancer for example, the main thing is not to kill mainly the cancer cells, a few months later they come back. Are we having a homogeneous cancer population or is there something like there are cells that are just becoming the cancer cells (people have been talking about cancer stem cells)? What is your opinion about that? Can it be solved just by using two targets at the same time? But really the main thing is still that, even if you kill many cells, we can see that it often comes back. We misunderstand how metastases form, how it happens. What is the research in there? What can selection do about that?

Richard Lerner: You are talking about custom therapy. Do you have time to make custom therapy for individuals in the antibody world? The answer is yes in some situations and no in others. In the case of lymphomas, every lymphoma has a different antibody on the surface. So you could in principle, take a sample of the lymphoma at the time of the first biopsy, figure out what the antibody is on the surface and make an armed antibody or killer antibody, custom therapy and charge less and less than the custom dendritic 'company' charges.

There is one custom therapy immunology, publicly traded company, that is Dendreon. They take your tumor, they build cells that don't like your tumor and put these back in the humans. That is not very effective. The problem is that if you do not pick a target that the cells requires, what happens is the cells will dump the target and then go on and metastasize and do all kinds of bad stuff. If you are going to do that drill, and I think it is possible in the future, if the payers pay for it, you have to find a target that the cells can't get rid of. And to find a target that is (A) a tumor marker and (B) the cells cannot get rid of it, then you have a chance to do exactly what you say. But a tremendous amount of effort is done on killer antibodies so to speak, against targets that can be missed by the tumor cells. Of course, the cells want to survive, the population dumps that and that is the end of it. (Schertler speaking about the fact that many cancers come back)

For the second part of the question, of course the Holy Grail is to get the cells to differentiate again, because the phenotype of cancer is dedifferentiation. If you had an agonist that would cause the dedifferentiated cell to redifferentiate, you would probably stop the cancer. A lot of cancers make the phenotype of the cells from which they come. For example, if you have an insulinoma in a patient and they are making brown fat (cachexia) by some mechanism, the tumor is hot, because you are making heat, not ATP. If you could make that tumor cold by dedifferentiation, it probably would not be a tumor anymore, it would be a β cell. There are precious few tumors that have the mature phenotype of the parental system. In other words: what the oncologists say is that sometimes you can recognize where they came from, so if you could, with dedifferentiation, make them go back to where they came from, you have got something really nice.

Sachdev Sidhu: I think that is a great point. We are really much thinking to kill cancer but perhaps what you need to do is remove most of it and kind of achieve "peace with honors"... And that is better done with agonist rather than antagonist. The cancer field is very heavily focused on antagonists and signal blocking. What is needed and is possible is to combine that with turning on differentiation signals.

Markus Aebi: We use penicillin since about 60 years. Now penicillin is no longer in use because there is resistance and *Fungi* have been using penicillin for 50 million years and it is still in use that we did not read the description how to use penicillin. We know the answer — we probably have to target multiple points in a cell, in a pathway. And then I wonder: Are there strategies, using combinatorial methods — as you are using very nicely — to identify those target pairs or target triplets, to make therapy much more efficient? We know from HIV treatment that this is the solution. You would use combinatorial methods to find the magic bullet but we know (at the same time) that this magic bullet does not exist because you are targeting evolving systems. We will never reach that goal.

Richard Lerner: That is exactly right. And so Christopher Walsh raised one Holy Grail, which does not address that point, and is making antibodies more toxic. The other thing that is a big deal in the field is to make bispecific antibodies that target two targets at once. Every big pharma and half the small pharma in the antibody space are making bispecifics. The nice thing about infectious antibody libraries is that you obtain bispecifics whether you want it or not, automatically you have a bispecific. And bispecifics would help get around the problem that you raised.

Markus Aebi: It might be if you go for a single molecule with two functions. I would rather love to see two different targets, with combinatorial methods, or I did not understand well your answer... (Richard Lerner interrupts to define what he means with bispecific: one antibody for two targets, with one head having the target A, and the other head targets B).

OK great! If you can come up with that system, that is fantastic. I think the chance is much higher to find two different targets and two individual antibodies, if you want to stay with antibodies. I am not an expert, but are there combinatorial methods then to get that?

Richard Lerner: Yes. Everybody has its own way of making such a system. Genentech is using knobs and holes. We use natural recombination at the protein level and so on and so forth.

Dario Neri: Briefly to integrate the answer. Of course, the history of cancer therapy is history of combination therapy, which has been very efficacious in certain types of leukemia and lymphoma but actually not very efficacious in the field of solid tumors. There are a few types in which chemotherapy was actually shown to eradicate cancer metastases of solid tumors but usually this doesn't work. I think we often forget that in order to try combination in the clinic, you need two agents that have typically gone into the clinic separately, show to have single agent activity, and maybe 5-6 years down the road, you can try the combination. This is actually limiting progress a lot.

At the same time, I think it is fair to say, If you looked at the last ASCI meeting, probably the biggest excitement was again from combination of two antibodies: anti-TLR4 and anti-PD1, where all of a sudden you see a response rate in metastatic melanoma, which is a deadly disease, and which has never really been reported before. So I think the direction is clearly that one. In the mouse, it is very easy to show cure with combination. What I have learned is that translation is very slow and we should maybe think of how to make it more efficient.

Markus Aebi: This is exactly what I was asking for. Are there tools to speed that up, you cannot do that in patients or in mice, it takes simply too long. Are there combinatorial methods in this respect?

Chris Walsh: This is really a follow up to an earlier question and Richard Lerner and his response around stem cells. You showed this amazing ability to find antibodies, which should differentiate CD34+ cells, why wouldn't you now go to the epithelium mesenchymal transition of cancer stem cells from different origins? However controversial that hypothesis is about whether they really drive metastasis, and ask can you readily find agonists antibodies which would work on different kinds of stem cells to block that transition?

Richard Lerner: That is right on, we are doing exactly that. The question is: What is the selection? And the selection we are using is: there is a breast cancer model in mice where you put human breast cancer tumors cells into the footpad and they all metastasize to the lung. Our selection is: you infect them with the antibody library and pull out those that do not metastasize. And then you do it again. And our guess is that those would be cells that cannot metastasize.

Lode Wyns: When I listen to the whole discussion it is obviously biological and medical science. But I think there are many other fields. I see many applications and we didn't say a word about it: agriculture and plant science for example that was not touched in this meeting. There are a few botanists here.

And then the other aspect (what Markus Aebi tried to raise a moment ago) is: to what extent do we want to provide these tools to the community? I am very positive about the idea of the anti-ome. That would be extremely efficient in many domains of science. On the other hand I think it will remain very individual in the future in a sense that if you want to study mechanisms, such as fibrillogenesis or if you are looking at GPCRs, then one antibody for one target is not enough. But we see that if we have ten or more antibodies, they each have their peculiar properties and use them for the fundamental aspects of the process.

Sachdev Sidhu: A comment on that. I think you made a great point. The interesting thing is that medicine has to be conservative and it might be that new frameworks, new ideas you can actually test a lot easier in agriculture or in industry where you are not worried about immunogenicity in patients. I think that it is a good area to get into, where new ideas can be explored with industrial applications that can then maybe make their way into medicine.

And the whole idea of distributing that I think is critical because they (antibodies) have to be used to figure out a funding structure and an impact where you can actually be doing something beyond protein engineering. Our hypothesis has been to "Act locally and distribute globally" and what I mean by that is: you build a network around a center and work on problems where you know you have to cross the wall/street, and use those to figure out what is working and then be able to distribute those to anybody else in the world who wants them. Otherwise it doesn't work. If you try to produce a lot of stuff that people don't even know about, you simply end up with freezers full of these things.

Richard Lerner: But Dev Sidhu you make one exception. I think you have been wildly successful and unusual. One thing you have to realize is that the world who wants an antibody is not as sophisticated a molecular biologist, as you are. If you want this to get attraction, you cannot only send the gene. You need to send them milligrams of the protein and the condition is if they want more, then "Here is the gene and you make more". But all these enterprises have failed because the world has many genes around and people don't know how to use how to express them. Send a milligram and then you will see traction.

Gebhard Schertler: Can I ask a question to Dario Neri? We have often toxicity when you want to deliver a toxin or something to a target. We have been very good at starting to get complementing systems for assays, where we actually make the proteins with two things come together. How far are we in payloads that assemble on the tumor cell?

Dario Neri: I think this the Holy Grail of protein engineering. How to engineer toxicity as a result of binding? To a certain extent this is what antibodies do. They swim in the body as naked IgG — they don't do anything but, if they are on the target cell, then they interact with the Fcγ receptors and do the job if accessory cells are around. That is the way we are engineering toxicity after binding. From my perspective, it is a very clear field for ingenuity and development but, on the other hand, we always have to ask if they are good to reach the target, and this is less trivial than one thinks. Typically, companies develop products without imaging, without really doing biodistribution studies in humans. That is going blind, that is assuming that the product goes to the target but maybe it does not. There is a beautiful paper of Genentech a few years ago in which they gave Herceptin (Trastuzumab monoclonal antibody) at very high dose to tumor bearing mice. They looked at the section of the tumor and you would have expected that every single tumor cell should be covered by the antibodies but actually only few perivascular tumor cells were reached. Again highlighting the fact that antibodies don't extravasate so well and the few that do extravasate find the antigen, bind and stay there. So even simple concepts, like reaching a target *in vivo*, are not that trivial.

Richard Lerner: There are a lot more important points that are not raised. Sometimes you get too much antibodies. Because, for example, if you take the "brown fat" antibody, it will work at 0.1 mg/ml. but if you go up to 1 mg/ml like your small molecules, then you suppress the system. So it could be in the case of Herceptin that you could actually give less, it might not downregulate the receptor and be better. But companies don't want to give less because to people you've been giving less, you can only charge less.

Markus Grütter: I have to close the discussion session. A lot of questions are still open and a lot needs to be done but there are great opportunities with these molecules. I thank all the panel members.

Session 6

Proteins in Supramolecular Machines

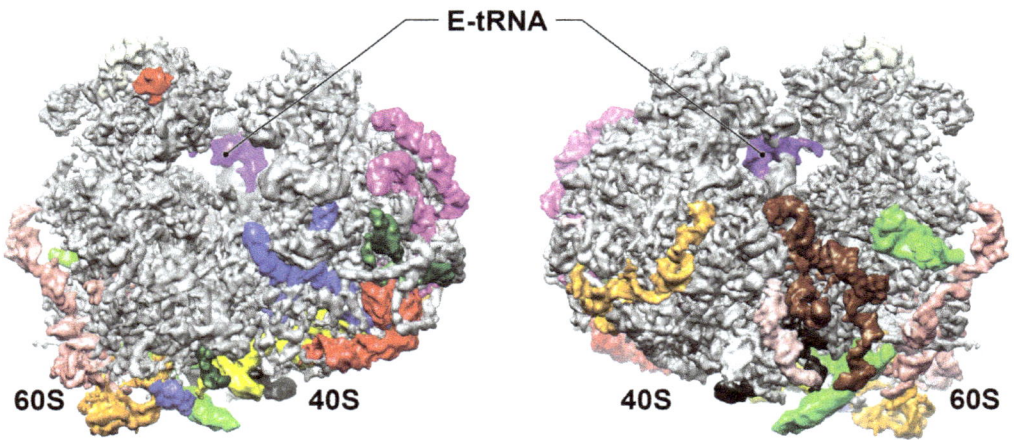

Cryo-EM density map of the 80S ribosome from *Trypanosoma brucei* at 5 Å resolution, reconstructed from over 200,000 particle projections. Left: front view, looking from the factor binding site into the intersubunit corridor, along the path of the tRNA. Right: back view. Ribosomal RNA expansion segments are rendered in different colors, and E-site tRNA is labeled as "E-tRNA."

Modification of figure from:
"High-resolution cryo-electron microscopy structure of the *Trypanosoma brucei* ribosome", Y. Hashem, A. des Georges, J. Fu, S. N. Buss, F. Jossinet, A. Jobe, Q. Zhang, H. Y. Liao, R. A. Grassucci, C. Bajaj, E. Westhof, S. Madison-Antenucci, J. Frank, *Nature* **494**, 385 (2013); Copyright held by Nature Publishing Corp.

ASSEMBLY OF FILAMENTOUS TYPE 1 PILI FROM UROPATHOGENIC *ESCHERICHIA COLI* STRAINS

RUDOLF GLOCKSHUBER

Department of Biology, ETH Zürich, Institute of Molecular Biology and Biophysics
Schafmattstrasse 20, CH-8093 Zürich, Switzerland

Understanding the assembly mechanism of supramolecular complexes and their molecular architecture at atomic resolution is one of the major challenges in biology. Despite the increasing number of solved three dimensional structures of large supramolecular assemblies and a large body of data on cellular factors required for their assembly *in vivo*, it is still extremely challenging to reconstitute supramolecular assemblies from all purified components *in vitro* and to obtain a complete picture of the assembly process that integrates structural, biochemical and *in vivo* data.

In the following, I will summarize recent results on one of the very few supramolecular assemblies that could be successfully reconstituted *in vitro*, of which all structural components and assembly factors are known, and where *in vitro* and *in vivo* data result in a uniform picture of the assembly process. The results are in line with Richard Feynman's famous statement "What I cannot create, I do not understand", in particular, because the reconstitution of the entire assembly process *in vitro* led to the discovery of a previously unknown family of protein folding catalysts and the identification of a membrane catalyst that accelerates the assembly of protein subunits independently of energy.

Assembly of filamentous type 1 pili from uropathogenic *Escherichia coli* strains

Type 1 pili of uropathogenic *Escherichia coli* strains are filamentous, 0.5–2 μm long surface organelles anchored to the outer bacterial membrane and are required for bacterial attachment host cells. They are composed of the four structural subunits FimH, FimG, FimF and FimA [1] (Fig. 1). The main structural subunit FimA assembles into a right-handed, helical quaternary structure and forms the pilus rod. FimH, FimG and FimF associate to a short tip fibrillum composed of a single copy of FimH at the tip of the fibrillum and one or several copies of FimF and FimG. The adhesin FimH is the first subunit that is incorporated into the pilus. It is composed of an N-terminal lectin domain and a C-terminal pilin domain that interacts with the following subunit FimG. The FimH lectin domain recognizes the mannose units of the glycoprotein receptor uroplacin 1a of epithelium cells of the urinary tract [2]. FimA, FimF, FimG and the pilin domain of FimH are structurally homologous proteins that share an immunoglobulin-like fold lacking the C-terminal β-strand

[3]. In addition, FimA, FimF and FimG possess an N-terminal extension of 15–20 residues, termed donor strand, which inserts into the preceding subunit in an anti-parallel orientation relative to its C-terminal β-strand and thereby completes its fold. This donor strand complementation mechanism is the common structural principle underlying quaternary structure formation and explains why the adhesin FimH can only be incorporated at the tip of the pilus [3].

Type 1 pili are assembled via the chaperone-usher pathway [4] (Fig. 1). This term refers to the fact that the periplasmic chaperone FimC and the assembly platform (usher) FimD in the outer membrane are essential for pilus biogenesis. FimC interacts with all pilus subunits in the periplasm and delivers them to FimD at the outer membrane. FimC-subunit interactions differ from subunit-subunit interactions in that an extended polypeptide segment of FimC is inserted in the opposite (parallel) orientation into the pilin fold of the subunit. FimD recognizes incoming FimC-subunit complexes, catalyzes subunit assembly (see below) and mediates the translocation of folded subunits to the extracellular space.

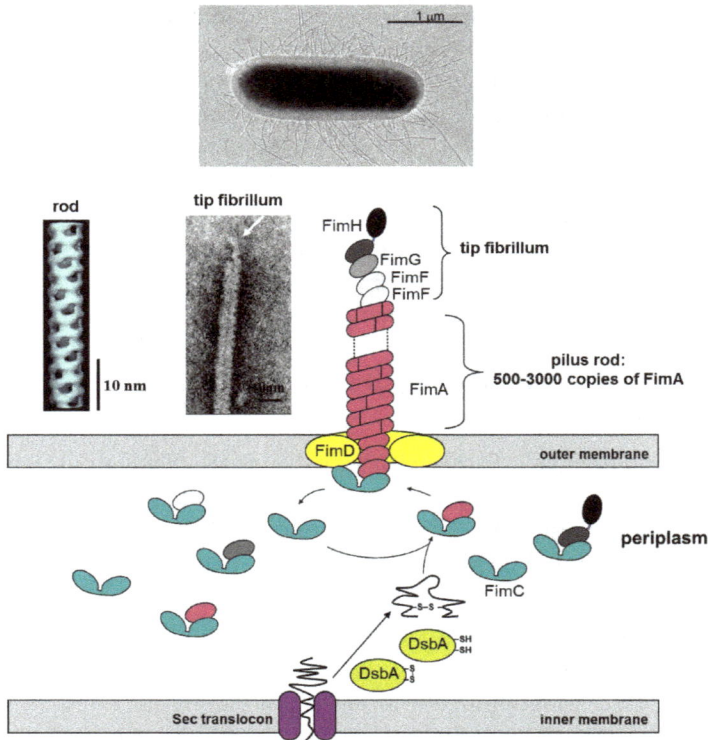

Fig. 1.   Subunit composition and biogenesis of type 1 pili from *Escherichia coli*. (Top) Electron micrograph of the *E. coli* K12 wild type strain W3110 bearing type 1 pili. (Bottom) Mechanism of DsbA-, FimC- and FimD-catalyzed assembly of type 1 pili, and quaternary structure of the pilus rod and the tip fibrillum (see text for details).

Type 1 pili are extraordinarily stable against unfolding and dissociation. To study the origin of this extraordinary stability, we constructed subunit variants that lacked the N-terminal donor strand and were artificially extended at the C-terminus by the donor strand of a neighboring subunit [5]. This enforced intramolecular, anti-parallel self-complementation and thus mimicked the conformational state of subunits in the pilus. All self-complemented variants proved to be extremely stable, with free energies of unfolding in the range of 67–85 kJ mol^{-1} [5]. In addition, they did not attain their folding equilibrium at physiological temperature due to extremely high activation energy barriers of unfolding and refolding, explaining the almost infinite stability of pili against dissociation [5, 6]. Surprisingly, bimolecular complexes between subunits and synthetic donor strand peptides show similar, high stabilities. An impressive example is the complex between FimG and the donor strand peptide of FimF (DsF). It is the most stable, non-covalent protein-ligand complex known to date, with a dissociation constant of 1.5 10^{-20} M and an extrapolated dissociation half-live of 4 billion years [5, 7]. This opens numerous possibilities for technical applications, *e.g.*, the use of the DsF peptide as tag in affinity chromatography or for the stable immobilization of proteins on surfaces [7].

All pilus subunits exhibit very low rates of spontaneous folding. Specifically, the main subunit FimA shows a folding half-life of 1.6 hours [8]. We identified the chaperone FimC as highly efficient and specific catalyst of subunit folding [8, 9]. The rate-limiting step in the catalysis is the binding of the unfolded subunit to FimC, while subunit folding on the surface of FimC occurs with half-lives of less than a second. FimC thus can accelerate folding of subunits by more than 10^4-fold [8].

All structural pilus subunits possess an invariant, structural disulfide bond, which is introduced into newly translocated subunits by the dithiol oxidase DsbA (Fig. 1). We found that FimC only recognized unfolded pilus subunits in which the disulfide bond was already formed [10]. As only FimC-bound subunits are assembly competent, this mechanism guarantees that only disulfide-intact subunits are incorporated into the pilus [10]. The kinetic analysis of FimC- and DsbA-catalyzed folding of FimA, together with the *in vivo* concentrations of both catalysts in the periplasm, predicted and *in vivo* half-life of FimA folding of approximately two seconds [10]. Thus, FimC and DsbA very efficiently overcome the kinetic bottleneck of the slow, spontaneous folding of pilus subunits.

The most fascinating component in the type 1 pilus assembly system is certainly the assembly platform FimD. It is composed of an N-terminal domain (FimD$_N$) that recognizes incoming FimC-subunit complexes [11], a β-barrel transmembrane domain forming the pore for subunit translocation, a plug domain and a C-terminal periplasmic domain [12, 13]. FimD exists in two distinct conformations: in its inactive state, the plug domain occupies the translocation pore. The reconstitution of pilus assembly from FimC-subunit complexes in the presence of catalytic amounts of FimD [14] showed that FimD can only be converted to its active state by the

FimH-FimC complex. The solved X-ray structures of both conformations demonstrated that the FimH lectin domain occupies the translocation pore in the ternary FimD-FimC-FimH complex and displaces the plug domain towards the periplasmic side of the membrane [12]. FimA polymers only form slowly from FimC-FimA complexes, with a half-life of about 10 hours at 37 °C [14]. FimD proved to be a very efficient catalyst of FimA polymerization, and pilus assembly was completed within 20 minutes in the presence of catalytic amounts of FimD (0.5 equivalents) [14]. FimD-catalyzed pilus assembly functions independently of energy and is exclusively driven by the higher stability of subunit-subunit complexes compared to FimC-subunit complexes [14]. The catalytic cycle of FimD starts from a state in which the last subunit at the growing end of the pilus is capped by FimC. The next FimC-subunit complex then binds to the N-terminal FimD domain (FimD$_N$). The donor strand of the incoming subunit then displaces the FimC molecule capping the pilus end. The last step of the cycle is a shift of the pilus by the length of a subunit towards the extracellular space, and the chaperone bound to the newly incorporated subunit caps the pilus [14]. A detailed analysis of the individual rate constants of the binding of FimC-subunits to FimD and the rates of subunit-subunit contact formation revealed that FimD determines the specificity of the assembly process in that contacts between natural pairs of subunits are formed dramatically faster than wrong subunit-subunit contacts [14, 15]. The successful reconstitution of pilus assembly should allow a quantitative prediction of the length distribution of pilus rods formed in the presence of FimD.

Development of antiadhesive drugs for treatment of urinary tract infections

The high frequency of urinary tract infections (UTIs), with about 50% of all women suffering from at least one UTI during their life [16], and the high rates of UTI recurrence after therapy with antibiotics, require new strategies for UTI treatment. Recurrent UTIs are most likely a consequence of intracellular bacterial communities that persist in epithelium cells of the urinary tract. As internalization of the bacteria by epithelium cells is type 1 pilus dependent, an obvious strategy to prevent recurrent UTIs is the development of antiadhesive drugs. Towards this goal, two different concepts have been followed. First, synthetic inhibitors of pilus assembly termed pilicides have been identified that inhibit *de novo* pilus assembly by binding to the assembly chaperone [18]. Second, a series of α-D-mannosides that mimick the natural glycan of the target receptor and bind to the isolated FimH lectin domain in the low nanomolar range has been developed [19]. The advantage of these FimH antagonists is a competitive inhibition of tissue adhesion of pili that have already formed. Both types of antiadhesive drugs may be used in the future in combination with classical antibiotics.

An intriguing property of type 1 piliated *E. coli* strains is their ability to increase their affinity to target surfaces under tensile mechanical force [20], *e.g.*, under flow conditions during excretion of the urine. This "catch bond" behavior most likely increases the life time of tissue-bound bacteria and the probability of pathogen uptake by urinary epithelium cells, thus promoting bacterial persistence inside host cells during therapy with antibiotics. The "catch bond" behavior has been attributed to two different conformational states of FimH observed in the three-dimensional structures of ligand-free FimH in the context of the tip fibrillum and the isolated FimH lectin domain with bound ligands [20]. While the FimH lectin domain in the tip fibrillum shows specific interactions with the FimH pilin domain and an open conformation of its carbohydrate binding pocket, the isolated lectin domain exhibits a more elongated conformation with a narrower binding pocket and strong conformational changes in the surface loops that interact with the pilin domain in full-length FimH. These data, together with the observation that the isolated lectin domain shows an about 100-fold higher affinity for carbohydrate ligands compared to full-length FimH in the tip fibrillum, have led to the hypothesis that ligand binding of FimH favors the separation between the lectin and the pilin domain and that, conversely, force-induced domain separation increases the affinity for ligands. This suggests that full-length FimH, rather than the isolated FimH lectin domain, is the physiologically relevant drug target.

Acknowledgments

This project was supported by the ETH Zürich and the Swiss National Science Foundation (SNF) within the framework of the NCCR Structural Biology program and by SNF grants 310030B-138657 and 31003A-122095 to R.G.

References

1. E. Hahn, P. Wild, U. Hermanns, P. Sebbel, R. Glockshuber *et al.*, *J. Mol. Biol.* **323**, 845 (2002).
2. G. Zhou, W.-J. Mo, P. Sebbel, T. A. Neubert, R. Glockshuber *et al.*, *J. Cell Sci.* **114**, 4095 (2001).
3. D. Choudhury, A. Thompson, V. Stojanoff, S. Langermann, J. Pinkner *et al.*, *Science* **285**, 1061 (1999).
4. A. Busch, G. Waksman, *Philos. Trans. R. Soc. Lond. B. Biol. Sci.* **365**, 1112 (2012).
5. C. Puorger, O. Eidam, D. Erilov, M. Vetsch, G. Capitani *et al.*, *Structure* **16**, 631 (2008).
6. D. Erilov, C. Puorger, R. Glockshuber, *J. Amer. Chem. Soc.* **129**, 8938 (2007).
7. C. Giese, F. Zosel, C. Puorger, R. Glockshuber, *Angew. Chem. Int. Ed.* **51**, 4474 (2012).
8. C. Puorger, M. Vetsch, G. Wider, R. Glockshuber, *J. Mol. Biol.* **412**, 520 (2011).
9. M. Vetsch, C. Puorger, T. Spirig, U. Grauschopf, E. U. Weber-Ban *et al.*, *Nature* **431**, 329 (2004).

10. M. D. Crespo, C. Puorger, M. A. Schärer, O. Eidam, M. G. Grütter *et al.*, *Nat. Chem. Biol.* **8**, 707 (2012).
11. M. Nishiyama, R. Horst, O. Eidam, T. Herrmann, O. Ignatov *et al.*, *EMBO J.* **24**, 2075 (2005).
12. G. Phan, H. Remaut, T. Wang, W. J. Allen, K. F. Pirker *et al.*, *Nature* **474**, 49 (2011).
13. S. Geibel, E. Procko, S. J. Hultgren, D. Baker, G. Waksman, *Nature* **496**, 243 (2013).
14. M. Nishiyama, T. Ishikawa, H. Rechsteiner, R. Glockshuber, *Science* **320**, 376 (2008).
15. M. Nishiyama, R. Glockshuber, *J. Mol. Biol.* **396**, 1 (2010).
16. B. Foxman, P. Brown, *Infect. Dis. Clin. North Am.* **17**, 227 (2003).
17. I. U. Mysorekar, S. J. Hultgren, *Proc. Natl. Acad. Sci. USA* **103**, 14170 (2006).
18. V. Aberg, F. Almquist, *Org. Biomol. Chem.* **5**, 1827 (2007).
19. M. Scharenberg, O. Schwardt, S. Rabbani, B. Ernst, *J. Med. Chem.* **55**, 9810 (2012).
20. I. Le Trong, P. Aprikian, B. A. Kidd, M. Forero-Shelton, V. Tchesnokova *et al.*, *Cell* **141**, 645 (2010).

HIV ENVELOPE AND INFLUENZA HEMAGGLUTININ FUSION GLYCOPROTEINS AND THE QUEST FOR A UNIVERSAL VACCINE

IAN A. WILSON

Department of Integrative Structural and Computational Biology, The Scripps Research Institute, 10550 North Torrey Pines Road, La Jolla, CA 92037, USA

My view of the present state of the structure and function of viral fusion glycoproteins, their recognition by the immune system, and universal vaccines

Human immunodeficiency virus type 1 (HIV-1) and influenza viruses remain major threats to human health that plague mankind with continual epidemics and sporadic pandemics. These viruses display surface glycoproteins for binding to receptors of the host cell and fusion machinery to enable the viral and cell membranes to fuse so that the viral genome can be deposited into the cell for initiation of viral replication. Blocking of these activities by small molecules and small proteins or by antibodies would neutralize the virus and prevent or ameliorate infection. Although neutralizing antibodies can readily be elicited against either virus, they usually target regions of the virus envelope with little functional relevance. Viral evolution can then easily outpace antibody adaptation, resulting in strain-specific antibodies with little or no broadly neutralizing properties. The development of vaccines that elicit broadly neutralizing antibodies as a means for universal protection against these two viruses, remains an illusive but important goal. The pathogenesis of HIV-1 and influenza differ significantly: whereas HIV-1 generates a persistent infection, influenza is typically cleared within several days. However, similarities can be drawn in their ability to evade protective immunity and the mechanisms they use to infect cells. The critical first step in viral infection is attachment of the virus to the host cell, followed by fusion of the viral and hosts membranes and viral entry into host cells. In the case of influenza and HIV-1, these complex processes are mediated by their respective trimeric Type 1 membrane fusion glycoproteins, the hemagglutinin (HA) for influenza and envelope (Env or gp160) for influenza and HIV-1, respectively.

HA-mediated entry process – The HA trimer is derived from a single polypeptide precursor, HA0, which is proteolysed extracellularly resulting in a metastable mature trimer, composed of two polypeptides, HA1 and HA2. Structurally, the trimer can be subdivided into a membrane-distal "head" and a membrane-proximal "stem" [1, 2]. HA1 comprises the head and contains the sialic acid receptor-binding site. The HA stem region consists primarily of HA2, in addition to parts of HA1, and composes the fusion machinery, which is triggered in the low pH environment of late endosomes. Upon acidification, the HA trimer undergoes major conformational rearrangements that directs insertion of its fusion peptide into the endosomal membranes and leads to subsequent fusion of host and virion membranes. Despite decades of work, the details of how these viral glycoproteins trigger have still not been determined nor how to adequately prevent fusion or receptor binding. Prior to 2008, identification of antibodies that broadly neutralize different strains and subtypes of influenza virus were scarce and only one mouse antibody that bound to different subtypes was known [3]. Interestingly, in a series of unforeseen discoveries, three independent groups identified human antibodies that cross-react with multiple influenza subtypes [4–6]. As opposed to almost all other known HA antibodies that typically bind to the head region of HA and likely prevent receptor binding, these newly discovered antibodies bind the less accessible stem portion of the HA that houses the fusion machinery. Crystallographic and biochemical studies indicated that the antibody neutralizes the virus by blocking conformational rearrangements associated with membrane fusion, thus interfering with the virus's ability to deliver its genetic material into host cells and initiate infection. These findings have opened the door to further explorations into blocking the viral fusion apparatus as means to engineer a universal flu vaccine [7, 8].

Env-mediated entry process – The HIV-1 membrane fusion process and subsequent viral entry into the host cell is even less well understood compared to HA-mediated entry [9, 10]. This has been largely due to the absence of an Env trimer structure, which has been thwarted by its lack of stability as a soluble protein and extensive N-linked glycosylation (~ 81 sites per trimer). During Env synthesis, the gp160 precursors trimerize and are subsequently cleaved by proteases of the furin family into gp120 and gp41 subunits, which associate non-covalently. However, attempts to isolate Env trimers have generally failed due to aggregation, instability and lack of proper folding of uncleaved trimers [11–14]. The current view of Env-mediated entry involves a two-step process that is initiated by the binding of gp120 to the CD4 receptor on T cells. This encounter induces conformational changes in Env that result in interactions with a co-receptor (generally CXCR4 or CCR5) that triggers further conformational changes in the gp41 transmembrane segment that leads to formation of an extended coil-coil pre-fusion intermediate [15]. Subsequently, the resulting N-terminal peptide of gp41 is inserted and anchored into host cell membrane. The gp41 then forms a six-helix bundle that brings the viral and host membranes together.

Thus, understanding the function of viral pathogens has been facilitated by structure determination of their surface antigens but it has been a challenge for HIV-1, HCV and until recently for RSV [16] and Ebola [17] for example. However, use of different expression systems, manipulation of the glycosylation, stabilization of the antigens, and forming complexes with antibodies has provided opportunities for high resolution structural studies. However, capturing fusion intermediates has not been possible or visualizing the molecular events that lead to fusion between the viral membrane and cell membrane.

My recent research contributions to structure and function of viral glycoproteins, the structural bass of neutralization, and universal vaccines

My laboratory focuses on understanding the interaction and neutralization of microbial pathogens by the immune system through high-resolution X-ray structural studies of antibodies complexes with viral antigens. Recently, we have focused on structural investigations of the key sites of vulnerability on influenza and HIV-1 viruses with the ultimate goal of developing sustainable cross-serotype immune responses [18]. Antibody-mediated neutralization of influenza and HIV-1 viruses is a complex combinatorial problem for the human immune system as it is presented with diverse, highly variable and constantly evolving viruses. While neutralizing antibodies against these viruses are traditionally regarded as being strain specific, studies have shown that a much broader response can be mounted. While these examples provide compelling evidence that the immune system is capable of mounting a sustained, cross-serotype response against these viruses, how to elicit broadly neutralizing antibodies by vaccination is poorly understood. In that optic, we are investigating the structures of the viral antigens and the basis of broad neutralization by antibodies. We have delineated sites of vulnerability on the HA and Env to enable development of novel vaccine scaffolds and even small molecule inhibitors that may ameliorate or prevent disease progression [7, 19–21]. To address this challenging task, we have used an integrative approach using combined biophysical and biochemical methodologies in close collaboration with the other laboratories at Scripps and elsewhere by investigating complexes of viral glycoproteins with broadly neutralizing antibodies. We employ state-of-the-art X-ray crystallography, electron microscopy and glycan array technologies to provide key insights into influenza virus and HIV-1 neutralization, tropism and pathogenesis, to reveal novel strategies to control and combat future pandemics. Most importantly, we have high quality reagents for soluble hemagglutinin [8] and HIV-1 Env trimer [11, 12, 14] and gp120 etc.

Structural characterization of both HIV-1 Env and influenza HA in complex with an extensive toolbox of diverse antibodies has been very informative at defining the key sites of vulnerability on these viruses [18, 22–24]. These analyses have clearly succeeded in uncovering novel modes and sites of antibody recognition, not anticipated from previous studies. The epitopes on HIV-1 are more diverse in type and number and some of the most potent sites of neutralization depend on both glycans and protein [18, 25–31]. We are exploring whether identification of these sites of vulnerability can be directly utilized in vaccine settings. Interestingly, bnAbs have been identified to both flu and HIV-1 that bind to key functional regions of each viral spike: the receptor binding site and the stem region proximal to the viral membrane involved in the fusion machinery. Antibody attachment to the receptor binding site most likely leads to potent neutralization by directly preventing receptor engagement on host cells, whereas binding to a region proximal to the viral membrane appears to prevent membrane fusion. The information derived from targeting these sites by different antibodies can now be used also for small molecule [18, 23, 32, 33] and small protein inhibitors [34].

Outlook for future developments on HIV and Influenza vaccines

While the challenges presented by HIV-1, influenza and other incredibly evasive pathogens are daunting, the presence of high titers of neutralizing antibodies, some which are broadly neutralizing, in subsets of infected individuals suggest that the repertoire of the human immune system are up to the challenge. Whether modern vaccine design can appropriately elicit and harness this potential remains to be seen. The current structural arsenal of antibodies that cover both influenza A and influenza B and the availability of multiple structures of antibodies in complex with HA that target the same functionally constrained epitopes in the receptor binding site and stem, indicate that design of a universal broad-spectrum vaccine in the near future seems very promising. For HIV-1, we are working on a better structural understanding of the HIV-1 Env trimer to guide the design of more appropriate immunogens that possess all of the characteristics and constraints necessary to present the epitopes of broadly neutralizing antibodies in a more native context. Scaffolded epitopes are also being designed for use as immunogens to elicit specific and focused responses against specific epitopes [25, 35–39]. The challenge still remains are to how to use these immunogens to stimulate potent and broad B cell responses that mimic natural infection and, in the case of HIV-1, to elicit a faster response than the two years it takes to generate a broad neutralizing response in humans after HIV-1 infection [40–42].

Acknowledgments

Marc Elsliger and Jean-Philippe Julien are gratefully acknowledged for help with preparation of this manuscript as well as the NIH, the International AIDS Vaccine Initiative Neutralizing Antibody Consortium and Center, and the CHAVI-ID for support.

References

1. I. A. Wilson, J. J. Skehel, D. C. Wiley, *Nature* **289**, 366 (1981).
2. J. J. Skehel, D. C. Wiley, *Annu. Rev. Biochem.* **69**, 531 (2000).
3. Y. Okuno, Y. Isegawa, F. Sasao, S. Ueda, *J. Virol.* **67**, 2552 (1993).

4. M. Throsby, E. van den Brink, M. Jongeneelen, L. L. Poon, P. Alard *et al.*, *PLoS One* **3**, e3942 (2008).

5. A. K. Kashyap, J. Steel, A. F. Oner, M. A. Dillon, R. E. Swale *et al.*, *Proc. Natl. Acad. Sci. USA* **105**, 5986 (2008).

6. J. Sui, W. C. Hwang, S. Perez, G. Wei, D. Aird *et al.*, *Nat. Struct. Mol. Biol.* **16**, 265 (2009).

7. D. C. Ekiert, I. A. Wilson, *Curr. Opin. Virol.* **2**, 134 (2012).

8. D. C. Ekiert, G. Bhabha, M. A. Elsliger, R. H. Friesen, M. Jongeneelen *et al.*, *Science* **324**, 246 (2009).

9. S. A. Gallo, C. M. Finnegan, M. Viard, Y. Raviv, A. Dimitrov *et al.*, *Biochim. Biophys. Acta* **1614**, 36 (2003).

10. I. Markovic, K. A. Clouse, *Curr. HIV Res.* **2**, 223 (2004).

11. R. Khayat, J. H. Lee, J. P. Julien, A. Cupo, P. J. Klasse *et al.*, *J. Virol.* **87**, 9865 (2013).

12. P. J. Klasse, R. S. Depetris, R. Pejchal, J. P. Julien, R. Khayat *et al.*, *J. Virol.* **87**, 9873 (2013).

13. J. M. Binley, R. W. Sanders, B. Clas, N. Schuelke, A. Master *et al.*, *J. Virol.* **74**, 627 (2000).

14. R. Ringe, R. W. Sanders, A. Yasmeen, H. J. Kim, J. H. Lee *et al.*, *Proc. Natl. Acad. Sci. USA* **110**, 18256 (2013).

15. I. Muñoz-Barroso, K. Salzwedel, E. Hunter, R. Blumenthal, *J. Virol.* **73**, 6089 (1999).

16. J. S. McLellan, M. Chen, S. Leung, K. W. Graepel, X. Du *et al.*, *Science* **340**, 1113 (2013).

17. J. E. Lee, M. L. Fusco, A. J. Hessell, W. B. Oswald, D. R. Burton *et al.*, *Nature* **454**, 177 (2008).

18. J. P. Julien, P. S. Lee, I. A. Wilson, *Immunol. Rev.* **250**, 180 (2012).

19. D. R. Burton, R. Ahmed, D. H. Barouch, S. T. Butera, S. Crotty *et al.*, *Cell Host Microbe* **12**, 396 (2012).

20. D. R. Burton, P. Poignard, R. L. Stanfield, I. A. Wilson *et al.*, *Science* **337**, 183 (2012).

21. M. J. van Gils, R. W. Sanders, *Virology* **435**, 46 (2013).

22. P. D. Kwong, I. A. Wilson, *Nat. Immunol.* **10**, 573 (2009).

23. M. Hong, P. S. Lee, R. M. Hoffman, X. Zhu, J. C. Krause *et al.*, *J. Virol.* **87**, 12471 (2013).

24. N. S. Laursen, I. A. Wilson, *Antiviral. Res.* **98**, 476 (2013).

25. J. Jardine, J. P. Julien, S. Menis, T. Ota, O. Kalyuzhniy *et al.*, *Science* **340**, 711 (2013).

26. J. P. Julien, J. H. Lee, A. Cupo, C. D. Murin, R. Derking *et al.*, *Proc. Natl. Acad. Sci. USA* **110**, 4351 (2013).

27. J. P. Julien, D. Sok, R. Khayat, J. H. Lee, K. J. Doores *et al.*, *PLoS Pathog.* **9**, e1003342 (2013).

28. L. Kong, J. H. Lee, K. J. Doores, C. D. Murin, J. P. Julien *et al.*, *Nat. Struct. Mol. Biol.* **20**, 796 (2013).

29. J. S. McLellan, M. Pancera, C. Carrico, J. Gorman, J. P. Julien *et al.*, *Nature* **480**, 336 (2011).

30. R. Pejchal, K. J. Doores, L. M. Walker, R. Khayat, P. S. Huang *et al.*, *Science* **334**, 1097 (2011).

31. L. J. Ross, M. M. Binns, J. Pastorek, *J. Gen. Virol.* **72**, 949 (1991).

32. R. Xu, J. C. Krause, R. McBride, J. C. Paulson, J. E. Crowe Jr. *et al.*, *Nat. Struct. Mol. Biol.* **20**, 363 (2013).

33. P. S. Lee, R. Yoshida, D. C. Ekiert, N. Sakai, Y. Suzuki *et al.*, *Proc. Natl. Acad. Sci. USA* **109**, 17040 (2012).

34. S. J. Fleishman, T. A. Whitehead, D. C. Ekiert, C. Dreyfus, J. E. Corn *et al.*, *Science* **332**, 816 (2011).

35. M. L. Azoitei, B. E. Correia, Y. E. Ban, C. Carrico, O. Kalyuzhniy *et al.*, *Science* **334**, 373 (2011).

36. B. E. Correia, Y. E. Ban, M. A. Holmes, H. Xu, K. Ellingson *et al.*, *Structure* **18**, 1116 (2010).

37. J. Guenaga, P. Dosenovic, G. Ofek, D. Baker, W. R. Schief *et al.*, *PLoS One* **6**, e16074 (2011).

38. G. Ofek, F. J. Guenaga, W. R. Schief, J. Skinner, D. Baker *et al.*, *Proc. Natl. Acad. Sci. USA* **107**, 17880 (2010).

39. R. L. Stanfield, J. P. Julien, R. Pejchal, J. S. Gach, M. B. Zwick *et al.*, *J. Mol. Biol.* **414**, 460 (2011).

40. N. A. Doria-Rose, R. M. Klein, M. M. Manion, S. O'Dell, A. Phogat *et al.*, *J. Virol.* **83**, 188 (2009).

41. D. N. Sather, J. Armann, L. K. Ching, A. Mavrantoni, G. Sellhorn *et al.*, *J. Virol.* **83**, 757 (2009).

42. M. D. Simek, W. Rida, F. H. Priddy, P. Pung, E. Carrow *et al.*, *J. Virol.* **83**, 7337 (2009).

DECONSTRUCTION OF ITERATIVE POLYKETIDE SYNTHASES

CRAIG A. TOWNSEND

Department of Chemistry, The Johns Hopkins University, 3400 N. Charles Street Baltimore, MD 21218-2685, USA

The enormous progress of the last 25 years in natural products biochemistry has been dominated by investigation of giant microbial enzymes comprised of organized catalytic domains that function according to two distinct paradigms. Each is "programmed" to sequentially assemble simple precursor molecules over multiple steps without diffusible intermediates to synthesize highly elaborated products. Both synthetic strategies rely upon carrier proteins to hold and shuttle extending intermediates among catalytic domains in an ordered fashion. Collectively these megaproteins are impressive "supramolecular machines" exemplifying protein catalysis far more sophisticated than the classical view of "one gene, one enzyme, one reaction" [1], and are of central importance to the creation of small-molecule drugs for human and animal health.

Assembly-line and iterative catalysis in natural product biosynthesis

The application of molecular biological techniques to natural product biosynthetic problems in the mid 1980's ushered in the modern era of research in the field. Seminal events include isolation of the penicillin gene cluster and characterization of the non-ribosomal peptide synthetase (NRPS) ACVS that assembles the classical Arnstein L,L,D-tripeptide precursor of the antibiotic [2], Fig. 1. Biochemical experiments on isolated wild-type NRPS enzymes had already established a linear mode of accreting ATP-dependent synthesis [3], but now isolation of the corresponding biosynthetic genes and their translation to domains of recognized function in loosely repeated sets, or "modules," underscored the exquisitely organized nature of the "assembly line" synthetic program executed by these large enzymes. The end of 1990 marked the watershed co-discovery of genes encoding three giant modular polyketide synthases (PKSs) that synthesize the 6-deoxyerythronolide core of the macrolide antibiotic erythromycin [4, 5]. Together deoxyerythronolide B synthases (DEBS)1-3, totaling *ca.* 10^6 Da, opened the door to a second paradigm of assembly line megaenzymes whose synthetic program could be "read" from the N- to C-terminus.

The parallel universes of NRPS and modular PKS enzymes in their canonical forms (there are many exceptions and variations on the theme) are extravagant extensions of the classical view of enzyme function enunciated 70 years ago [1]; that is, each domain of the assembly line carries out a *single* chemical reaction and the carrier protein delivers the extending intermediates to the next programmed domain and so on to the final product release step. Among the superfamily of PKSs, however, the modular, assembly line enzymes are but one subtype. As shown in Table 1, four other families of Type I, or multidomain proteins, are known emblematic of the second programmatic strategy of "iterative" catalysis. Iterative catalysis fundamentally differs from the assembly line paradigm where a smaller set, often the minimal set, of catalytic domains is present, but their active sites are reused a fixed, or "programmed", number of times.

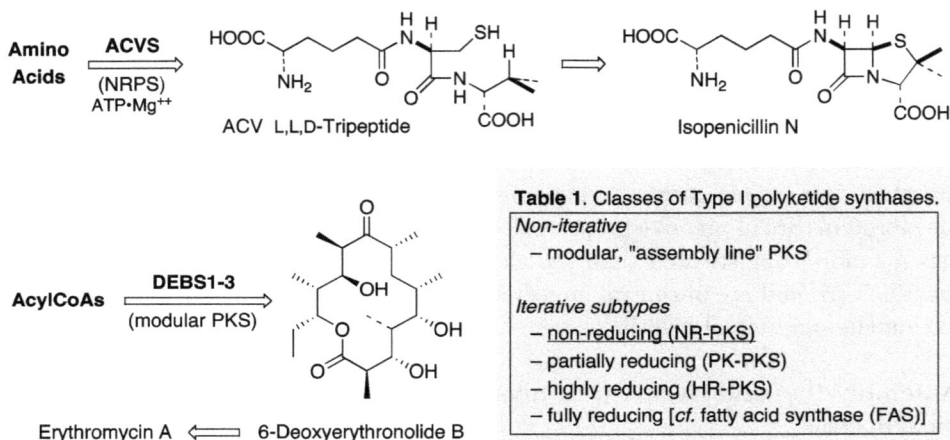

Table 1. Classes of Type I polyketide synthases.

Non-iterative
– modular, "assembly line" PKS
Iterative subtypes
– non-reducing (NR-PKS)
– partially reducing (PK-PKS)
– highly reducing (HR-PKS)
– fully reducing [*cf.* fatty acid synthase (FAS)]

Fig. 1. Representative "assembly line" natural products. The non-ribosomal peptide synthetase (NRPS) Arnstein tripeptide from ACV synthetase and that from the modular polyketide synthases (PKSs) deoxyerythronolide B synthases (DEBS) 1-3.

Iterative polyketide synthases as supramolecular machines

Fatty acid synthases (FASs) are effectively one subgroup of the Type I, iterative megaenzymes in Table 1 that share with the PKSs the use of acylCoA precursors and identical mechanisms of two-carbon chain extension, but differ in the extent of β-carbon processing. At one extreme are the FASs where reductions and dehydration occur at every β-carbonyl to produce a fully saturated fatty acid product. At the other lie the fully non-reducing polyketide synthases (NR-PKSs) where the same chain elongation machinery exists, but no further modifications occur at the β-carbon to generate a poly-β-ketone intermediate of "programmed" length, which is cyclized in specialized domains to aromatic products, Fig. 2 [6]. The aromatic and polycyclic aromatic products, pigments and toxins of fungi are some of the oldest natural products known. Ironically the function of modular PKSs, despite the greater structural complexity of their products, could be readily established by either inactivation of β-processing in specific modules or by inserting the chain-terminating thioesterase domain at the ends of internal modules to yield predictably truncated products. Iterative PKSs, however, remained shrouded in mystery and technical barriers. Inactivation of a synthetic domain in an iterative system stops all synthesis and next to nothing can be learned. A new approach was needed.

A series of contributions from my laboratory over the last dozen years has opened iterative PKSs to systematic investigation using methods that can be more widely applied. Our focus has been on the NR-PKS PksA, which is central to the biosynthesis of the potent environmental toxin and carcinogen aflatoxin B_1. First, we developed a bioinformatics algorithm, UMA, that can identify linker regions in a group of related polydomain proteins [7]. Application of UMA to PksA yielded the surprising result that unlike the four catalytic domains readily identified by conventional sequence analysis, a β-ketoacyl synthase (KS) domain, a malonyl acyltransferase (MAT) domain, and acyl-carrier protein (ACP) and a C-terminal thioesterase (TE) domain, a large, well defined N-terminal domain was clearly visible, and an unknown internal domain placed between the MAT and ACP domains. Were these functional in some way, or merely structural?

Using UMA as a guide, PksA at the gene level was cut (dissection) to yield mono-, di- and tridomains as soluble, functional proteins in *E. coli* that could be recombined (reconstitution) to deduce, or "deconstruct" [8], the synthetic role of each domain. The N-terminal domain we discovered is a "starter-unit acyltransferase" (SAT) domain [9], which in most NR-PKSs accepts acetylCoA from cellular primary metabolism and loads the PKS for synthesis [10]. In the case of PksA, however, a more specialized situation obtains that PksA has a closely associated yeast-like pair of fatty acid synthase subunits that prepare a hexanoyl starter unit that is transferred directly to the PksA SAT domain in a $\sim 1.3 \times 10^6$ Da $\alpha_2\beta_2\gamma_2$ complex to prime polyketide extension (Fig. 2) [11, 12].

Like the SAT domain, which bore no compelling sequence similarity to any other known protein (except other similarly placed domains in established and suspected NR-PKSs), the well-defined internal domain also showed no useful sequence homology. Deconstruction experiments, however, decisively demonstrated that it binds the fully elongated C_{20}-poly-β-ketone and controls the pluripotent self-condensation of this reactive intermediate to a *single* bicyclic product and, thus, was dubbed a "product template" (PT) domain. This bicyclic intermediate was then transferred from the ACP to the TE domain, not for typical hydrolytic release, but for a third, Dieckmann, cyclization to simultaneously release the product norsolorinic acid anthrone (Fig. 2), the first dedicated intermediate en route to aflatoxin [8]. X-ray crystal structures of the PksA PT and TE domains, the product determining cyclization catalysts, and molecular modeling have led to hypotheses to account for how these key cyclization steps occur [13, 14].

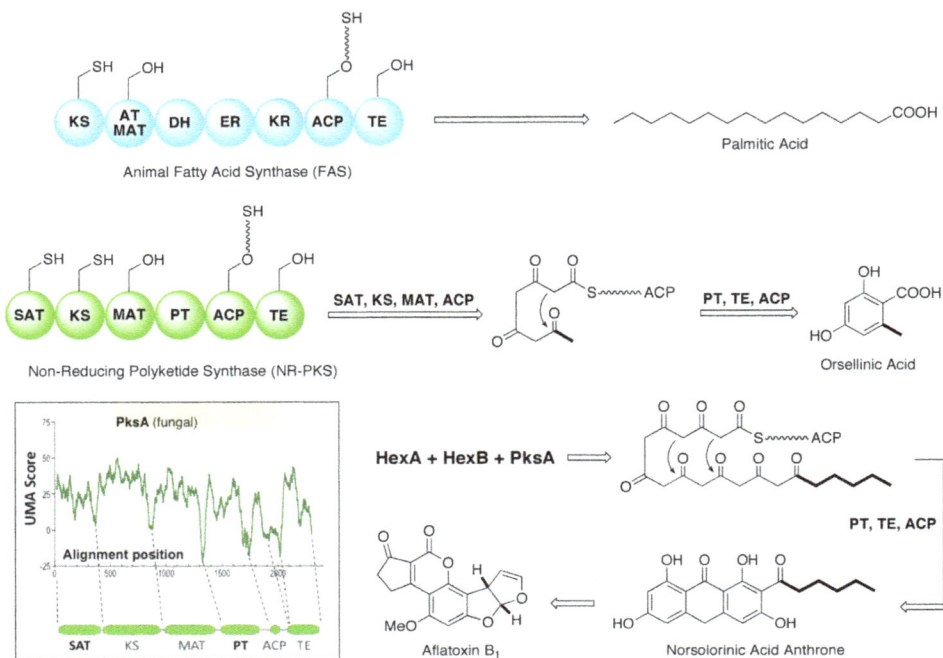

Fig. 2.   Comparison of animal fatty acid synthases and non-reducing polyketide synthases, which lack any β-processing domains but contain unique starter unit transacylase (SAT) and product template (PT) domains originally discovered by UMA analysis of PksA. The highly complex aflatoxin biosynthetic pathway is initiated by a complex of yeast-like fatty acid synthase subunits HexA and HexB, which form a complex with PksA and carry out the synthesis of norsolorinic acid anthrone, the first committed intermediate to the mycotoxin.

Outlook

• Fundamental issues of "programming" control in iterative systems of chain length, control of β-carbon processing during polyketide elongation, and control of cyclization events are in most respects unknown.

• The quarternary structures of these large, believed dimeric proteins remain important unsolved problems, although fragmentary information exists for mono- and didomains. As Professor Ban will address, the work of his and the Steitz groups on animal and yeast fatty acid synthases are pioneering successes and conceptual models for the field.

• Fragmentary information exists that domain•domain interactions affect "programming". For example, recent observations in PksA suggest that the KS domain "waits" for the correct starter, a hexanoyl group as opposed to acetyl, to be bound at the active site cysteine before polyketide extension is initiated, despite the demonstrable presence of acetyl groups on the ACP during synthesis [8, 15].

• The orchestration of 20, 30, 50 or more reactions without release of intermediates, but the total synthesis of a discrete product is an impressive hallmark of iterative catalysis where active sites are reused multiple times and each, therefore, must accommodate a series of structurally distinct intermediates as the polyketide chain extends. The "editing" behavior of TE domains is key to faithful and efficient synthesis [15, 16]. So too are balanced internal kinetics of synthesis and transfer among active sites to maintain synthesis in the forward direction [15]. The roles of protein dynamics, inter-domain cooperative effects (positive and negative), and the possibility of "triggering" initiated by substrate (the correct substrate) binding to advance correctly edited intermediates onward to product all present daunting problems for study.

• Necessary formation of higher order catalytic complexes is increasingly recognized in natural product biosynthesis where enzymes require "auxiliary" or "helper" proteins to function efficiently, if at all. The recently elucidated role of the MbtH superfamily activation of some, but not all, adenylation domains in NRPSs is a shining example of their importance [17]. *In trans* interactions of this sort will be found to play newly appreciated roles in the activities of NRPS and PKS enzymes.

• Finally, engineering and directed evolution of biosynthetic enzymes is in its infancy. Iterative systems are especially attractive because of the already promiscuous nature for all participating synthetic domains. Unlike modular systems where active sites typically function once during chain elongation, iterative catalytic domains are used multiple times to faithfully shepherd substrates having a remarkable breadth of molecular sizes to defined products. Possibilities can be envisioned where their intrinsic substrate flexibility can be exploited. What is lost in binding affinity may well be compensated for by high effective concentration provided by carrier-protein mediated intraprotein transfer [18].

Acknowledgments

The National Institutes of Health are gratefully acknowledged for their sustained support.

References

1. G. W. Beadle, E. L. Tatum, *Proc. Natl. Acad. Sci. USA* **27**, 499 (1941).
2. N. M. Sampson, R. Belagaje, D. T. Blankenship, J. L. Chapman, D. Perry *et al.*, *Nature* **318**, 191 (1985).
3. F. Lipmann, *Acc. Chem. Res.* **6**, 361 (1973).
4. J. Cortes, S. F. Haydock, G. A. Roberts, D. J. Bevitt, P. F. Leadlay, *Nature* **348**, 176 (1990).
5. S. Donadio, M. J. Staver, J. B. McAlpine, S. J. Swanson, L. Katz, *Science* **252**, 675 (1991).
6. J. M. Crawford, C. A. Townsend, *Nat. Rev. Microbiol.* **8**, 879 (2010).
7. D. W. Udwary, M. Merski, C. A. Townsend, *J. Mol. Biol.* **323**, 585 (2002).
8. J. M. Crawford, P. M. Thomas, J. R. Scheerer, A. L. Vagstad, N. L. Kelleher *et al.*, *Science* **320**, 243 (2008).

9. J. M. Crawford, B. C. R. Dancy, E. A. Hill, D. W. Udwary, C. A. Townsend, *Proc. Natl. Acad. Sci. USA* **103**, 16728 (2006).

10. J. M. Crawford, A. L. Vagstad, K. R. Whitworth, K. C. Ehrlich, C. A. Townsend, *ChemBioChem* **9**, 1019 (2008).

11. C. M. H. Watanabe, C. A. Townsend, *Chem. Biol.* **9**, 981 (2002).

12. J. Foulke-Abel, C. A. Townsend, *ChemBioChem* **13**, 1880 (2012).

13. J. M. Crawford, T. P. Korman, J. W. Labonte, A. L. Vagstad, E. A. Hill *et al.*, *Nature* **461**, 1139 (2009).

14. T. P. Korman, J. M. Crawford, J. W. Labonte, A. G. Newman, J. Wong *et al.*, *Proc. Natl. Acad. Sci. USA* **107**, 6246 (2010).

15. A. L. Vagstad, S. B. Bumpus, K. Belecki, N. L. Kelleher, C. A. Townsend, *J. Amer. Chem. Soc.* **134**, 6865 (2012).

16. A. G. Newman, A. L. Vagstad, K. Belecki, J. R. Scheerer, C. A. Townsend, *Chem. Commun.* **48**, 11772 (2012).

17. E. A. Felnagle, J. J. Barkei, H. Park, A. M. Podevels, M. D. McMahon *et al.*, *Biochemistry* **49**, 8815 (2010).

18. A. L. Vagstad, A. G. Newman, P. A. Storm, K. Belecki, J. M. Crawford *et al.*, *Angew. Chem. Int. Ed.* **51**, 1 (2013).

REGULATING RIBOSOME PAUSING DURING TRANSLATION

MARINA V. RODNINA

Department of Physical Biochemistry, Max Planck Institute for Biophysical Chemistry
Am Fassberg 11, Goettingen, 37077, Germany

Ribosome processivity and translational pauses

In bacteria, the average rate of translation elongation is 10–20 amino acids per second and the error frequency ranges between 10^{-4}-10^{-3} [1, 2]. The overall rate of translation elongation can be influenced by many factors, such as the rate of the codon-specific delivery of cognate aminoacyl-tRNAs (aa-tRNAs) to the A site of the ribosome, the abundance of the respective tRNAs, the chemistry of peptide bond formation, secondary structure elements in the mRNA, mRNA context, controlled ribosome stalling, collisions between individual ribosomes in polysomes, and the cooperation between translating ribosomes and the RNA polymerase machinery. As a result of this complex interplay, some codons are translated faster or slower than the average, *i.e.*, the rate of translation is not uniform along an mRNA. Due to the limited information, the rules that define the local speed and accuracy of translation in different regions of a given mRNA are not fully understood [3]. Usually, speed and accuracy of protein synthesis are tightly controlled throughout the translation process; however, the consequences of changes in translational efficiency on the cellular level, and the long-term evolutionary responses to such changes, are not known [4]. For a given amino acid sequence of a protein, multiple degrees of freedom still remain that may allow evolution to tune efficiency and fidelity of gene expression under various conditions [3]. Thus, the sequences on the mRNA and the tRNA pool provide an additional, rich informational framework that may define the proteome of the cell beyond the rules of the genetic code.

Stalling at proline sequences is alleviated by EF-P

Kinetic measurements suggested that peptide bond formation with P-site peptidyl-tRNAs ending with proline or with Pro-tRNA^{Pro} in the decoding site of the ribosome turned out to be surprisingly slow [5, 6], which might lead to stalling of translation as a whole. In fact, translating ribosomes tend to be stalled on mRNA sequences coding for two consecutive prolines followed by Asn, Asp, Glu, Gly, Pro, or Trp [7]. We discovered that unlike peptide bond formation with the other amino acids, translation of those sequences is accelerated by a specialized translation factor, elongation factor P (EF-P) [5, 8]. Due to the presence of the pyrrolidine ring in proline, the torsion angle for the N–Cα bond restricts the number of accessible conformations and imposes structural constraints on Pro positioning in the peptidyl transferase center of the ribosome, thereby sterically hindering peptide bond formation. Of more than 4000 annotated proteins in *Escherichia coli*, > 300 contain motifs of three or more consecutive prolines or X-Pro-Pro-X motifs, where X is one of the stalling-promoting amino acids [9]. Among those proteins metabolic enzymes, transporters, and regulatory transcription factors are frequent, explaining the pleiotropic phenotypes caused by deletions of the genes coding for EF-P or its modification enzymes, including effects on virulence and bacterial fitness [10]. Given that the synthesis of a peptide bond usually does not require an auxiliary factor, EF-P appears to be a recent evolutionary addition to the repertoire of translation factors, possibly required to properly balance the cellular amounts of regulatory proteins containing proline repeats.

EF-P carries a unique post-translational modification which requires the action of three enzymes, YjeK, YjeA, and YcfM [10–13]. The low catalytic proficiency of unmodified EF-P may explain why deletions of the *yjeA* or *yjeK* genes that code for the modification enzymes lead to phenotypes that are similar to, or only somewhat milder than, the deletion of *efp*, the gene coding for EF-P [14, 15]. On the other hand, even in the absence of active EF-P, small amounts of proteins with proline stretches are slowly formed; this may be sufficient to support the viability of strains in which the *efp*, *yieA*, and *yieK* genes are deleted. The modification enzymes, which are lacking in higher eukaryotes, provide attractive new targets for the development of new, highly specific antimicrobials.

EF-P and evolution

Of the 21 proteinogenic amino acids, Pro (together with Ala, Glu, and Gly) tends to be lost during the evolution [16], possibly because it stalls the translation. The emergence of a factor that specifically accelerated Pro incorporation might result from an evolutionary trade-off between the crucial role of Pro clusters in some proteins and the requirement for rapid, processive protein synthesis. One can envisage that the specific requirements of eukaryotic cells, particularly with respect to translational control, compartmentalization of protein synthesis, and larger protein domains, might lead to emergence of additional, yet uncharacterized proteins that modify the functions of the ribosome or of the universal translation factors [17]. Identification and understanding of this auxiliary machinery would be a very exciting avenue for future research.

Acknowledgments

The work was funded by the Max Planck Society and by a grant of the Deutsche Forschungsgemeinschaft.

References

1. M. V. Rodnina, *Ribosomes: Structure, Function, and Dynamics*, eds. W. Wintermeyer, R. Green (Springer, Wien, New York), 199 (2011).
2. I. Wohlgemuth, C. Pohl, J. Mittelstaet, A. L. Konevega, M. V. Rodnina, *Philos. Trans. R. Soc. Lond. B Biol. Sci.* **366**, 2979 (2011).
3. H. Gingold, Y. Pilpel, *Mol. Syst. Biol.* **7**, 481 (2011).
4. D. A. Drummond, C. O. Wilke, *Nat. Rev. Genet.* **10**, 715 (2009).
5. L. K. Doerfel, I. Wohlgemuth, C. Kothe, F. Peske, H. Urlaub *et al.*, *Science* **339**, 85 (2013).
6. I. Wohlgemuth, S. Brenner, M. Beringer, M. V. Rodnina, *J. Biol. Chem.* **283**, 32229 (2008).
7. C. J. Woolstenhulme, S. Parajuli, D. W. Healey, D. P. Valverde, E. N. Petersen *et al.*, *Proc. Natl. Acad. Sci. USA* **110**, E878 (2013).
8. S. Ude, J. Lassak, A. L. Starosta, T. Kraxenberger, D. N. Wilson *et al.*, *Science* **339**, 82 (2013).
9. L. Peil, A. L. Starosta, J. Lassak, G. C. Atkinson, K. Virumae *et al.*, *Proc. Natl. Acad. Sci. USA* **110**, 15265 (2013).
10. W. W. Navarre, S. B. Zou, H. Roy, J. L. Xie, A. Savchenko *et al.*, *Mol. Cell* **39**, 209 (2010).
11. L. Peil, A. L. Starosta, K. Virumae, G. C. Atkinson, T. Tenson *et al.*, *Nat. Chem. Biol.* **8**, 695 (2012).
12. H. Roy, S. B. Zou, T. J. Bullwinkle, B. S. Wolfe, M. S. Gilreath *et al.*, *Nat. Chem. Biol.* **7**, 667 (2011).
13. T. Yanagisawa, T. Sumida, R. Ishii, C. Takemoto, S. Yokoyama, *Nat. Struct. Mol. Biol.* **17**, 1136 (2010).
14. S. B. Zou, S. J. Hersch, H. Roy, J. B. Wiggers, A. S. Leung *et al.*, *J. Bacteriol.* **194**, 413 (2012).
15. S. B. Zou, H. Roy, M. Ibba, W. W. Navarre, *Virulence* **2**, 147 (2011).
16. I. K. Jordan, F. A. Kondrashov, I. A. Adzhubei, Y. I. Wolf, E. V. Koonin *et al.*, *Nature* **433**, 633 (2005).
17. L. K. Doerfel, M. V. Rodnina, *Biopolymers* **99**, 837 (2013).

THE MOLECULAR MECHANICS OF THE RIBOSOME

JIE ZHOU, LAURA LANCASTER, ZHUOJUN GUO,

JOHN PAUL DONOHUE and HARRY F. NOLLER

Department of Molecular, Cell and Developmental Biology and Center for Molecular Biology of RNA, University of California at Santa Cruz, Santa Cruz, CA 95064, USA

Research on the ribosome as a supramolecular machine

Ribosomes, which are responsible for synthesis of proteins in all organisms, are among the largest and most complex of biological supramolecular machines, comprising several dozens of protein and RNA molecules. During the past two decades, our understanding of the structure of the ribosome has made spectacular progress, first from cryo-EM and then from X-ray crystallography. With all-atom structures for examples of both bacterial and eukaryotic ribosomes in hand, the stage is set for understanding the mechanisms of protein synthesis at the atomic level. Fundamental questions that have gone unanswered for several decades are now beginning to be resolved. Protein synthesis involves translation of the nucleotide sequences of messenger RNAs into proteins via the genetic code. Although the detailed mechanisms of initiation and termination of translation differ between bacterial, archaeal and eukaryal systems, the mechanism of elongation is similar for all organisms.

Each round of the elongation cycle takes place in three steps: (i) selection of aminoacyl-tRNA, (ii) catalysis of peptide bond formation and (iii) coupled translocation of mRNA and tRNA. Each of these steps is believed to be based mainly on the properties of ribosomal RNA (rRNA), rather than the ribosomal proteins, in keeping with the notion that the first ribosomes had to be made only of RNA. Aminoacyl-tRNA selection, based on codon-anticodon base pairing, is monitored by interactions with universally conserved bases of 16S rRNA in the decoding site of the small ribosomal subunit [1]. Peptide bond formation is catalyzed by peptidyl transferase, an enzymatic activity of 23S rRNA in the large ribosomal subunit [2]. Although translocation of mRNA and tRNA is catalyzed by the protein elongation factor EF-G, it seems likely that its fundamental mechanism is again based on ribosomal RNA.

Following peptide bond formation, the mRNA must be moved through the ribosome by one codon. In addition, the deacylated P-site tRNA and peptidyl A-site tRNA must be moved from the A and P sites to the P and E sites, respectively. These represent large-scale movements of 20 Å to 70 Å, which take place on a time scale of about 50 ms. Chemical probing experiments showed that tRNA movement occurs in two steps, involving hybrid binding states [3]. In the first step, the

A-site peptidyl-tRNA moves into the A/P hybrid state and the deacylated P-site tRNA moves into the P/E hybrid state. In the second step, the A/P tRNA moves into the P/P state (the classical P site) and the P/E tRNA moves into the E/E state (the classical E site). The first step is reversible and can occur spontaneously, while the second, rate-limiting step depends on catalysis by EF-G.

Studying ribosome mechanics by crystallography and FRET

Cryo-EM studies by Frank and Agrawal showed that when EF-G binds to the ribosome, the small ribosomal subunit rotates by 5–10° counterclockwise relative to the large subunit [4]. In a more recent cryo-EM study from the Spahn laboratory, a ribosome complex trapped with fusidic acid in an intermediate state of translocation, contained a tRNA bound in a position in between the P/E and E/E states [5]. In this complex, the head of the 30S subunit was rotated by about 18°, suggesting that the rate-limiting step of translocation, in which the mRNA and anticodon stem-loop moieties are moved through the small subunit, is accomplished by rotation of the 30S head. We sought to (a) obtain direct evidence for rotation of the 30S head during translocation and (b) to solve the crystal structure of a trapped translocation intermediate.

We used ensemble FRET methods to study 30S head rotation in real time [7]. We labeled specific positions of ribosomal proteins in the head and body of the 30S subunit by *in vitro* reconstitution with fluorophore-labeled proteins. The in-

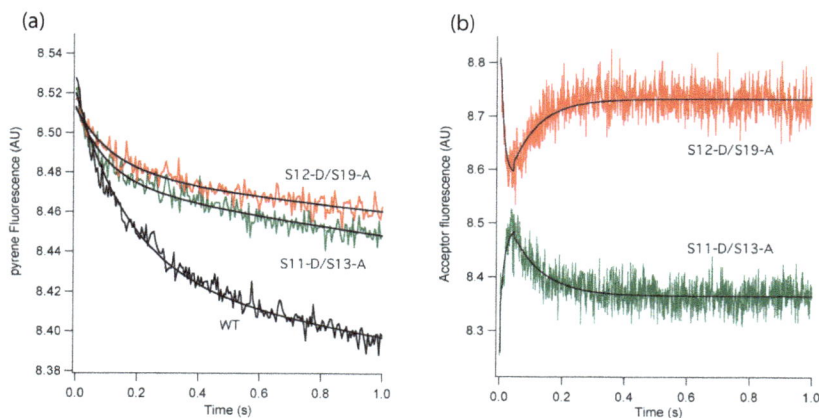

Fig. 1. Rotation of the 30S subunit head during EF-G-catalyzed translocation [6]. (a) mRNA translocation, as followed by quenching of fluorescence of a pyrene dye attached to position +9 of mRNA. (b) Rotation of 30S subunit head, followed by changes in FRET efficiency between donor Alexa 488 and acceptor Alexa 568 probes attached to position 101 of S11 and position 11 of S13 for the S11-D/S13-A pair (green), and position 108 of S12 and position 56 of S19 for the S12-D/S19-A pair (red). Pre-translocation 70S ribosome complexes were rapidly mixed with EF-G-GTP at time = 0 and pyrene fluorescence (a) or Alexa 568 emission (b) were recorded. Each trace represents the average of ten reactions. WT: wild-type, unlabeled ribosomes; AU: arbitrary units.

termolecular distance between proteins S19 (head) and S12 (body) was predicted to increase, and the distance between S13 (head) and S11 (body) was predicted to decrease during translocation. These predictions were confirmed (Fig. 1) using a rapid mixing stopped-flow approach. The forward rate of head rotation was slower than the forward rate of 30S subunit body rotation [8], and the reverse rate of head rotation was roughly similar to the reverse rate of 30S body rotation. Our results are consistent with coupling of 30S head rotation to the second step of translocation.

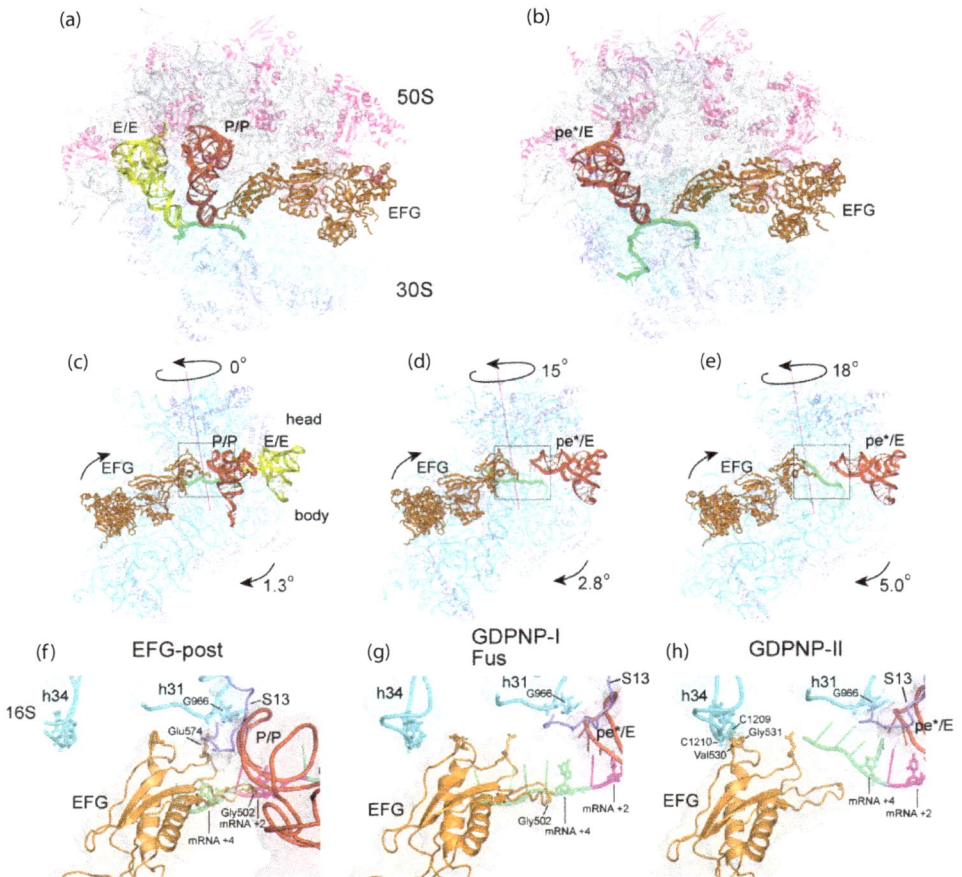

Fig. 2. Structures of trapped 70S ribosome·EF-G complexes [6]. (a,b) Overall views of the (a) non-rotated 70S·EF-G-post complex [10] and (b) Fus 70S·EF-G complex with tRNA bound in the pe*/E state. (c–e) Interface views showing 30S subunit body and head rotation in the (c) EF-G-post state; (d) GDPNP-I and Fus complex; and (e) GDPNP-II complex. (f–h) Close-up views of EF-G domain IV interactions with the 30S subunit head in the (f) EF-G-post complex; (g) GDPNP-I and Fus complex; and (h) GDPNP-II complex. 16S rRNA, cyan; 30S proteins, blue; 23S rRNA, grey; 50S proteins, magenta; mRNA, green; P/P tRNA, red; E/E tRNA, yellow; pe*/E tRNA, red; EF-G, orange.

To determine the structures of rotated translocation intermediates, we trapped the ribosome either with fusidic acid (which allows hydrolysis of GTP but prevents release of EF-G) or the non-hydrolyzable GTP analogue GDPNP. Three different trapped complexes were crystallized (one with fusidic acid and two with GDPNP) and their structures solved at resolutions between 3.5 Å and 4.1 Å (Fig. 2; [6]). These structures provide numerous insights into the mechanism of translation, including conformational changes that take place in the ribosome, tRNA and EF-G, as well as details of the many changes in intermolecular contacts that occur during translocation. The fusidic acid complex and one of the GDPNP complexes (GDPNP-I) shows 30S head rotation of 15°, while the GDPNP-II complex shows a head rotation value of 18°, similar to that observed in the Spahn cryo-EM reconstruction [5]. The tRNA is trapped in intermediate states between the P/E and E/E states (the pe*/E state), in which the anticodon stem-loop (ASL) of the tRNA is bound to the P site of the 30S head, but also to a position near the E site of the 30S body. (We term these states "chimeric hybrid states" following the terminology of Greek mythology in which the beast called *chimera* has the head of a lion and the body of a goat.) Most intriguingly, we observe that two universally conserved bases of 16S rRNA intercalate between bases of the mRNA, potentially acting as "pawls" of a translocational ratchet. The conserved and functionally important "switch loop I" in the GTPase active site of EF-G is found to be ordered in the GDPNP structures for the first time. Comparison with structure of the fusidic acid-stabilized complex suggests that fusidic acid acts by mimicking the structure of the ordered switch loop I.

A working model for the mechanism of translocation

Our findings, together with previous results from many other laboratories, suggest a model for translocation involving a sequence of large-scale structural changes in the ribosome. (i) Following peptide bond formation, the tRNAs move into the A/P and P/E hybrid states. This movement is accompanied by forward (counterclockwise) intersubunit rotation, and is promoted by, but not necessarily dependent on, binding of EF-G·GTP. (ii) Forward (counterclockwise) rotation of the head of the 30S subunit advances the mRNA and the ASLs of the tRNAs and their associated mRNA codons relative to the body of the 30S subunit, into the ap/P and pe/E chimeric hybrid states. This step is dependent on binding EF-G, but does not appear to require GTP hydrolysis. (iii) Reverse (clockwise) rotation of the 30S head while disengaged from the ASLs completes movement of the tRNAs into their P/P and E/E states. This may be accompanied by, or followed by, reverse (clockwise) rotation of the 30S subunit body, returning the ribosome to its classical, non-rotated conformation. Hydrolysis of GTP is required for release of EF-G, but also appears to be necessary for some other step(s) in the translocation process [9].

Outlook for the future of research on ribosome translocation

In spite of decades of research, often using ingenious and powerful approaches, our understanding of the mechanism of ribosomal translocation remains incomplete. Among the many remaining mysteries are: (i) Is the mRNA moved actively or passively during translocation? (ii) How are mRNA and tRNA movement coupled, so as to prevent frame-shifting? (iii) What is the role of EF-G in translocation? (iv) What does GTP hydrolysis accomplish? (v) How can EF-G bind when the A and P sites are occupied by tRNA? (vi) What is "the ratchet"? (vii) What is the pawl of the ratchet? (viii) How are the movements of the 30S body and head, and other dynamic features, coordinated?

Acknowledgments

We thank the U.S. National Institutes of Health, the Agouron Foundation and the California Institute for Quantitative Biosciences (QB3) for support of this work.

References

1. J. M. Ogle, D. E. Brodersen, W. M. Clemons Jr., M. J. Tarry, A. P. Carter *et al.*, *Science* **292**, 897 (2001).
2. P. Nissen, J. Hansen, N. Ban, P. B. Moore, T. A. Steitz, *Science* **289**, 920 (2000).
3. D. Moazed, H. F. Noller, *Nature* **342**, 142 (1989).
4. J. Frank, R. K. Agrawal, *Nature* **406**, 318 (2000).
5. A. H. Ratje, J. Loerke, A. Mikolajka, M. Brünner, P. W. Hildebrand, *et al.*, *Nature* **468**, 713 (2010).
6. J. Zhu, A. Korostelev, D. A. Costantino, J. P. Donohue, H. F. Noller *et al.*, *Proc. Natl. Acad. Sci. USA* **108**, 1839 (2010).
7. Z. Guo, H. F. Noller, *Proc. Natl. Acad. Sci. USA* **109**, 20391 (2012).
8. D. N. Ermolenko, H. F. Noller, *Nat. Struct. Mol. Biol.* **18**, 457 (2011).
9. M. V. Rodnina, A. Savelsbergh, V. I. Katunin, W. Wintermeyer, *Nature* **385**, 37 (1997).
10. Y. G. Gao, M. Selmer, C. M. Dunham, A. Weixlbaumer, A. C. Kelley *et al.*, *Science* **326**, 694 (2009).

EXPLORING THE DYNAMICS OF SUPRAMOLECULAR MACHINES WITH CRYO-ELECTRON MICROSCOPY

JOACHIM FRANK

Howard Hughes Medical Institute, Department of Biochemistry and Molecular Biophysics
650 W. 168ᵗʰ Street, New York, NY 10032, USA

Present state of research

Supramolecular machines perform their work in the cell by going through many different states, distinguished by different conformations and free-energy levels. Ideally, in order to find out how these machines work, we would create a suitable *in vitro* environment containing all components including energy supply that allows the machine to function. We would then aim to take a "movie," capturing their structure at highest resolution in a continuous fashion. Keeping within that film analogy, we might consider taking a large number of "snapshots" in equal small time intervals, each short enough, as in the macroscopic world, to eliminate jarring transitions. However, we would find out that this project has flaws both on the conceptual and the practical level. Conceptually, it is incorrect to equate a molecular machine's progress to the workings of a macroscopic machine in motion since the states are not ordered in sequence of time but are visited in a stochastic manner, with occasional irreversible events such as NTP hydrolysis as the only mark of progress. In practical terms, there are in fact two problems, one affecting the way data for any given state can be captured, the other affecting the ability to obtain coverage of states in a continuum.

First of all, the visualization of a structure requires some form of radiation which imparts energy on the molecule and changes it in the very process. Minimization of these damaging interactions calls for a radiation dose so low that averaging over a sufficiently large population is required for visualization. This requirement, in turn, translates into a complicated way each snapshot must be taken: the structural information in every state has to be gathered from many different copies of the molecule. While in forming such an average, crystallographic approaches — X-ray and electron crystallography — are able to take advantage of the regular arrangement of molecules in a crystal, single-particle electron microscopy must first determine the precise orientation of each molecule from its projection image [1]. The second hurdle interfering with the idea of making a movie is that only a limited number of states are sufficiently populated to allow a three-dimensional structure to be determined. Thus the "movie frames" in between these states remain empty, undetermined.

The difference between crystallographic and single-molecule approaches has opposite consequences for resolution and functional relevance: the high order achievable in a crystal makes it possible to obtain very high resolution, but the conformational state that the molecule is trapped in may not be relevant to its function. When, on the other hand, the structure is obtained from multiple images of free-standing "single" molecules as they are engaged in their work, functional relevance is guaranteed (*e.g.*, [2]), yet until recently atomic resolution has not been achieved, except for viruses and other molecules with high symmetry [3, 4].

However, the introduction, in the past year, of direct-detection cameras [5, 6], some of which have single-electron counting and super-resolution capabilities, has radically changed this situation. Recent accomplishments obtained with the help of such cameras portray a field in rapid transition, with the ultimate claim to occupy a position in structural biology matching the one held until now by X-ray crystallography. Three-dimensional density maps (reflecting the reconstructed Coulomb potential in a 3D array) obtained in the 3 Å range for particles with high symmetry [7, 8] and now even entirely asymmetric assemblies [9] enable *de novo* chain tracing and the construction of accurate atomic models without the aid of fitting existing structures. This of course pertains not only to the protein parts of a molecular machine, but to nucleic acid components, as well.

Another aspect to be emphasized in single-particle electron microscopy is the ability to obtain an entire inventory of co-existing states of a macromolecule from a single sample. Recent development of powerful software using maximum-likelihood methods has made it possible to extract multiple structures, one for each state, from a heterogeneous mixture [10]. This capability means that to some extent, with help from other techniques such as single-molecule FRET, inferences can be drawn on the dynamics of the supramolecular machine ("Story in a Sample" [11]).

So far I have spoken of a reductionist, *in vitro* approach by electron microscopy toward study of supramolecular machines, whose aim is a portrayal of their structure and dynamics at atomic resolution. Another approach, *electron tomography*, is the attempt to visualize the machine within the context of the cell. The progress with this approach in recent years, with automated tilt data collection having become routine, has been quite impressive [12]. In rare cases the interesting part of a cell is thin enough to be penetrated by the electron beam. For thicker samples, high-pressure freezing must be used followed by sectioning with diamond knife or Focused Ion Beam (FIB) milling. While FIB milling of a frozen sample requires very specialized equipment, it is becoming the preferred method of sectioning of biological material as it is virtually free of artifacts associated with cutting [13, 14].

My lab's recent research contributions

My lab studies the mechanism of protein biosynthesis in both eubacteria and eukaryotes, using single-particle cryo-electron microscopy of ribosomes that are in various states of translation. These states are characterized by different conformations of the ribosome itself and different binding configurations of mRNA, tRNA, as well as a variety of translation factors (in the case of bacteria, EF-G, EF-Tu, etc.). From density maps reconstructed, such as in ref. [2], atomic models are built by docking and flexible fitting of structures in the Protein Data Bank. Of special interest to us has been translocation, or the mechanism by which the ribosome transports mRNA and tRNAs bound to it to the next codon in each cycle of the polypeptide elongation. We early recognized that this movement is facilitated by a ratchet-like reorganization of the ribosome [15]. More recently we discovered [16] that the ribosome at the point after peptide bond formation exists in multiple conformational states characterized by different intersubunit rotations and tRNA binding configurations. Thanks to the above-mentioned development of novel classification software these states could be extracted and independently reconstructed [17].

Most recent contributions by my group to the field of biosynthesis include the 5-Å resolution reconstruction of the ribosome from *Trypanosoma brucei*, a eukaryotic parasite causing Sleeping Sickness [18], and the determination of the structure of the mammalian translation pre-initiation complex [19]. At the date of preparation of this manuscript, the *T. brucei* ribosome reconstruction represented one of the highest-resolution structures of an asymmetric molecule, though still employing the old technology of recording on film (Fig. 1). Compared to the ribosomes of other eukaryotes, the *T. brucei* ribosome possesses unusual features, such as its large RNA expansion segments which may be docking platforms for *T. brucei*-specific protein factors, possibly associated with the need of the parasite to rapidly adapt to hosts with widely different body temperatures.

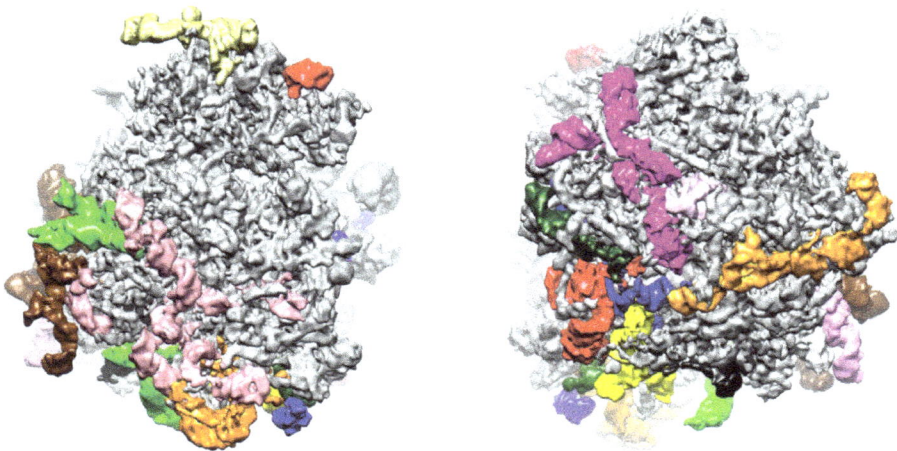

Fig. 1. Cryo-EM density map of the ribosome from *Trypanosoma brucei* at 5 Å resolution. (Left) The ribosome (grey) viewed from the solvent side of the small subunit; (right) viewed from the solvent side of the large subunit. Ribosomal RNA expansion segments are rendered in different colors.

The visualization of the mammalian 43S pre-initiation complex presents an example for a highly heterogeneous specimen that could nevertheless be characterized by a series of reconstructions which depict different combinations of factors attached to the small ribosomal subunit. Of these, only one representing 4.4% of the whole dataset contained the initiator tRNA and most factors that are important in setting up the 43S initiation complex and the staging for the scanning of the mRNA for the start codon (Fig. 2).

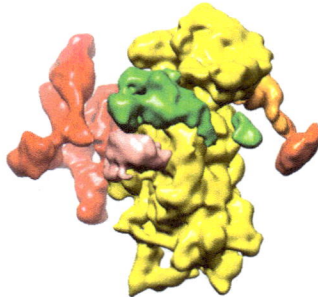

Fig. 2.   43S translation pre-initiation complex at 11.6 Å resolution, reconstructed from a subpopulation of 29,000 cryo-EM particle images. The components of this supramolecular complex are: 40S ribosomal subunit (yellow), eIF3 (red), DHX29 (green), initiator tRNA (yellow-orange), and eIF2 (orange, attached to initiator tRNA).

Outlook to future developments of research

As noted in the introduction, the new generation of detectors employed in the transmission electron microscope is currently revolutionizing cryo-electron microscopy as a tool in structural research, in particular in the study of supramolecular complexes. Due the gain in contrast and resolution achievable, it will soon be possible to reconstruct supramolecular machines in their entirety — proteins along with the nucleic acid components — in multiple, functionally relevant states. The novel technology has a bearing on both different experimental approaches outlined in the beginning: single-particle reconstruction and electron tomography of cell sections. Both techniques will be transformed because of the gains in resolution. Determination of atomic structures by single-particle reconstruction will become routine for "well-behaved" supramolecular complexes, *i.e.*, those that occur only in a few well-populated states. High-throughput methods will be available similar to those now found in X-ray crystallography. At the same time, electron tomography will reach the range of resolutions that allows the signatures of conformational states of a molecule to be recognized within the context of a cell, enabling the tracking of information relevant for the description of its functional state *in situ*. In this way, it will be possible to directly link localized processes in the cell to the dynamical behavior of atomic structures determined by a combination of single-particle cryo-EM, X-ray crystallography, and single-molecule FRET.

Acknowledgments

Funding has been provided by Howard Hughes Medical Institute and grants NIH R01 GM29169 and GM55440. I thank Yaser Hashem for the preparation of the figures and for a critical reading of the manuscript.

References

1. J. Frank, *Three-dimensional Electron Microscopy of Macromolecular Assemblies* (Oxford University Press, 2006).
2. E. Villa, J. Sengupta, L. G. Trabuco, J. LeBarron, W. T. Baxter *et al.*, *Proc. Natl. Acad. Sci. USA* **106**, 1063 (2009).
3. X. Zhang, L. Jin, Q. Fang, W. H. Hui, Z. H. Zhou, *Cell* **141**, 472 (2010).
4. N. Grigorieff, S. Harrison, *Curr. Opin. Struct. Biol.* **21**, 265 (2011).
5. A. R. Faruqi, *J. Phys. Condens. Matter* **21**, 314004 (2009).
6. X. Li, S. Zheng, C. R. Booth, M. B. Braunfeld, S. Gubbens *et al.*, *Nat. Meth.* **10**, 584 (2013).
7. B. E. Bammes, R. H. Rochat, J. Jakana, D.-H. Chen, W. Chiu, *J. Struct. Biol.* **177**, 589 (2012).
8. D. Veesler, T.-S. Ng, A. K. Sendamarai, B. J. Eilers, C. M. Lawrence *et al.*, *Proc. Natl. Acad. Sci. USA* **110**, 5504 (2013).
9. X.-C. Bai, I. S. Fernandez, G. McMullan, S. H. W. Scheres, *eLife* **2**:e00461 (2013).
10. S. H. Scheres, *J. Mol. Biol.* **415**, 406 (2012).
11. J. Frank, *Biopolymers* **99**, 832 (2013).
12. V. Lučié, A. Rigort, W. Baumeister, *J. Cell Biol.* **202**, 407 (2013).
13. M. Marko, C. Hsieh, R. Schalek, J. Frank, C. Mannella, *Nat. Meth.* **4**, 215 (2007).
14. K. Wang, K. Strunk, G. Zhao, J. L. Gray, P. Zhang, *J. Struct. Biol.* **180**, 318 (2012).
15. J. Frank, R. K. Agrawal, *Nature* **406**, 318 (2000).
16. X. Agirrezabala, J. Lei, J. L. Brunelle, R. F. Ortiz-Meoz, R. R. Green *et al.*, *Mol. Cell* **32**, 190 (2008).
17. X. Agirrezabala, H. Liao, E. Schreiner, J. Fu, R. F. Ortiz-Meoz *et al.*, *Proc. Natl. Acad. Sci. USA* **109**, 6094 (2012).
18. Y. Hashem, A. des Georges, J. Fu, S. N. Buss, F. Jossinet *et al.*, *Nature* **494**, 385 (2013).
19. Y. Hashem, A. des Georges, V. Dhote, R. Langlois, H. L. Liao *et al.*, *Cell* **153**, 1108 (2013).

CRYSTALLOGRAPHIC STUDIES OF EUKARYOTIC RIBOSOMES AND FUNCTIONAL INSIGHTS

NENAD BAN

Institute of Molecular Biology and Biophysics,
ETH Zürich (Swiss Federal Institute of Technology), Zürich, 8093, Switzerland

State of research (proteins in supremolecular machines)

Recent developments in the technology and methods for crystallographic and electron microscopic data collection and structure determination have made it possible to approach particularly complex problems in structural biology [1, 2]. Some of these improvements include; i) the development of methods to rapidly freeze samples in the native hydrated state for crystallographic and electron microscopic experiments, ii) the availability of high-intensity synchrotrons and beamlines with a tunable wavelength for crystallographic experiments and the availability of a new generation of very stable electron microscopes with possibilities for automatic data acquisition, iii) the use of extremely sensitive direct detection devices, iv) improved software for data integration and phasing of the measured amplitudes in crystallography and for particle alignment and 3D reconstruction in cryo-electron microscopy combined with an increase in computer speed [3]. These developments are now propelling the studies of large macromolecular assemblies to a new level and several new structures have emerged in the past few years that reveal interesting architectural features of protein components in large cellular complexes and their functional roles, for some examples see: [1, 4–7]. The section below outlines our contributions to the field using as an example our structural studies of eukaryotic ribosomes, multi-megadalton ribonucleoprotein complexes responsible for protein synthesis.

Recent research contributions (proteins in supramolecular machines)

The ribosome is a large cellular assembly that plays a central role in the process of protein synthesis in all organisms [8]. Although basic aspects of protein synthesis are preserved in all kingdoms of life, eukaryotic ribosomes are much more complex than their bacterial counterparts, require a large number of assembly and maturation factors during their biogenesis, use numerous initiation factors, and are subjected to extensive regulation [9, 10].

Recently we obtained detailed structural information on eukaryotic ribosomes by determining the first complete structures of both eukaryotic ribosomal subunits each in complex with an initiation factor [9, 11]. These results are of fundamental importance for understanding protein synthesis in eukaryotes and are the prerequisite for understanding the structural basis of the regulation of protein synthesis in normal cells and how it is perturbed in various diseases.

Crystal structure of the eukaryotic 40S ribosomal subunit in complex with initiation factors 1 and 1A

The eukaryotic 40S ribosomal subunit plays a central role in the initiation of protein synthesis during which it interacts with more than a dozen eukaryotic translation initiation factors (three in bacteria) [12]. These functional differences are reflected in the size and complexity of the eukaryotic 40S subunit; it is 500 kDa larger than the 30S subunit with 18 eukaryotic proteins that have no homologues in bacteria and it contains a considerably longer rRNA. Furthermore, due to its role in translation initiation, which differs substantially between prokaryotes and eukaryotes, the eukaryotic 40S subunit is the target of regulation in a number of cellular processes including development, differentiation, stress response and neuronal function.

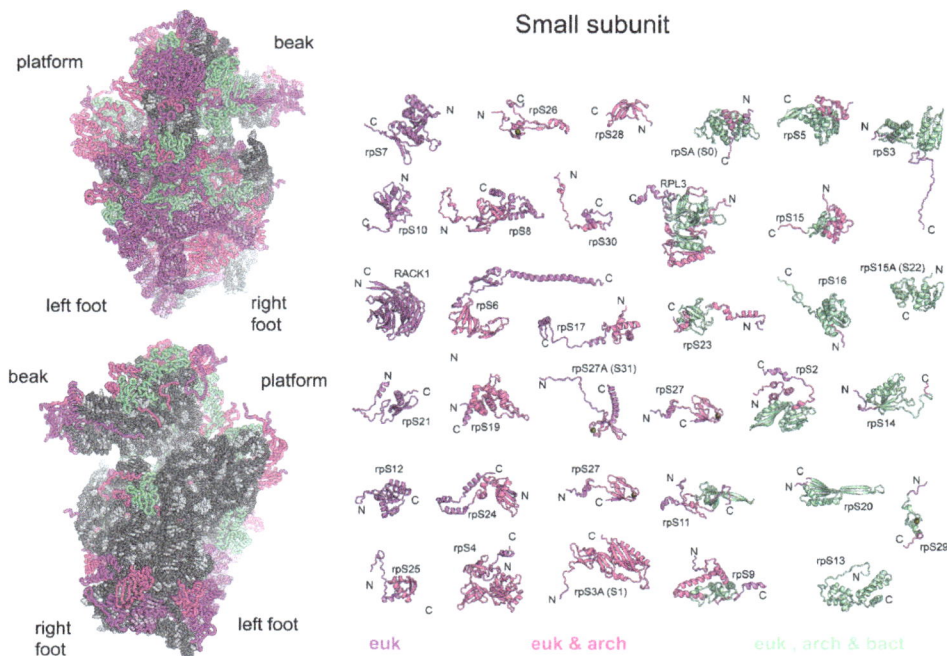

Fig. 1. Two views of the small 40S subunit of the eukaryotic ribosome. Proteins in this ribonucleoprotein complex the colored based on conservation in bacteria, archaea and eukaryotes. Structures of individual proteins forming the complex are shown as a collage on the right. Their evolutionary origin is indicated by the coloring scheme.

To gain insight into the functioning of the eukaryotic ribosome and the process of translation initiation in eukaryotes, we determined the crystal structure of the eukaryotic 40S ribosomal subunit in complex with eukaryotic initiation factors 1 and 1A [5, 13]. The structure defines the locations and the folds of all 33 eukaryotic ribosomal proteins (Fig. 1) and reveals the fold of the 18S rRNA. Furthermore, it provides information on proteins responsible for ribosomal functions in eukaryotic-specific signaling and regulation, and it permits mapping of mutations that lead to genetic diseases. Finally, the interactions with the two initiation factors eIF1 and eIF1A provide a basis for understanding their roles during eukaryotic translation initiation and the role of mutations that affect the start codon recognition [14].

Crystal structure of the eukaryotic 60S ribosomal subunit in complex with initiation factor 6

The eukaryotic 60S ribosomal subunit catalyzes peptide bond formation and provides a platform for binding of various factors involved in nascent polypeptide processing, folding, targeting to the membranes and membrane insertion. The 60S subunit is considerably larger and more complex than its bacterial counterpart [15], with many additional eukaryotic-specific functions connected to the regulation of protein synthesis or ribosome assembly and maturation. Furthermore, the large ribosomal subunit is a major target for various antibiotics, and therefore revealing

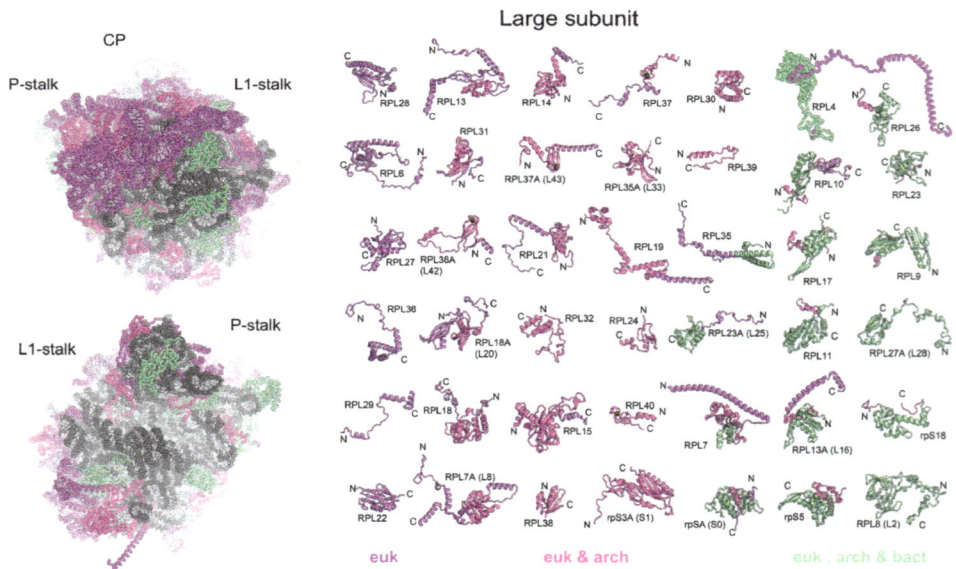

Fig. 2.   Two views of the 60S subunit of the eukaryotic ribosome. Proteins in this ribonucleoprotein complex are colored based on conservation in bacteria, archaea and eukaryotes. Structures of individual proteins forming the complex are shown as a collage on the right. Their evolutionary origin is indicated by the coloring scheme.

the differences between the eukaryotic 60S subunit and the bacterial 50S subunit is important for our understanding of antibiotic specificity of compounds that target the bacterial ribosome and those that specifically inhibit the eukaryotic ribosome.

Recently we described a complete atomic structure of the 60S ribosomal subunit in complex with the eukaryotic initiation factor 6 [6]. The structure of the eukaryotic large ribosomal subunit reveals a network of interactions between eukaryotic-specific ribosomal proteins and ribosomal RNA expansion segments, demonstrating that novel eukaryotic-specific architectural solutions provide long-distance tertiary stabilization of this ribonucleoprotein complex (Fig. 2). It uncovers the roles of eukaryotic ribosomal protein elements in the stabilization of the active site and the extent of eukaryotic-specific differences in other functional regions of the subunit. The structure also offers insights into the regulatory mechanisms of the eukaryotic ribosome, the maturation of the large ribosomal subunit, the principles of antibiotic specificity, and the structural basis of extra-ribosomal roles of various eukaryotic-specific ribosomal proteins. Finally, it elucidates the molecular basis of the interaction with eukaryotic initiation factor 6, involved in both initiation of protein synthesis and ribosome maturation; The release of eIF6 from the ribosome has been connected with the Shwachman-Bodian-Diamond syndrome, an inherited disorder with a wide range of abnormalities and symptoms.

Outlook and future developments of research (proteins in supramolecular machines)

Based on the established methodology and obtained structural information we are now pursuing studies of various functional complexes of eukaryotic, archaeal and bacterial ribosomes to better understand initiation of translation in eukaryotes and ribosomal evolution and assembly in general. Future developments in this field are likely to stem from the above described developments in the field of cryo-electron microscopy combined with single particle reconstruction techniques. This is emerging as one of the most powerful methods in structural biology with huge potential for future development in the following years. Over the next few years atomic resolution reconstructions using cryo-electron microscopy will be in reach for a number of interesting biological systems.

Acknowledgments

The author thanks F. Voigts-Hoffmann for generating the figures, D. Boehringer, S. Klinge, M. Leibundgut, F. Voigts-Hoffmann and J. Rabl for discussions. Research described here was supported by the Swiss National Science Foundation (SNSF), the National Center of Excellence in Research (NCCR) Structural Biology program of the SNSF, and European Research Council grant 250071 under the European Community's Seventh Framework Programme (N.B.), and the Max Roessler prize to N.B.

References

1. X. C. Bai, I. S. Fernandez, G. McMullan, S. H. Scheres, *eLife* **2**, e00461 (2013).
2. M. Mueller, S. Jenni, N. Ban, *Curr. Opin. Struct. Biol.* **17**, 579 (2007).
3. J. P. Abrahams, N. Ban, *Meth. Enzymol.* **374**, 163 (2003).
4. A. Ben-Shem, N. Garreau de Loubresse, S. Melnikov, L. Jenner, G. Yusupova *et al.*, *Science* **334**, 1524 (2011).
5. J. Rabl, M. Leibundgut, S. F. Ataide, A. Haag, N. Ban, *Science* **331**, 730 (2011).
6. S. Klinge, F. Voigts-Hoffmann, M. Leibundgut, S. Arpagaus, N. Ban, *Science* **334**, 941 (2011).
7. D. L. Makino, M. Baumgartner, E. Conti, *Nature* **495**, 70 (2013).
8. T. M. Schmeing, V. Ramakrishnan, *Nature* **461**, 1234 (2009).
9. S. Klinge, F. Voigts-Hoffmann, M. Leibundgut, N. Ban, *Trends Biochem. Sci.* **37**, 189 (2012).
10. S. Melnikov, A. Ben-Shem, N. Garreau de Loubresse, L. Jenner, G. Yusupova *et al.*, *Nat. Struct. Mol. Biol.* **19**, 560 (2012).
11. F. Voigts-Hoffmann, S. Klinge, N. Ban, *Curr. Opin. Struct. Biol.* **22**, 768 (2012).
12. J. R. Lorsch, T. E. Dever, *J. Biol. Chem.* **285**, 21203 (2010).
13. M. Weisser, F. Voigts-Hoffmann, J. Rabl, M. Leibundgut, N. Ban, *Nat. Struct. Mol. Biol.* **20**, 1015 (2013).
14. C. E. Aitken, J. R. Lorsch, *Nat. Struct. Mol. Biol.* **19**, 568 (2012).
15. N. Ban, P. Nissen, J. Hansen, P. B. Moore, T. A. Steitz, *Science* **289**, 905 (2000).

SESSION 6: PROTEINS IN SUPRAMOLECULAR MACHINES

CHAIR: RUDOLF GLOCKSHUBER

AUDITORS: J.-F. COLLET[1], H. REMAUT[2]

(1) Welbio, Avenue Hippocrate 75, 1200 Brussels, Belgium and de Duve Institute,
Université catholique de Louvain, Avenue Hippocrate 75, 1200 Brussels, Belgium
(2) Structural and Molecular Microbiology, VIB Department of Structural Biology, VIB,
and Structural Biology Brussels, Vrije Universiteit Brussel, Pleinlaan 2, 1050 Brussels, Belgium

Discussion among panel members

Rudi Glockshuber: I have a question to Craig Townsend. When you do this polyketide synthesis *in vitro*, for example with purified polyproteins and all the substrates, what are the best yields, which have so far been reported? How many turnovers did you get and how pure is the end-product that you get?

Craig Townsend: The end-product is surprisingly pure. The interesting thing is that one can combine these proteins and they will actually still work together despite having been taken apart. They even faithfully do the chemistry that you expect them to do. I think that one of the disadvantages of taking apart, deconstructing them, is that the efficiency of the process is worse probably than when all proteins are together. So this is an important fundamental question and we have thought about it a lot. The amounts we are making are small, but I think that the comparison that will be useful to make is to have a fully intact protein and measure an overall flux without and with the various components. We have done a combination of two parts of three domains apiece, we have done three plus two plus one and three plus one plus one plus one. We can see just qualitatively by HPLC that the efficiency drops down. I think it can be reasonably efficient.

Rudi Glockshuber: And what is the maximum number of turnovers which has been observed?

Craig Townsend: I don't know the answer yet but I think we can run these reactions over periods of hours and still see production. I think the proteins are fairly robust if you cut them correctly, if you disassemble them correctly. The individual pieces are fairly stable and they will associate sufficiently well to be pretty good little machines.

Nenad Ban: I have a question for Ian Wilson. I noticed that with the HIV trimer structural interpretation, that you have been able to obtain at about 5 Å, you had, sometimes, atomic models underlying this structure. The question is: it is in a way very tricky to make interpretations at 5 Å, so I presume that in some cases you had previously solved domains. But do you think that, in some cases, you can actually trace the fold? We have done a lot of interpretations at this resolution range and it seems extremely difficult, let say, to either use molecular replacement procedures or perform very accurate fold traces at that resolution.

Ian Wilson: We have a lot of models for about 50% of this structure. And we also have the model of the antibody. So the antibody can be placed really pretty accurately. We have a large part of the top domain, which is gp120. So it is missing some of the hypervariable loops. However, there has been multiple structures solved of that, and so they can be really fitted quite well in the structure solved. For the molecular replacement, we have about 50% of the mass that we were using to get the solution. We haven't a lot of the pieces that we did not have for helices, which we can see quite well below 5 Å. And so we have been fairly conservative about building in only the helical regions. We have density, which can account for all of the rest of it but we are cautious about the interpretation of that at the moment. We have both the cryo-EM structure and the X-ray structure and we do not have to deal with molecular replacement for the cryo-EM structure. These were independently solved and, basically, all of the main features, all of the main new features agreed, as well as the fitting of the previously known structures.

Joachim Frank: This is a follow-up question (to Ian Wilson). You have this 5.7 Å structure. How does it compare to the controversial structure that was published by Mao *et al.* in PNAS who claimed a very similar resolution?

Ian Wilson: There is almost no correlation at all of the new regions. Similar to what you described, by cryo-EM, we have multiple states in the reconstruction, in the sample. Andrew Ward's group used an antibody for which there were some unliganded complexes, one-liganded, two-liganded and three-liganded. They have done a reconstruction of each of these. We have looked at the limit at which we can actually do a reconstruction for the unliganded trimer and I think the best you can do without adding features such antibodies is around 12 Å. That would be very discrepant with the reported 6 Å structure. We have chosen not to do any comparison with that structure because I think it is highly controversial, but we see no features in common with our structure at all. And it does not look like a type I membrane protein.

Rudi Glockshuber: I have also a question to Ian Wilson. How heterogeneous is your HIV protein with respect to the glycans?

Ian Wilson: So for the crystallography, (note that there are 81 glycans on the structure), we basically expressed them in cells that produce high mannose. We then bind the antibody. The antibody binds with high affinity. And so we could actually take PNGase F and actually clean off the rest of it down to single sugars. So that really enabled us to actually get a very nice homogenous preparation that could crystallize and get us to that resolution. For the EM, it is actually much more interesting because we can actually express the glycans in 293T cells. And so we actually have much more native-like glycans and those glycans can actually be observed in the EM density. Clearly there will be some heterogeneity there but we can already see at least the core of the glycans, which is the same for almost all of the glycans, appearing in the EM density. So this is really very exciting because you are not able to really define the epitopes unless you actually have the problem of glycosylation on there because many of these antibodies for example that we are actually looking at require binding to both high mannose and to complex glycans. So when one looks at the epitopes as defined on the trimer structure as distinct from engineered smaller constructs, one actually finds that the epitopes are much larger, includes more glycans, and includes more surface, including the protomers, adjacent protomers, that you don't see when you are looking at monomer units.

Marina Rodnina: I have a question to Rudi Glockshuber. You have this very high affinity, incredibly high, right? Can you understand that in molecular terms, so trace that high affinity back to hydrogen bonds, to salt-bridges and so on?

Rudi Glockshuber: The very slow off-rate is most likely caused by simultaneous formation of twenty inter-molecular β-sheet hydrogen bonds, because the bound peptide is inserted between the neighboring strands of the protein and it is a contiguous hydrogen bond network of twenty simultaneous hydrogen bonds which you would have to break to release the peptide. The complex really behaves like a single polypeptide chain. The peptide only falls off when the protein unfolds simultaneously. In addition, every second residue points towards the inside, because it is a β-strand, and there is also very good surface complementarity.

Marina Rodnina: Isn't it a problem for drug design because then you have to find drugs which are competing with this extremely tight binding?

Rudi Glockshuber: I personally believe that pilus assembly inhibitors are not a good idea. The main reason is not so much the stability but rather that an assembly inhibitor does not destroy the pili which are already present when an infection has started. I think that a much more promising strategy is towards adhesion inhibitors which compete, like mannose analogs for example, with receptors for binding of the pilus, in combination with antibiotics therapy.

Nenad Ban: Another question for Rudi Glockshuber. Now that you are saying this, it reminds me of the fact that, maybe there is nothing to it, but people sometimes say you should drink cranberry juice when you have an urinary tract infection. Could it be that in cranberry juice there is some compounds that actually might be mannose analogs or something like this?

Rudi Glockshuber: Yes, this is definitively true. I think some of the compounds have even been identified. There is another quite interesting aspect of the pili because they increase apparent affinity during sheer forces. For example, under flow conditions during excretion of the urine, they increase their apparent affinity. The reason is because there is most likely a domain separation between the pilin domain and the lectin domain under sheer forces. And this induces a conformational change in the lectin domain of the adhesin, which has a higher affinity and about five orders of magnitude lower off rate.

Rudi Glockshuber: If there are no further questions to the first three talks, I would like to suggest that we maybe start quickly with a couple of questions on the ribosome talks and then make a break and continue afterwards.

Marina Rodnina: I have a question to Joachim Frank. Which is about the time-resolved cryo-EM, so where are we there?

Joachim Frank: I didn't have an opportunity to talk about the experimental design. But there is a technology that employs a microchip with two inlets in a reaction chamber and then the compound it forms is spread directly on the grid, and the grid is plunged down into the cryogen. So one is able to get a time resolution down to ten milliseconds with this device. We tested this device, which was manufactured at RPI, using ribosome subunit association as a sort of test case. We were able to identify through different time points the formation of bridges, so we have intermediate states in the formation of bridges. So the entire thing really works and calls for applications that target kinetics of, for instance, the decoding process and so on.

Rudi Glockshuber: I also have a question to Joachim Frank. How dangerous, how big is the likelihood that you identified too many conformations? I mean some conformations that are maybe not physiologically relevant.

Joachim Frank: Yes, very dangerous. I think one really has to analyze the crop that comes from the classification very, very carefully. What we do, at the resolution that we achieved, we can really identify individual factors by fitting X-ray structures convincingly. I think there is a certain resolution threshold that has to be reached before one can confidently do this and below that threshold (*i.e.*, when the resolution is not good enough) one simply cannot.

Nenad Ban: Harry, I have a question for you (Harry Noller). I think it is very beautiful the way you described it and the results that you've achieved of understanding the rotation of the head of the small subunit with respect to repositioning of the messenger RNA and the tRNAs and I remember this is different from what was initially thought based on some early structures of the small subunit where, I think, it was frequently shown that the head has to open completely to let the tRNA and the codon stem loop to move from P to E site and from A to P site. In your current understanding of how this repositioning occurs, you have stepwise events that include maintaining contacts with the head and then separately with the body of the small subunit. Did you ever look at, let's say, the extent of contacts, or maybe, some energetic contributions, because obviously, this way you managed to break the contacts one at the time relative to the head or relative to the body. How would that agree with, let say, either the sequence of steps, or whether they are balanced, these affinities?

Harry Noller: I think it is very early days still in this understanding of this process. For example we can see the rotation of the head and the tRNA following the rotation of the head. But now we know the head has to return. So it must somehow release its contacts with the tRNA during the return. This would be some kind of ratchet mechanism where the pawl is released. I think we have no idea how this happens at present but it could involve something like what you first mentioned, maybe an orthogonal rotation of the head upwards, or maybe a deformation, for example, of helix 34 of 16S RNA that is sort of a scaffold for the things that contact the tRNA. So I think we are just at the beginning, we see the outlines finally of what the mechanism is for the two mains steps of translocation but the details that you are eluding to, I think, are going to get to the essence of the mechanism, which again is going to be a RNA mechanism, no doubt.

Discussion among all attendees

Rudi Glockshuber: I propose to continue our discussion on the specific issues on the ribosome talks and after that have a more general discussion on the main future challenges of structural biology of supra-molecular complexes.

Kurt Wüthrich: I was very intrigued by the stops in the translation that you described. You alluded to the possibility that these are related to the folding of the nascent polypeptide chain. Do you have any evidence or even data to show for this?

Marina Rodnina: We don't have our own evidence but there is evidence in the literature which suggest that if you mutate a rare codon of the mRNA to a more frequently used codon, there are cases that, thereby, you change the rate of translation and then you see apparently different folding which is manifested through a different function of the protein. There is, in the meantime, I think, three or four reports in this direction. One of them was published, let's say five years ago while two or three are very recent. I have to say that so far we don't have our own evidence and we are working on that. I also have to admit that these previous papers have a number of issues. So it is not at all clear whether it is really the case or not. But there is an interesting indication, and I think this is an interesting way to go for. Also, this is definitively not true for all proteins, because from what we have seen for a number of proteins, for example, β-barrel proteins, which are very likely to fold post-translationally, they don't have stops at all. But we're just starting. So it is just the α-helical and the proteins of mixed α and β structures, which show these stops. Even we cannot exclude that this one particular topology which we tend to study all the time, so this Rossmann-fold basically, makes all the stops. We still have to see that.

Kurt Wüthrich: Are the stops also related to proline or to multi-proline sequences, which definitively have a special role in protein folding?

Marina Rodnina: Not in these proteins. We have analyzed that very carefully; these proteins do not have double prolines. They have single prolines, which do not cause stopping. What we have seen is that there are some stops, which are definitively due to rare codons and rare codon clusters. There are some stops, which are related to the Shine-Dalgarno-like sequences, very likely. But about 50% of stops which we have seen, we don't know what is that.

Nenad Ban: Because of these observations that Marina Rodnina mentioned, it was also suspected that, perhaps, this is partially the reason why occasionally eukaryotic proteins are difficult to express in bacteria because if we optimize codon usage, then we get fast read-through through potential pause sequences which are important for gradual domain-wise folding of large eukaryotic proteins. Now since it is possible to optimize codons for bacterial expression of eukaryotic proteins, it is also attempted to simulate a little bit the distribution of codon as it is observed in the eukaryotic system.

Jason Chin: Marina, I was curious about the EF-P experiment. There is an interesting thing going on there both with the modification of EF-P and with what is happening mechanistically at the peptidyl-transferase center. Specifically, I was curious, if you like, on the substrate side, whether you have been able to dissect out when EF-P is necessary? If you do the experiment, can you do the experiment, for example, with N-methylated but non-cyclic amino acids to dissect out the effect of N-methylation versus the geometry constraint? What do you think mechanistically is happening in the peptidyl-transferase center in turn on EF-P, do you actually think that there is facilitated RNA catalysis or whether the protein is participating directly in facilitating peptide-bond formation?

Marina Rodnina: Good question and very little answers. First of all, it is difficult to say how EF-P is contributing to catalysis because in the first place we don't understand why proteins are so slow. Apparently, when two prolines are there, there is something in the structure of the catalytic center which is not positioned properly. So why proline does not do that, we simply don't know. Of course we also do not know what EF-P is doing. All these other experiments are all planned and should be done. The point is that, if you look at the exact contribution, of for example, modification which is looking in the peptidyl-transferase center, this is only a factor of four to catalysis. This might mean that there are two contributions. One of them is that EF-P is doing in fact something to the t-RNA, somehow pushing it into proper position and then there is additional catalytic effect.

Sachdev Sidhu: What about relationship between pausing and secretion? Is there any influence on what happens in the ER with glycosylation, etc.?

Marina Rodnina: This would be all in eukaryotic systems. We know that there is pausing. But this is usually not necessarily by the pausing which we are talking about during translation, because there is specialized pausing, for example translational arrest. I can imagine there are more tools there. I suppose this is type of pausing which is much slower than the pausing which I am talking about. One should not forget that this pausing during translation introduces small pauses. They would not be sufficient to put it somewhere in a different compartment.

Sachdev Sidhu: When it is already going into the compartment is there influence on how it folds in the ER if something is stalled?

Marina Rodnina: This is a good question. We don't know how this pausing is affected by anything from the outside. Also not by all these proteins which are interacting with the exit tunnel of the ribosome, and there are many of them (Nenad mentioned some of them). For example, we know that SRP, in bacteria, does not affect it. But in eukaryotes it does affect it, very strongly. Trigger factor does not seem to affect it, from what we know. But for the rest, we don't know such as if targeting to the translocon affects the rate of translation, this is really not known.

Chi-Huey Wong: It is exciting to see the combination of EM and X-ray. I have a question about glycosylation of membrane proteins. We know in eukaryotic system, it is co-translational. The question is how would the system do that? And how would the system ensure that after glycosylation and before folding, the sugar chain would be on the surface instead of the interior of the protein. Is there a correction system there to ensure this process?

Marina Rodnina: This is a very complex question, so I'll try to give a very short answer to that. This is probably an interplay between many components in the system. The first step in this process will be targeting of the ribosome with the nascent peptide to the translocon. And then this peptide continues to be synthesized while being in the translocon or starting to exit the translocon. These helices, which are to be in the membrane at the end, are synthesized one by one. They have to find their proper topology. How they find their proper topology is not exactly clear. There are models that this occurs spontaneously due to physical properties of the trans-membrane segments themselves and the membrane. But I don't think we know more.

<u>Ray Stevens</u>: We are reaching the end of the meeting, so it is time for sort of crazy questions. Each of you showed these beautiful supra-molecular assemblies. The next level of complexity is the cell. When are we going to see sort of the structure of the cell and what technology for prokaryotes and eukaryote? Any comments on that?

<u>Joachim Frank</u>: Well there is, as you might know, technology available already with electron microscopy, so-called electron tomography. In this case, one actually actively tilts a cell in all possible directions and then forms a three-dimensional image. The application of this technology is severely restricted by the limited penetration range of the electrons, so one can only work with 0.2-0.3 microns. One either has to use very thin protrusions of the cell or, if interested in the entire cell, one has to use some kind of slicing method. There is the "FIB" (Focused Ion Beam) milling technology which is now being increasingly applied in biology. It is possible to actually look at selected sections of a cell and reconstruct these in their entirety. But of course the resolution is limited, and one does not get into the sub-nanometer range that we are now used to with the single-particle reconstruction methods.

<u>Ray Stevens</u>: Sort of a follow-up on that, it seems that one of the limiting factors is going to be the hybrid approach... we are going to use multiple technologies again, like each of you sort of demonstrated, but computational is probably going to be one of the biggest, I think, limitations to really advance this kind of dreams or visions. Any comments on what computationally, from anybody else, what the breakthrough is going to be, to imagine this type of dream?

<u>Gebhard Schertler</u>: Can I first say something to experimental? There is soft X-ray imaging which has made a lot of progress, particularly in Berlin and in the US. Recently I have seen a first tomogram where we can see actually tubulin subunits in a soft tomogram of a full prokaryotic cell. If you want to go to high resolution and even thickest slices (here you can go to 10 microns of range in one tomogram), you have to use hard X-rays. You can then go to 50 microns most likely. That covers roughly a cell or a cell-cell interphase. In the very long run, this is not the very close future, the hard X-ray imaging will come along with a few techniques. I think it will fill in a range above the EM tomography. When we can look for example at membrane topology in very great details and then go into a certain section and do an EM tomogram. In this way, we kind of close a blind gap which is now very difficult to access. It might also become faster than, for example, these methods at the moment where we do block phase imaging, which are roughly doing the same, but very slowly.

<u>Chris Walsh</u>: Given that the questions changed, I think maybe it is time to hear about RNA and the big-bang theory from Harry Noller.

<u>Harry Noller</u>: Sounds like we really ran out of topics.

<u>Chris Walsh</u>: Well, that's the next level!

<u>Harry Noller</u>: Well, it starts with a question: How do you evolve the ribosome? Unless you are a creationist, you look at this thing and you say "this is impossible". So how to develop? If you accept the fairly widely held idea that there was something like a RNA world that preceded the protein-DNA-RNA world and which was based on the functional properties of ribosomes and other kind of functional RNAs which could then self-replicate to create new copies of themselves, then you had a genetic system that could go forward and evolve. One problem, that I'm sure everyone is aware of, is that RNA is wonderful, especially compared to DNA, in its capabilities for folding into interesting three-dimensional structures with tertiary interactions, and so on, and evolved catalytic properties that have been shown by the ribosome people, but it is severely restricted in terms of its folding capabilities compared with proteins. And also severely restricted with respect to its functional groups. So you could imagine that one early version of translation was simply to make peptides that bound to RNA and stabilized novel or only weakly-sampled conformations and there are ample examples of this. For example, the HIV Tat-TAR interaction: the Tat protein binds to the TAR element of the HIV RNA to a simple bulged helix with the hairpin loop at the end and induces formation of a complex tertiary interaction, a base triple, and closes the two co-axial helices to tilt by 90°. Anyway, it is a quite dramatic alteration in structure and Frankel and Puglisi and their collaborators showed a long time ago that you could mimic this with a peptide, which is an arginine-riche peptide from the Tat protein, and then remarkably, they showed that you could even do this with arginine peptide at millimolar concentrations. So even just binding an amino acid at millimolar concentration could do this but the advantage of a peptide is that you can get away with lower concentrations. And so maybe what happened initially is that you started making peptides that greatly enlarged the structure space of RNA and therefore its functional capabilities and its catalytic and other kind of efficiencies. Then, you would imagine that this would of course select for a peptidyl transferase-like activity and ancillary activity. This will then lead to larger and more interesting peptides and eventually these peptides would be sort of prosthetic groups on the RNA, contributing catalytic functions. Eventually the RNA could shrink and we ultimately end-up with the reverse, the RNA being the prosthethic group of a protein for example. An idea which people think may be the remnant of an early ribosome. That is the big bang theory.

<u>Marina Rodnina</u>: I can continue on the big-bang a little bit. Because I think it is also interesting to see how evolution developed after the big-bang. An interesting thing is that apparently the peptidyl-transferase center of the ribosome remains as unspecific as it was at the beginning, probably. This is the reason why Jason and Hiro are able to make their tricks with unnatural amino acids, because the ribosome

just does not care what comes. This is probably the remainder, the fossil, if you wish, of this thing. But then with the time the evolution apparently went into the direction of speed and fidelity and this led to the development of elongation factors, which in fact only are necessary to make it all faster. The more recent evolution apparently leads to all these proteins which are (right now we have found them as orphan proteins) often of unknown functions. And which probably one by one will be shown to do something with translation, like EF-P, EIF5A. Joachim has just told me that there is another protein which is binding to the same place, which is also one of these orphan protein which we don't know what it is doing. It seems that the perspective is to make proteins which are better in helping the ribosome than changing the ribosome itself. This is our fossil of the big-bang.

Nenad Ban: It is just a short comment on the fact that it seems that at the beginning there was RNA and there were lipid bilayer vesicles. It is interesting to note that today protein targeting machinery consists of an RNA element, the signal recognition particle that maybe is due to the fact that this process is also very ancient and may have occurred at the time of the "big-bang" Harry was referring to.

Jason Chin: I have two questions. One relates to how we extrapolate from small number of cases to generality, I'm thinking on how Harry showed some work from Venki's lab on the decoding center where at least, as I understand it, there is a small number of structures showing coding-decoding interactions in the decoding center. More recently, for example, there was a structure of pseudouridylated tRNA with quite different decoding properties. We don't, for example, as far as I know, have structure of every cognate codon-anticodon interactions to see to what extent the rules that Venki' illustrates very nicely explain in principle those kind of structures, the genetic code wobbling, how that is a generally played out. My question is to what extent should we elucidate structurally every single case and to what extent do we just take it on faith this is the right principle that has been uncovered? I have another question but I'll stop.

Harry Noller: So I was careful in my talk not to say...

Jason Chin: It wasn't a criticism to what you said in your talk, I was trying to understand where we should go.

Harry Noller: No, it's a point well taken. This is a very interesting turn of events to see these unusual structures from Venki's lab and from Marat Yusupov recently that sort of change how we think about this. I would just say that the presence of the ribosomal RNA at the scene of the crime as it were interacting intimately with these base-pairing tells us that RNA is involved intimately in this discrimination mechanism. How it plays out may turn out to be different from what we are thinking right now.

Marina Rodnina: I would like to add to that. Crystal structures and structural work is not alone. There is a lot of biochemical and biophysical work from different groups...

Jason Chin: Including from you?

Marina Rodnina: ...which somehow test at least part of these suggestions, which come from structural models. And if they fit, it is fine. If not, something else comes. We have to reconsider that and we have to design new experiments and test it. This is a normal way in our field as we do that and I think this is one of the reasons for the progress in the last ten years.

Jason Chin: Marina, do you think it's worth getting structures of every single tRNA codon-anticodon interaction, would that be a useful activity?

Marina Rodnina: I would suggest to check a couple of them. From the biochemical evidence, we know which steps are the same for each type of codon-anticodon base pairing, because with biochemistry you can simply test it, it is much easier. Then you can define precise experiments where you would like to test that because biochemistry tells you there is something interesting. There I would definitely make crystal structures or cryo-EM. For example, regarding the cryo-EM of the near-cognate, we have only one structure from Joachim Frank's lab. So I would say there is still some place, some room for improvements for looking at the near cognate. This is still an unfinished story.

Harry Noller: I have a question to Nenad. Now that we have a detailed three-dimensional structure of the 80S ribosome. What is all the extra stuff for? Can we generalize about that or even address a few specific topics?

Nenad Ban: Obviously, there are some hints and some of those additional functions specifically related to eukaryotic protein synthesis, I mentioned in the talk. However, I think this is still mechanistically poorly understood and the structure hopefully will provide tools to both interpret and devise genetic or biochemical experiments to better understand these functions. That said, obviously only some new portions of the eukaryotic ribosome structures will have specific functional roles. Many of these extra features will be purely architectural. Maybe some of these features are critical during the assembly process of eukaryotic ribosomes in the nucleus. What is interesting, just as an example, is that the exit site of the ribosome that interacts with the membranes during insertion of membrane protein seems to be very flat and there is not that much added additional proteins or RNA mass in this region. However, eukaryotic ribosomes have one of these RNA expansion segments that would sit right above this exit and this expansion segment has to change conformation when ribosome synthesizes membrane proteins, as it was seen in the EM reconstruction of the translocon for example in the Beckman lab. We see that it has to move out the way, and deletion of this helix is again lethal. So obviously it has some important role but it is not clear which. It is tempting to assume that maybe this helix could serve some sort of a regulatory role or facilitate, let's say, binding of certain factors that are involved during folding of proteins. And then it moves out the way when the ribosomes have to interact with the membrane. There are many questions of that sort that will be interesting to look at in the future.

Lode Wyns: Talking about peculiar features, Joachim Frank, you showed us *Trypanosome* ribosomes. Could you say what is peculiar in there?

Joachim Frank: It is very intriguing. Some of these RNA expansion segments are much larger than in any other eukaryote that we know. Unfortunately, nothing is known about these. We speculate that it has something to do with initiation and also the fact that *Trypanosomes* have to switch hosts. Essentially, they have to jump-start in the new host immediately in order not to perish. We see these kind of expansion segments in every trypanosomal ribosome that we have looked at. So far, we cannot assign a function to these.

Rudi Glockshuber: I have a question to Nenad Ban, regarding the trigger factor. In your movie, it looked as if for a multidomain protein with say two or three sequentially arranged domains, that trigger factor could only support the folding of the N-terminal domain, and once the synthesis proceeds, the trigger factor falls off but can no longer rebind. Is this true?

Nenad Ban: Yes, indeed. We can assume this might be the case, based on the structural features of the ribosome-bond trigger factor. But more importantly, evidence for this type of sequential folding of individual domains and multidomains proteins aided by trigger factor that cycles on and off the ribosome in a fashion that I have shown only for one domain, has been proposed and demonstrated biochemically with FRET studies in Ulrich Hartl's lab and Bernd Bukau's lab, based on the structural model that we have initially proposed.

Jason Chin: This a question for Nenad Ban. Is this phenomenon that has been reported as the translational ramping at the N-terminal region during expression. Is that related to the trigger factor?

Nenad Ban: I'm not entirely sure what you are referring to, could you elaborate a little bit?

Jason Chin: There is an increase in protein synthesis rate from the N-terminus, depending on who you talk to, over the first 20 to 30 amino acids or something like this, the speed of translation ramps up.

<u>Nenad Ban</u>: Ok I wasn't aware of that.

<u>Jason Chin</u>: So the question is does this somehow correlate with the trigger factor working on the N-terminus and then being released?

<u>Nenad Ban</u>: Interesting idea, but I suspect that this might not be the case because the trigger factor is known to re-associate with the next segment of unfolded polypeptide. It is not present just initially.

<u>Ian Wilson</u>: What is really the mechanism for the differentiation between membrane proteins and soluble proteins. It seems hard to understand how you can actually recognize, as it is coming out, whether this is going to fold up into an α-helix, whether it's going to be hydrophobic enough, or is just a few hydrophobic residues?

<u>Nenad Ban</u>: This is a very interesting question, and here we also open the question of how the activities of these various factors at the exit of the ribosomal channel are spacially and temporally regulated, because you may have noticed that several of them bind virtually to identical sites at the exit of the ribosomal tunnel. Obviously, there is some sort of an interplay that is currently poorly understood. I think there is a lot of research happening in this area. To better understand for example why trigger factor initially interacts with proteins and how it is displaced, or maybe for a short time, we have simultaneous binding of signal recognition particle and trigger factor in a distorted conformation, because in their energetically favored state they would clearly clash one with another, even though their direct binding sites do not overlap. It is a very good question and I think it is a matter of affinities. It is also clear that these factors do not have very high affinities for the ribosome. They bind with micromolar affinities, so they obviously cycle and they have some residence time on the ribosome. The bottom line is that with respect to future membrane proteins, it is not clear entirely how the signal recognition particle recognizes it on the ribosome. It is known, for example the Wolfgang Wintermeyer lab has reported that for a nascent chains that carry a signal sequence, already before the signal sequence exits, the ribosome increases somehow the affinity for the signal recognition particle. But mechanistically, it is not clear how that could happen. In the end, obviously, there are other factors that later have to step into the game. The translocon for example, has to also bind to the same place where the signal recognition particle binds. And only recently we understand how the GTP-controlled conformational change exposes simultaneously the binding site for the translocon and allows signal sequence hand over from the signal recognition particle to the translocon.

<u>Rudi Glockshuber</u>: I have a question to all ribosome experts. I think the antibiotics which are known to bind to the ribosomes all bind to the intact ribosome, the intact quaternary structure. But there are no antibiotics that inhibit ribosome assembly. Eukaryotic ribosomes have about two hundreds assembly factors, or something like

that and so one could easily think of assembly factors as drug targets. Why have such kind of drugs not been identified?

Harry Noller: It has been suggested by some assembly people that ribosomal assembly is so hugely cooperative that the binding affinity of an antibiotic might simply be overwhelmed by the massive cooperative interactions of all of the proteins and assembly factors.

Marina Rodnina: I have a comment on that, a very short one. The comment is that I don't think that anybody has ever been looking for drugs against eukaryotic ribosome assembly, because I don't know why you should do that?

Rudi Glockshuber: I of course meant drugs against bacterial ribosome assembly.

Marina Rodnina: The answer is very simple. This is too fast. You have to try to develop a lead compound against something which takes a long time. Elongation for example takes a long time. There is a good possibility this drug would come and will bind there. The ribosomes spends most time during the elongation. Therefore the large majority of antibiotics which we have right now, are against peptidyl-transferase or against decoding (*e.g.*, aminoglycosides).

Tom Muir: I want to come back to this proline effect. As I understand it, at least one of the problems is that the peptidyl-tRNA ending in proline is a slightly slower reactant. I have a comment that might relate to this. It turns out that peptides ending in proline, peptide esters or thioesters ending in proline, are intrinsically less reactive as acylating agents compared to the other amino acids. It is quite a dramatic effect actually. It used to be thought that was related to the conformational effect with the pyrrolidine ring. It turns out that it is not the case, it rather relates to an anti-π^* interaction involving the ester carbonyl that deactives as an acylating agent. That is maybe related to what you are seeing.

Marina Rodnina: Well, maybe. The only thing what we have, we can compare that to the model studies which have been done for peptide-bond formation including effects of different amino acids on the A-site substrate and P-site substrate. And from these studies, proline was not particularly different from anything else. There were more effects coming from charges. For example positive charges on the terminal amino acid led to favorable effects and a negative charge was unfavorable. Proline was unremarkable under these studies and so it still remains to be seen.

Markus Aebi: I have a question to Ian Wilson. You showed this beautiful machine of these fusion proteins. How is the cell or the virus directing the synthesis of such a loaded spring that it maintains in a high-energy state. It is synthesized in a high-energy state and only reacts when it encounters its ligand. Have you any idea on how this is done?

Ian Wilson: Yes. Clearly we are used to proteins folding in their lowest energy state, and quite clearly these fusion proteins are metastable proteins. But in a sense they are in a lower energy in as much as the environment in which they are produced. Each of these fusion proteins in order to adopt a different state actually has to bind something, or to be subjected to some different chemical environment. And that varies from fusion protein to fusion protein. For influenza virus, binding on the surface of cells does not trigger any conformational changes. The virus gets taken into the endosomes and in that compartment it's acidic. It is this acidity that actually is the trigger. Now you are in a low pH environment and that triggers the conformational change. For HIV it is a different story, you've got two receptors and each one of them actually triggers a conformational change. When the envelope protein binds to CD4, it actually is in a pretty closed state. When CD4 binds, the heads start to move apart and rotate around. That exposes and helps form the co-receptor binding-site, and only when CD4 binds, do you actually start to change the conformation sufficiently that the protein goes into its fusion-active form. This occurs on the membrane surface and it is not activated by pH. The second trigger is not pH, it is the binding of an additional co-receptor. So they are in a sense in a semi-stable state, they are reasonably stable proteins in a particular environment and it is only through binding ligands or change in chemical environment that you come to these dramatic changes that lead to fusion.

Don Hilvert: Can I just follow up on that? Since you undergo these very large conformational changes, is it possible to elicit antibodies to the intermediates using truncated constructs, to trap the fusion transition that occurs?

Ian Wilson: I'm sure people have tried to look at these and to look at mutations that actually help stabilize some of these intermediates. At the moment we have not really trapped many of the intermediates because once it goes, it goes very quickly. That movie I showed you was actually a representation of the beginning state and the end state, and you need some sort of fantasy of the in between as to how it actually goes. It's just that we have not managed to trap an intermediate state. It may be possible for things like HIV, where you may be going into some sort of intermediate state, to actually trap that, and you in fact can trap this stage with antibodies.

Kurt Wüthrich: Is there some protein-related new chemistry in these macromolecular assemblies? What I'm looking for is perhaps best illustrated with work that Don Hilvert has been involved with some time ago, work on enzymatically active antibodies, which have not been discussed at all here. What impressed me at the time, and actually I don't quite understand why this is not carried on and emphasized more clearly, is that organic synthetic reactions that would otherwise only happen in extremely dry non-aqueous solvents, run in aqueous media in these enzymatically active antibodies. Do you have something similar to report about chemistry in these macromolecular assemblies which would be unlikely to happen outside of these "machines"?

Judith Klinman: I just want to say many enzyme active sites have very low dielectric constants, in which case you don't really need a machine. You just need to control the local dielectric constant to drive reactions that normally would go in an organic solvent.

Kurt Wüthrich: Is it not surprising to you that the Diels-Alder reaction would run in an enzymatically active antibody in aqueous solution?

Judith Klinman: Well, it is a surprising reaction. I just think the degree of hydration and the way in which the active site residues control the dielectric, the proximity for sure is a factor, but no I don't find it that astonishing.

Marina Rodnina: Perhaps I want to answer this question in a little bit different way. We are dealing with normal translation, but we know that people are introducing amino acids and making products which have not been there before. This is more a question to Jason [Chin] and to Hiro [Suga], who make all these fantastic products. This is all was ribosome does. It just can take a number of substrates and unnatural substrates which have never been there before.

Don Hilvert: There are still amide bonds.

Nenad Ban: Maybe I can just ask Craig Townsend to comment a little bit on the strategies to develop potentially new natural products with polyketide synthases, are there "surgical" procedures to redesign the order of their active sites or to design new types of multi-enzymes.

Craig Townsend: Right, well, for the last twenty years people have had a great deal of interest in doing this with modular systems. I think the difficulty there is that as you proceed through each individual domain that has been optimized evolutionary, it is hard to get a good overall yield. Our experience is limited, I did not have time to talk about this, but it is possible to make hetero-combinations of these individual catalytic domains that have a similar function and the remarkable thing is that they will work together. They will also carry out synthesis in a way that — we call programming because we don't understand exactly what controls these events — if a particular ring system is made by a PT-domain for example, it can be supplied with a polyketide precursor that is shorter or longer, and amazingly it still binds and does the cyclization chemistry faithfully. So it would be wonderful to have structural information of these individual pieces to have some idea of what controls these events, but I think the intrinsic flexibility of binding can be taken advantage off. I am sure it can be optimized, there is doubtlessly downstream evolutionary kind of things that one would want to do to make them efficient, but the intrinsic activity is quite good.

Jason Chin: I have a question for Craig Townsend and maybe for Nenad Ban about assembly line biosynthesis. Where are we with understanding at the mechanistic level, by which I don't mean chemical mechanism, I mean structure and dynamics of the protein and how that relates to catalytic steps or transfer steps in term of both dynamics and structure. Is that the missing information for moving this forward or is there something else that would be more important? What is your view?

Craig Townsend: This is a complicated question because the intrinsic fact for those iterative and modular systems is that the intermediates are all bound, there are no free intermediates, so a lot of the standard kinds of biochemical tests that you would like to be able to apply aren't so easy to do. There are a lot of reactions, if you count up, just for making a fatty acid where things are very nicely orchestrated to be repeated every single cycle, it's fifty reactions that take place. The individual steps, rates and flux and so forth, people have worked at this, but it's complicated.

Jason Chin: Are there biochemical ways to trap these intermediates?

Nenad Ban: I would like to comment on one other aspect. Currently, the structural information on the fungal, mammalian type of fatty acid synthases gives us a relatively good idea of the basic architecture of a single module of a polyketide synthase. However, we have relatively limited information on how these modules assemble into complete polyketide synthases that have to shuttle substrates from one module to another for each step of the series of reactions. I think this is definitely a very interesting future direction of research in understanding both the structure and the dynamics of these machines.

Wilfred van der Donk: I saw David Sherman give a talk. In that case it was I think a PKS or NRPS, I don't remember. They had done cryo-EM. It was really quite different from the fatty acid synthase. Even the way the individual domains were organized, I was surprised. Instead of being out, in kind of like a butterfly type, they moved back in and they looked very different. I think there are still a lot of things to be understood there.

Chris Walsh: This polyketide synthase, would there be a simple trick that it can make a complex polymer or it could be another processing enzyme? There is a very big interest in different materials with different properties as well, but often to make them solid actually, it is good to have them in a polymer state. Here you will be able to have kind of a polymer with a certain sequence of chemistry.

Craig Townsend: That is possibly true. There is a limitation in the iterative systems in that there are essentially four ring types that are made, right now. You can make fused rings, up to six for sure, maybe four fused together. Then you will be at the end of what is currently possible to do in a product template domain. I think it will be difficult to make an extended condensed polymer. Maybe pieces linked together individually, groups of three and four perhaps, that might be possible to do.

The other problem is the intrinsic reactivity of the poly-β-ketone. It is remarkable, it is not something that you would be easily able to do in a laboratory, to make something like that. The trick is a kinetic one, one thing that I would have liked to have said but did not say, is that the populations, in an iterative system, of the individual active sites are essentially zero. Except for the malonyl transferase. So the syntheses that initiate chain extension are very rapid. You can actually make a poly β-ketone sufficiently quickly. That is, it hasn't had a chance to actually react with itself, and that's then controlled in a template domain by being bound and cyclized in a fairly rigid and defined way. So it takes quite a bit of engineering to change that beyond about four rings.

Don Hilvert: I might add in that context, that there are enzymes that polymerize activated acids to make poly-esters, poly-hydroxybutyrate for example. You can imagine adapting such systems or copying that strategy to make artificial polymerases.

Nenad Ban: I have one more question to Rudi (Glockshuber). It is interesting to see that these pathogenic bacteria attach to the cell walls using the pili. It is curious to think of what sort of forces would be necessary to detach the pili. Has anybody done maybe some, let's say, optical tweezers experiments to see whether the subunits will unfold or whether they will break or whether they will pop out the anchor site on the bacteria. What is the weakest point?

Rudi Glockshuber: The first observation is that you have this apparent increase in binding affinity to the receptor upon application of sheer force. The hypothesis is that this is related to the domain separation in the adhesin and a conformational change in the lectin domain that then has a lower off rate for the ligand. If you continue pulling, then you see the unwinding of the helical pilus rod. And a gain in length by about a factor of two. And only then, they fall apart.

Nend Ban: Presumably then unfolding?

Rudi Glockshuber: Or by breaking donor strand domain interactions, that is not entirely clear.

Don Hilvert: I also have a question to Rudi (Glockshuber). How sensitive is this F-tag to mutations and can you use the tag itself as an inhibitor for pilus formation?

Rudi Glockshuber: This is a very good question. We have not tested the peptide for inhibitory activity. One could add it to the periplasm, by just adding it to the medium and it would maybe diffuse to the periplasm. What we have done: the FimG-dsF peptide complex is the most stable non-covalent protein complex known to date. But it has actually quite a bad on-rate of only 300 $M^{-1}sec^{-1}$. We have tried to increase charge complementarity between peptide and protein by introducing mutations in the protein next to the peptide binding-site and also by mutating the peptide residues which point towards the surface. We have now, it is not published yet, a new combination of FimG variant/DsF variant which has 10^{-22} molar dissociation constant, and an about one hundred times better on-rate.

Don Hilvert: That could be very useful then for competitive inhibition with the natural system.

Rudi Glockshuber: Yes, but we have not tested it yet.

Chris Walsh: I would like to go to another question. It is about electron microscopy, but also about the pili. So in 3D electron microscopy, we can do this with the ribosome because it is very big. We hope that we will get smaller but we will have a problem to get below 200,000. The question I have is, can we use these pili filaments to insert relatively small molecules? And then have a helical architecture that we can look at in EM, which is nicely presented as a helical structure. Is there any insertion point we could use in the pili?

Rudi Glockshuber: I believe that it is difficult. Because the translocation pore of the assembly platform is just wide enough to allow translocation of an individual folded subunit. One can also reconstitute the pili after dissociation and let them reassemble. The pili become a little bit shorter, or significantly shorter, but still fifty nanometers or so long. This is maybe possible but very likely not with the assembly platform.

Chris Walsh: Joachim, how far are people with using scaffolds for electron microscopy?

Joachim Frank: I am not aware of any use right now. Long time ago, we used the ribosome itself in order to find the position of an rRNA by inserting a tRNA genetically, so we found the tRNA stuck at the outside. I think people have been trying to do similar experiments. One with the group II intron. We have a first reconstruction of a group II intron in collaboration with Marlene Belfort. But we need to identify the components since the reconstruction is at low resolution. One idea is to make a similar kind of insertion.

Rudi Glockshuber: We still have about ten minutes. There are so many exciting developments in electron microscopy, we also have free electron laser technology and so many experts are here. I would like to propose that we maybe spend the next ten minutes on the future of structural biology. What are your thoughts where structural biology, in particular of supra-molecular structures, will stand in ten or twenty years from now? Do you agree with this?

Bernard Henrissat: I would agree if I can ask a last question before? This is on the pili. I would like to know what is the conservation of this machinery across bacterial species? Do we expect the same properties always or is there diversity there? Second, are there any interactions between the pili and other surface components of the cell such as peptidoglycans?

Rudi Glockshuber: With respect to peptidoglycan, I do not know any reports. Similarity: the pair-wise sequence identity within type 1 pilus subunit is not more than 35% or so. Other pili from Gram-negative bacteria which are also assembled according to the same mechanism share sometimes even less than 20% sequence identity with the corresponding subunit from other species. But structurally they are all identical. There have been a couple of experiments, or reports where it was tried to replace for example the chaperone from one pilus system from a certain species by the chaperone from another species. Sometimes it worked, sometimes it did not. Overall, I think the general mechanism that I described is the same for this kind of pili.

Jason Chin: I have a question. We don't really have a representative of the other side of the argument here, but I guess I'm curious what the panel think about the relative merits of optical methods, of subdiffraction imaging, so-called super-resolution, and where that's going to go and how that is or is not useful and might complement electron microscopy? I guess Joachim [Frank] might be the best person to start.

Joachim Frank: I really don't know the range that is still possible in the light microscopy field, but I'd like to mention something that has not been mentioned before, which is the power of correlative methods, where LM is combined with cryo-EM. It is possible now, through the development of certain stages, to follow a live process in detail and then freeze it at a particular moment. And then keep track of the coordinates and transfer everything into the electron microscope. Even now a stage is being developed, in which everything stays in the electron microscope for the duration of both LM and EM imaging. This is a very exciting future of an approach in which one can study something that is happening *in vivo* and then immediately take a look at a much higher magnification.

Harry Noller: I also had a question, probably for Joachim [Frank]. Can one combine electron tomography with the direct electron detection technology to now solve the cell and beginning of course with the shallow?

Joachim Frank: Yes, absolutely. The first results that we are obtaining show that we have a much better resolution in tomography. It is very exciting, but it is also very humbling because we are now dealing with incredibly large volumes. We have to sort of disentangle these. The computational efforts are really mind boggling there. One of the nice things with these direct detection cameras is that they allow you to do multiple frames. So, the K2 summit camera from Gatan catches forty different frames per second. In this case, any motions of the specimen that might be beam-induced or are due to instability of the instrument can be compensated. This is part of the reason for the improvement in resolution. So absolutely, electron tomography is one of the big winners in this game.

Gebhard Schertler: So we had a huge development in X-ray diffraction, what a free electron laser can do is give you tremendous intensities, so a huge number of photons in a very short time. Here you see, we have femtoseconds up there. What this diagram here shows, is actually a very important simulation done in 2000, where actually it was proposed that every atom would maybe absorb an X-ray photon that is not the diffracting photon but then electrons leave the molecule and actually there is a Coulomb explosion from the residual there. How can that be useful? If you make the X-ray pulse short enough, like 2 femtoseconds in this case, we can actually see the molecule before it explodes.

Here you see the second important thing how this was practically made useful. We are injecting now little crystals or particles, or we could do ribosomes, into the machine with a liquid jet and the little light you see there is actually the destructive power of the free electron laser but we can record these patterns, and now, at room temperature, structures have been solved to 1.5 Å resolution of various things, new molecules and also very recently of membrane proteins with a slightly modified technique. So we really look at a very new technique where we can also do spectroscopy and other methods in completely different timeframes. Another important thing about this technique is that we have now laser like X-ray light, so holographic techniques, or phase reconstruction and retrieval will really change dramatically over time, so I think we have a fantastic development but actually it is working already, it's not that we need years to develop, we can apply it to biological molecules now.

Rudi Glockshuber: Maybe one last comment and then we have to finish.

Ian Wilson: So looking at fantasy, but may be it isn't so much fantasy, I mean doing single molecules. That's the ultimate, you don't have to get a crystal anymore and you can do everything off a single molecule. Whether that's actually going to pan out, at least that's the way people are thinking now, that would really revolutionize what we can do, if we can get high resolution information from single molecules.

Gebhard Schertler: But there is a big debate whether EM isn't better at it, so I think we have to find the right niche for every method. I am not convinced that actually free electron lasers in the first generation will be very good for single particles, they are very good actually for difficult membrane proteins, difficult crystallization projects, in giving you high resolution but at the moment they are not a single molecule method.

Joachim Frank: To comment on this, as I pointed out when I spoke, we are really on the verge of atomic-level resolution for several big enough macromolecules and we will see this reflected in the literature in the next two or three years; we will soon have an entire promenade of structures captured in actual 'live' situations; in other words macromolecular machines 'caught in the act' and so this will be a very exciting prospect in future research.

Gebhard Schertler: I think the other exciting thing is really the combination and spread, what we need is methodology that spans the whole timeframe, the whole spatial frame, and one single technique most likely cannot do that.

Rudi Glockshuber: Ok, then this is I think quite a good final statement on this part of the discussion. I would like to thank all the speakers of this morning session, and those of you who contributed to the discussion for their excellent contributions, thank you very much.

Index